Deep-Water Fisheries of the North Atlantic Oceanic Slope

edited by

Alan G. Hopper

formerly Technical Director
of the Sea Fish Industry Authority,
Hull, U.K.

Kluwer Academic Publishers

Dordrecht / Boston / London

Published in cooperation with NATO Scientific Affairs Division

Proceedings of the NATO Advanced Research Workshop on
Deep-Water Fisheries of the North Atlantic Oceanic Slope
Hull, U.K.
March 1–4, 1994

A C.I.P. Catalogue record for this book is available from the Library of Congress

ISBN 978-90-481-4563-8

Published by Kluwer Academic Publishers,
P.O. Box 17, 3300 AA Dordrecht, The Netherlands.

Kluwer Academic Publishers incorporates the publishing programmes of
D. Reidel, Martinus Nijhoff, Dr W. Junk and MTP Press.

Sold and distributed in the U.S.A. and Canada
by Kluwer Academic Publishers,
101 Philip Drive, Norwell, MA 02061, U.S.A.

In all other countries, sold and distributed
by Kluwer Academic Publishers Group,
P.O. Box 322, 3300 AH Dordrecht, The Netherlands.

Printed on acid-free paper

CONTENTS

PREFACE

During the next decade mankind is likely to turn more and more to the exploitation of the resources of the deep ocean. These efforts will be directed at oil and gas reserves, mineral wealth and the living resources, especially fish. At the same time the deep ocean could become more widely used for the disposal of industrial waste. These trends are particularly apparent in the North Atlantic around which most of the world's major industries are concentrated and where the fish stocks of the continental shelf are heavily depleted.

In spite of more than a century of research in oceanography and marine biology we still lack sufficient knowledge to appreciate fully how to manage our use of these resources. There seems little doubt that the deep water marine ecosystem is very fragile, and that ill-considered exploitation by fishermen, or the secondary effects of oil, gas and mineral extraction could cause rapid and irreparable damage. This may or may not be important in world economics but it would seem most sensible to assess these risks before plunging into deep waters.

The Workshop which provided the material for this book was first conceived in 1992 when it was realised that there was a considerable amount of contemporary research of the deep water fisheries of the North Atlantic. Much of it was being carried out independently and very little was being made available to fishermen or the administrators who at some time in the future may have to exercise control of these new fisheries. The experience from New Zealand, where the commercial fishery for orange roughy in deep water had alarmingly depleted these stocks in the space of 10 years could not be ignored. The Workshop brought together scientists, fisheries experts, fishermen and administrators to listen to papers and to assess the current state of knowledge and discuss how to proceed in the future. The book is aimed at all these people and has the simple message that we have only a little time to make sure that we exploit these resources sensibly and avoid putting them at risk.

The planning of the Workshop took almost two years and it was attended by 45 scientists, engineers, fishermen and administrators from 15 countries. The book is divided into four sections;- scientific papers which have been refereed in accordance with scientific protocol, summary papers describing current or past research activity, technical papers and finally a synopsis of the workshop sessions which formed an important part of the programme and from which recommendations for future research and technical development have been made.

The Workshop was partly funded by NATO and by the EU DGXII (Fisheries). It was organised jointly by the Sea Fish Industry Authority and the Scottish Association for Marine Science. The organisers are grateful to the two funding agencies, and to the delegates who gave freely of the time to share their knowledge and experience.

INTRODUCTION

Until recently the fishing grounds of the continental shelf zones in the North Atlantic have provided adequate fish stocks for the needs of commercial fishermen. Today most of these stocks are described as fully exploited, and on which no further increase of fishing effort should be permitted, or they are over exploited and therefore in decline. In these circumstances the commercial fisheries are severely restricted and inevitably this has led to an increased interest in the deep-water species of the Atlantic oceanic slope (circa 500 m. to 2000 m.).

The biology of the fishes of the deeper water has been extensively studied by scientists for more than a century, but very little work has been done to estimate the stock sizes. In general there is a shortage of relevant information to give the fishermen, or the legislators who must exercise some degree of control on these resources if the past mistakes arising from too much effort on the shelf fisheries are to be avoided in deep water. At the same time there has not been an occasion at which current scientific and technical knowledge has been shared. This was the justification of the NATO Advanced Research Workshop held in Hull in March 1994 the proceedings of which are presented in this volume.

Up until now there has only been a limited amount of fishing effort on the oceanic slope zones and directed at only a few species such as roundnose grenadier (*Coryphaenoides rupestris*), roughhead grenadier (*Macrourus berglax*), blue ling (*Molva dypterygia*), orange roughy (*Hoplostethus atlanticus*), black scabbard fish (*Aphanopus carbo*) and Greenland halibut (*Reinhardtius hippoglossoides*). The cost of fishing in deep water, the high engine power needed and the lack of a market demand for many of the unfamiliar species are the limiting factors.

Most of the deep-sea stocks discussed in this volume are regarded as virgin or accumulated stocks. Many too are slow growing, with low inherent productivity. Scientists agree that the ecosystems in which these species live are different from the shelf ecosystem and that an all-out and ill considered exploitation of these deep water stocks will put the future of these important resources at risk.

Whilst there is a considerable and encouraging measure of agreement between scientists in their understanding of these deep-sea stocks, there are some differences of opinion on the hypotheses of stock structure. This is understandable in view of the complexity of the subject and the need for extensive time series data. The differences are highlighted in the continuing debate on the theories of the stock structure of the roundnose grenadier. Russian research mainly from the period 1960 to the mid 1980s suggested a single stock throughout the whole of the North Atlantic, (Lisovsky and Troyanovsky, this volume). More recent research and a re-assessment of earlier data now suggests that this hypothesis is no longer tenable, (Atkinson, this volume), and there is more than one stock. Roundnose grenadier is one of the most studied of all the deep-water fish and the lack of a definitive conclusion on the stock structure reinforces the need for international collaboration and a full and comprehensive sharing of data, if we are to understand this complex deep sea ecosystem and manage it effectively.

The planning of the Workshop took almost two years and it was attended by 45 scientists, engineers, fishermen and administrators from 15 countries. The book is divided into four sections;- scientific papers which have been refereed in accordance with scientific protocol, summary papers describing current or past research activity, technical papers and finally a synopsis of the workshop sessions which formed an important part of the programme and from which recommendations for future research and technical development have been made.

The Workshop was partly funded by NATO and by the EU DGXII (Fisheries). It was organised jointly by the Sea Fish Industry Authority and the Scottish Association for Marine Science.

Acknowledgement

The editor gratefully acknowledges the invaluable and patient help given by Miss Louise Allen for secretarial support and keyboard skills, Miss Christine Drewery in for the graphics and Mrs Jean Dorsett for her organisational back up from the time the project was first conceived to the completion of the manuscript. All are employed by the Sea Fish Industry Authority.

GLOSSARY OF TERMS

Abyss The deep ocean floor beyond the continental slope (see sketch).

ANOVA A statistical method to test whether two or more mean values differ significantly.

Assemblage A group of species that regularly occur together in a particular place.

Benthos The plants and animals that live in or on the sea bottom.

Biomass The total weight of living matter present per unit area and per unit time.

Boreal Pertaining to the north, as in the boreal region that lies between the temperate regions and the Arctic.

Directed fishery A fishery in which a particular species of fish is the target.

Epipelagic The upper region of the open ocean, usually about 200 m. deep from the surface where sufficient light penetrates to support photosynthesis.

Longline A method of fishing consisting of a main line to which there are attached several hundred short branch lines which carry the baited hooks. Longlines may be set in the mid water zone or on the bottom. They are used in some deep-water commercial fisheries.

Macrozooplankton The larger planktonic animals such as krill, jellyfish and small fishes

Nekton Those swimming animals of the ocean that are capable of moving against the currents.

Otolith The small bone found in the ears of many fishes from which annual growth can be determined by counting the concentric rings in the otolith.

Plankton The small plants and animals of the ocean that drift with the currents and are incapable of swimming against them.

Population The subset of individuals within a species that occupy the same range and regularly interbreed.

Rise The area of the ocean bottom that lies between the bottom of the slope and the abyssal ocean floor.

Slope The edge of the continental shelf where the bottom descends steeply to the rise.

Scattering layer. A layer in the upper 1000 m. of the ocean detected acoustically by echo sounders and composed of macrozooplankton and smaller nekton.

Teleostei (teleost fishes) Fishes with a swim bladder and a rayfin bone structure.

Trawl The most common method of fishing in which a conical netting bag is towed by a trawler at speeds up to 4 knots. Fish in the path of the trawl are herded into the net. There are many design variations including trawls for mid-water fishing and for bottom trawling.

Trophic Pertaining to nutrition. The trophic levels in a food web are plants, herbivores and carnivores

ABBREVIATIONS

DFO Department of Fisheries and Oceans (Canada)
NAFO North Atlantic Fisheries Organisation
NATO North Atlantic Treaty Organisation
CPUE Catch per unit effort
TAC Total allowable catch
MSY Maximum sustainable yield
ITQ Individual transferable quota
MDS Multi-dimensional scaling
M Natural mortality factor
F Fishing mortality factor
ICNAF International Commission of the Northwest Atlantic Fisheries
INIP Instituto Nacional de Investigação des Pescas
IPMAR Instituto Português de Investigação Maritima
Int Integer (Jørgensen, this volume)
GRT Gross registered tonnage
GDR (former) German Democratic Republic
FRG (former) Federal Republic of Germany

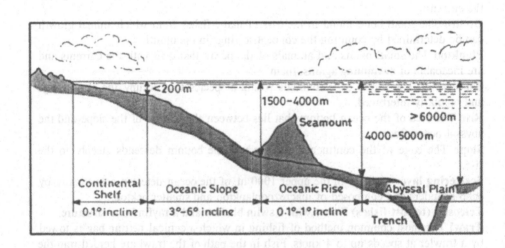

SECTION 1   Scientific papers

ENVIRONMENTAL AND BIOLOGICAL ASPECTS OF SLOPE-DWELLING FISHES OF THE NORTH ATLANTIC.

JOHN D.M. GORDON[1,] NIGEL R. MERRETT[2] & RICHARD L. HAEDRICH[3]

[1]*Scottish Association for Marine Science*
PO Box No. 3, Oban, Argyll PA34 4AD, Great Britain

[2]*Department of Zoology, The Natural History Museum*
Cromwell Road, London SW7 5BD, Great Britain

[3]*Department of Biology, Memorial University*
St. John's, Newfoundland A1B 5S7, Canada

ABSTRACT

The environmental and biological aspects of slope dwelling fishes of the North Atlantic are discussed in the form of answers to the following questions. (1) How do the physical features of the continental slope and shelf compare? (2) How does the physical environment of the slope differ from that of the shelf? (3) How do the demersal species assemblages on the shelf and slope differ from one another and are the latter basically different from the pelagic oceanic assemblage? (5) How do the basic distribution patterns (vertical and horizontal) of slope-dwellers compare with their shelf-dwelling counterparts? (6) How does the vertical distribution pattern of fish biomass correlate with the trophic input to the oceanic environment? (7) What is known about deep-sea fish population structure and breeding biology?

The continental slope is the true edge of the deep-sea and, because of the relatively steep gradients is a physically and biologically layered environment. The adaptations of the slope fishes stem mostly from their food supply, which decreases strongly with depth, and their growth, population regeneration, and time for recovery will all be very slow. Deep-water fish populations cannot be expected to support the levels of exploitation that have been applied to shelf populations.

1

A. G. Hopper (ed.), Deep-Water Fisheries of the North Atlantic Oceanic Slope, 1–26.
© *1995 Kluwer Academic Publishers.*

1. Introduction.

Fisheries for deep-water demersal species in the North Atlantic have increased and diversified quite considerably in recent years (Reinert, this volume; Troyanovsky and Lisovsky, this volume; McCormick, this volume; Thorsteinsson and Valdimarsson, this volume). If these fisheries are to provide a sustainable resource the stocks must be assessed and biological information necessary for management must be assembled. This is not an easy task, and even the long established fishery for roundnose grenadier *(Coryphaenoides rupestris)* in the western Atlantic is managed by precautionary quotas based, in part, on previous landings (Atkinson, this volume). This is because the Northwest Atlantic Fisheries Organisation (NAFO) considers that the basic catch and effort data are inadequate and that there are too many important gaps in our knowledge of the biology of the species to enable an assessment to be made. The orange roughy *(Hoplostethus atlanticus)* fishery of New Zealand began in the 1970s but it is only comparatively recently that adequate stock assessments have been made, and for some stocks these have come too late to prevent them being overexploited (Clark, this volume).

The basic requirements for a stock assessment are reliable estimates of biomass and catch and effort data. The estimation of both these parameters is more difficult in the deep-sea than on the shelf, but the technical difficulties can be overcome. While a number of approaches are possible, it is most likely that random stratified trawl surveys will be the best option for biomass estimation because of the difficulty of carrying out acoustic surveys or egg and larval surveys in deep-water. Considerable improvements can be made to the assessment by incorporating biological information into the models. Examples are vertical and horizontal distribution, age and growth, length-frequency composition, reproduction and the relationships, for example through food webs, to other species which co-occur with target species.

The environmental and biological differences between the slope and the shelf can be quite large, and, as a consequence, the slope fish fauna is rather different from the fauna of the shallower shelf. In this paper we explore some of these differences by means of a series of questions and answers. The perspectives we bring to bear on these questions converge from studies of taxonomy and systematics, biological oceanography, deep ocean species and community ecology, and fisheries. Many of the specific examples we make come from long-term studies of the Rockall Trough (Mauchline 1986, 1990) and the Porcupine Seabight (Rice et al. 1992). These regions are, in fact, among the very few where full and broad-scale multi-disciplinary investigations of the ocean have been carried out from shallow areas to the very greatest depths and over several seasons and years. But we are also able to call on a sufficient body of information from other parts of the Atlantic (e.g. Haedrich et al., 1975; Haedrich and Krefft, 1978, Merrett and Marshall, 1981; Merrett and Domanski, 1985) so that we are not uncomfortable with drawing general conclusions.

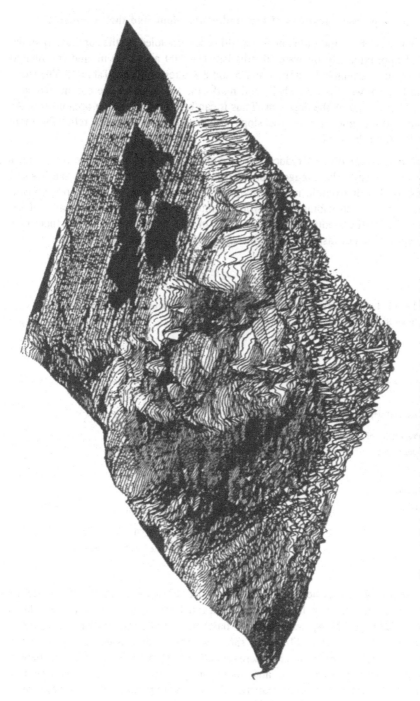

Figure 1. Bathymetric chart of the Rockall Trough region.(modified from a chart by the British Oceanographic Data Centre)

2. How do the physical features of the continental slope and shelf compare?

The proportions of the total area of the world ocean occupied by the continental slope, delineated approximately by water depths between 200 and 2000 m, and continental shelf (<200 m) are relatively similar at 7.5 and 8.8% respectively (Table 1). The slope is generally narrower than the shelf, and marks the true edge of the continental land masses and the edge of the deep-sea. Thus it can be said to form the oceanic rim. As the name implies, gradients on the slope are much steeper and the relief far more accentuated than the shelf.

These properties combine to reduce the accessibility of bottom trawl fishing grounds on the slope. Despite the exaggeration of the vertical scale, the chart of the Rockall region in the North Atlantic makes this point forcibly (Figure 1). The topography is varied and steep, with submarine canyons common features. The fishable area of the slope can be further reduced because many of the target species have their own discrete vertical depth distributions that may occupy only a part of the slope.

TABLE 1. Comparisons of some physical features of the continental shelf and slope (from Sunderman 1986)

	SHELF	SLOPE
Nominal depth distribution	0-200 m	200-2000 m
Proportion of total world ocean area	7.5%	8.8% (200-1000 m:4.4% 1000-2000 m:4.4%)
Gradient	<1:1000	>1:40 (30-60)
Width	Few>300 km.	Few to 150 km.
Relief	20 m	Locally 2000 m. (associated with canyons)

Twenty years ago fishermen would have avoided trawling on many of the steeper areas of the slope where both the gradient and relief are far more pronounced than on the shelf (Table 1). Now, with satellite navigation and track plotters, it has been possible to build up a chart library of safe tows that enable fishermen to repeatedly avoid all the known obstacles. This increased efficiency, with less time being spent on searching, repairing trawls etc. and more on productive fishing, has important implications for catch and effort statistics that are so important for stock assessment. Because the technology is changing fast, care must be taken in interpreting apparent changes in stock size which could result as easily from this increased efficiency as from biological causes.

For those slope species which can be attracted to bait, longline or trap fisheries are an alternative to trawling on difficult ground (Hareide, this volume; Jørgensen, this volume). In the eastern Atlantic longlines are already widely used on the upper slope to catch species such as ling (*Molva molva*), blue ling (*Molva dypterygia*) and tusk (*Brosme brosme*). But this technique is very selective and even in close relatives, for example roughhead (*Macrourus berglax*) and roundnose (*Coryphaenoides rupestris*) grenadiers, one species will take a bait while the other will not. Deep-water trap fisheries are not well developed in the Atlantic but there is a potential for the development of specialised fisheries for species such as the deep-water red crab (*Chaceon affinis*, formerly *Geryon affinis*) (Reinert, this volume). Another static gear which can be used on ground unsuitable for bottom trawling is the gillnet. These nets have been used on the Portuguese slope to catch deep-water sharks and by Spain in the deep water to the west of Hatton Bank to catch monkfish (*Lophius piscatorius*).

3. How does the physical environment of the slope differ from that of the shelf?

Because most studies of the continental slope have been initiated by oceanographers with general and varied interests in that area (e.g. Rowe and Haedrich, 1979), there is a considerable amount of information on the environment and on the processes. In particular, one of the longest time series of hydrographic studies in slope areas has taken place in the Rockall Trough, northeastern Atlantic (Ellett et al., 1986), and in this section we cite the Rockall results as broadly representative of a deep-sea area.

As a general rule temperature decreases with depth but the rate of change can vary from area to area. There is generally a well defined summer thermocline to depths of about 500 m which breaks down during the winter and early spring (Figure 2). This deep, winter mixing causes a redistribution of nutrients that are of vital importance to the phytoplankton bloom in the spring. In most higher latitude parts of the ocean, the burst of primary production, the so-called spring bloom, can be very dramatic (Prasad and Haedrich, 1993, 1994) and the growth achieved at that time then drives the system, including the fisheries, for the rest of the year.

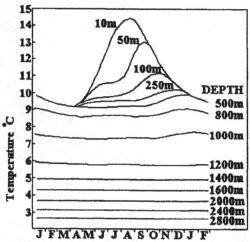

Figure 2. Temperature-depth profiles, winter and spring, Rockall (after Mauchline, 1990)

6

Almost all the food energy reaching the seabed on the slope is derived from this surface production.

Below 500 m the water temperature continues to decline gradually but the annual variation in most places is negligible (<0.5°C). The Rockall Trough is separated from the Norwegian Basin by a number of underwater ridges, such as the Wyville Thomson Ridge between Shetland and the Faroe Islands (Figure 1). At depths down to the top of the ridge the temperature regime is similar on either side of the ridge. Below the sill depth of the ridge the temperature decreases gradually to the west but to the east the temperature decreases rapidly to below 0°C (Figure 3). Not unexpectedly, these differing temperature regimes have a profound effect on the composition, abundance and biomass of the fish fauna. In the Rockall Trough species diversity and biomass peak at between 1000 and 1500 m whereas in the Norwegian Basin diversity, abundance and biomass decrease rapidly below about 500 m (cf. Figure 7 below). In one study off the coast of Norway the biomass at 1000 m decreased to 1% of that on

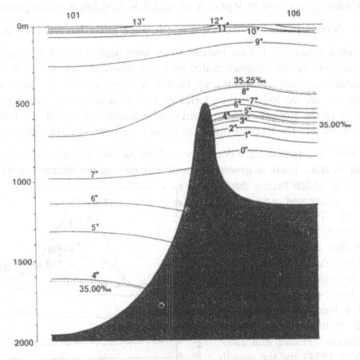

Figure 3. Temperature structure either side of the Wyville-Thomson Ridge.

the upper slope (400 to 500 m) and the numbers of species decreased from 10 to 3 (Bakken et al.,1975). There was also a change with depth from a boreal fauna similar to that in the Rockall Trough to a boreal-arctic/arctic fauna (Bergstad and Isaksen, 1975); a similar clear faunal transition occurs across the ridge at the Davis Strait between the Arctic Ocean and the Irminger Sea (Haedrich and Krefft, 1978).

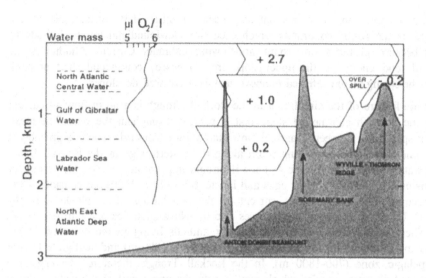

Figure 4. Oxygen profile, water masses and circulation patterns, Rockall (after Mauchline, 1990).

Annual changes in salinity on the slope are generally small and are not likely to affect the fishes living there. Long term changes in the surface salinity of the Rockall Trough are well documented and can be related to changes in the distributions of the water masses that constitute the Atlantic Ocean (Ellett et al., 1986). One such change was the so called 'Great Salinity Anomaly' which was characterised by a period of unusually low surface salinity in the central Rockall Trough in the mid 1970s. Although this anomaly was described for surface waters its effect can be detected in the deeper water (Ellett, 1993).

In general, oxygen levels are close to saturation in continental slope waters. Oxygen concentrations change with depth in the Rockall Trough (Figure 4) and the lowest values at about 1000 m are indicative of Gibraltar or Mediterranean Water, which originates from the Straits of Gibraltar and, because of its high salinity and low oxygen, is detectable over a wide area of the Atlantic. Distinctive water types such as this can help to describe the distribution of certain oceanic animals. The black scabbard fish (*Aphanopus carbo*), for example, seems to be associated with the Mediterranean Water in the deep Atlantic (Martins and Ferreira, this volume).

The patterns of temperature, salinity, and oxygen distribution in the ocean result from movements by the oceanic currents and circulation. The gradients so established, combined with the depth-related diminution of light and increase of pressure, are the main environmental characteristics of the continental slope. Because of the linkages established by the circulation of water, conditions at any one place can be understood only by reference to situations that may occur at other, quite distant, places. An example is the Arctic conditions which occur off Newfoundland (at the latitude of Paris) because of the strong southward-flowing Labrador Current. Another is the input of food energy to the upper slope from a nearby continental shelf or bank where the spring bloom referred to above may be most well-developed.

The main feature of the circulation in the Rockall Trough is a north flowing current along the Scottish continental slope that is present throughout the year. The mean current speeds are between 3 and 30 cm/s increasing northwards. There is probably a consistent counter current going south along the eastern edge of the Rockall Bank. Mass water movements (Figure 4) have important implications for the life history of organisms that produce pelagic eggs and larvae. Mauchline (1990) has considered the influence of the Scottish slope current on the populations of zooplankton in the Rockall Trough and estimated the losses into the Norwegian Sea and northern North Sea. These were considered to be large for animals living in the epipelagic zone (0-500 m), moderate for the mesopelagic zone (300 - 1000 m) and negligible for the bathypelagic zone (700-1800 m). In the Rockall Trough, therefore, the eggs and larvae of deep-water species which may occupy the upper part of the water column will have a greater chance of being dispersed into areas that may be unsuitable for their further development than those remaining close to the adult living depth.

The deep-sea, at least at depths below about 500 m, is an environment of relatively broad physical constancy and yet many of the animals, both fish and invertebrates, appear to have seasonal cycles of reproduction or growth (Tyler, 1988). Two parameters which might be implicated in the synchronisation of these cycles are diurnal tidal variations or annual variations in deep-ocean currents. The seasonal signal from the sinking of the spring bloom must play an important role.

4. How do the demersal species assemblages on the shelf and slope differ from one another and are the latter basically different from the pelagic oceanic assemblage?

Almost 100 years ago Woodward (1898) wrote of the antiquity of the deep-sea fish fauna, declaring that 'those out-of-date forms of life which can no longer compete with the vigorous shore-dwelling races, are compelled to retreat to the freshwaters on the one hand, or to the deep-sea on the other'. While the subject has been addressed in general terms subsequently (e.g. Andriyashev, 1953), recent work has indicated that enzyme function under the high pressure and low temperature of the deep-sea environment may preclude shelf species from extending beyond depths of about 500 m (Somero et al., 1983). Thus, while permanent residents of the upper 400 m or so of the open ocean share the same ancestral affinities as shelf-dwelling fishes, the true deep-sea fish fauna has a quite separate composition (Merrett, 1994). This work has shown that the North Atlantic Basin overall contains at least 1094 deep-sea species, representing 143 families and 25 orders. There are rather more pelagic species (589) than demersal (505), distributed among 93 pelagic and 72 demersal families.

Dividing the pelagic ichthyofauna into two zones on the basis of their ancestral affinities (see above, 0-400 m and 400+ m) and considering the demersal zone as one, there is very little overlap in family distribution among them (Table 2). More than three-quarters of the families are restricted to a single zone, with only the Gempylidae (scabbard-like fishes) represented in all. Members of this family are often pseudoceanic (e.g. see Merrett 1986) and therefore concentrated around the oceanic rim and oceanic islands (Nakamura and Parin 1993), where zonal intermixing is most likely. Three families occur in both the upper pelagic and the demersal zones, while 13 are common to the lower pelagic and demersal. There is a striking contrast in species representation by order between the lower pelagic and demersal assemblages which, taken together, constitute the true deep-sea fauna (Figure 5a).

TABLE 2. Vertical distribution of oceanic fishes by family and species in the North Atlantic Basin: (a) subdivided by major zones with the proportion of families restricted to each indicated (b) subdivisions of the demersal zone. (From Merrett, 1994)

ZONE	DEPTH (m)	TOTAL No. OF FAMILIES	TOTAL No. OF SPECIES	PROPORTION RESTRICTED TO ZONE
(a)				
Pelagic	0-399	28	80	89%
Pelagic	>400	66	509	79%
Demersal	All	72	505	78%
(b)				
Demersal	200-399	30	74	
	400-1999	58	347	
	2000-3999	14	64	
	>4000	7	20	

Species of the Myctophiformes (lantern fishes) (order no.12; 17% zonal frequency) are the only representatives of the lower pelagic assemblage to have been commercially exploited hitherto (Gjøsæter and Kawaguchi, 1980). A far greater diversity among orders represented is evident from the demersal assemblage when compared with the lower pelagic assemblage (22 vs 13). Commercially exploited species are more widespread among the demersal assemblage also, but still remain a small proportion of the total species richness. They are represented by the squaliform (no.3; 6% - e.g. Squalidae: Portuguese dogfish (*Centroscymnus coelolepis*)), rajiform (no.5; 6% - e.g. Rajidae: Arctic skate (*Raja hyperborea*), gadiform (no.13; 19% - e.g. Macrouridae: roundnose grenadier (*Coryphaenoides rupestris*), roughhead grenadier (*Macrourus berglax*); Gadidae: blue ling (*Molva dypterygia*), greater forkbeard (*Phycis blennoides*); Moridae: mora (*Mora moro*)), lophiiform (no.15; 1% - e.g. Lophiidae: monkfish, (*Lophius piscatorius*)), beryciform (no.18; 3% - e.g. Trachichthyidae: orange roughy (*Hoplostethus atlanticus*)), scorpaeniform (no.21; 8% - e.g. Scorpaenidae: redfish (*Sebastes mentella*)) and pleuronectiform (no.23; 4% - e.g. Pleuronectiformes: Greenland halibut (*Reinhardtius hippoglossoides*)) species.

Differences in species representation by order in the demersal assemblage increases with depth (Figure 5b). The frequency of ordinal representation increases from 15 orders in the upper slope zone (200-399 m - a quasi-shelf zone) to 21 orders in the mid-slope zone (400-1999 m). Representation is then much reduced in the rise (2000-3999 m; 11 orders) and abyssal zones (4000+ m; 6 orders). Some of this reduction arises from the evident absence of the Agnatha (no.1) and chondrichthyans (no.2-6, with the exception of a small representation of Rajiformes (benthic - no.5) on the continental rise) at the two lower levels. Furthermore no demersal Anguilliformes (no.7) nor Notacanthiformes (no.8) are found predominantly at abyssal depths. Species' representation within orders is also variable among zones. The percentages of the benthic orders, Rajiformes (no.5; 12.5%), Scorpaeniformes (no.21; 16.7%) and Pleuronectiformes (no.23; 15.2%) are higher on the upper slope than elsewhere. The Squaliformes (no.3; 8.1%) are concentrated in the mid-slope zone, where the Anguilliformes (no.7; 7.6%) and the Gadiformes (no.13; 22.1%) peak also. The Salmoniformes (no.9; 10.9%) reach their highest proportion in the rise zone, where the Ophidiiformes (no.14; 35.9%) and the Scorpaeniformes (no.21; 14.1% - all family Liparididae) are also important. While the numbers of species centred in the abyssal zone are low (20), the Ophidiiformes (no. 14; 50%), Aulopiformes (benthic - no.11; 20%) and Beryciformes (no.18; 10%) all assume their highest proportions at this level.

The higher species diversity found on the continental slope has an important implication for fish community structure and ultimately, therefore, for the fisheries there. By and large, commercially important species tend to occur in low diversity communities where they make up a significant proportion of the total community biomass (Haedrich, this volume). This is the case in many continental shelf communities, especially in higher latitudes. But when more species are present, as is the situation on the continental slope, dominance by one or a very few species is reduced. Unless the gears applied are very selective, any fishery in that area will therefore take a mixture of species as a matter of course. Processing time will no doubt be increased and discards are likely to be high.

5. How do the basic distribution patterns (vertical and horizontal) of slope-dwellers compare with their shelf-dwelling counterparts?

There is considerable variation in the vertical range of distribution of slope species. Some like the deep-sea eel *Synaphobranchus kaupi* in the Rockall Trough have a wide

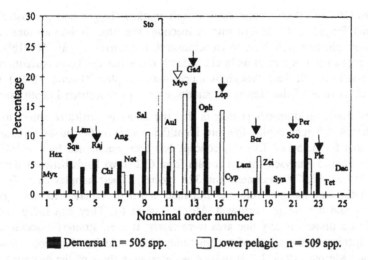

Figure 5a. Species representation by order in the deep Atlantic- lower pelagic vs demersal orders

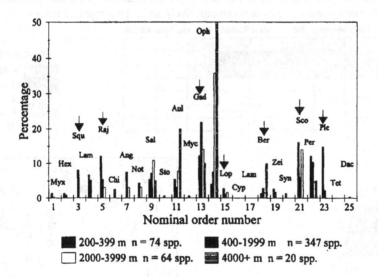

Figure 5b. Species representation by order in the deep Atlantic- demersal orders.

Arrows indicate orders in which exploited species occur. 1, Myxiniformes; 2, Hexanchiformes; 3, Squaliformes; 4, Lamniformes; 5, Rajiformes; 6, Chimaeriformes; 7, Anguilliformes; 8, Notacanthiforme s; 9, Salmoniformes; 10, Stomiiformes; 11, Aulopiformes; 12, Myctophiformes; 13, Gadiformes; 14, Ophidiifor mes; 15, Lophiiformes; 16, Cyprinodontiformes; 17, Lampriformes; 18, Beryciformes; 19, Zeiformes; 20, Syngnathiformes; 21, Scorpaeniformes; 22, Perciformes; 23, Pleuronectiformes; 24, Tetraodontiformes; 25, Dactylopteriformes. (From Merrett, in press.)

12

depth range (500 to 2000 m) and show a well marked "bigger deeper" distribution (Gordon and Bergstad, 1992). (In this connection, the broader implications of the bigger-deeper phenomenon have been discussed in Merrett et al. (1991b)). The roundnose grenadier has almost as broad a depth range but the size distribution with depth, at least in the Rockall Trough, is much more complex (Gordon, 1979a). Other species, such as some of the deep-water sharks, have a more restricted depth range.

A recurring theme in deep-sea studies is that the fauna is vertically zoned to form communities at different depths that are identifiable over rather broad geographical areas. At least for demersal fish this concept has been questioned by Haedrich and Merrett (1990) who analysed the catch data from 692 trawl hauls between depths of 204 and 5345 m (96,779 specimens and 325 species). Their samples came from the Atlantic continental margin from the Bahamas north to Canada, Iceland and the British Isles and around to northwest Africa (Figure 6). They concluded that the demersal fishes present in any one area were rarely, if ever, strongly associated with any other species in the sense of a community that could be identified elsewhere. Haedrich and Merrett (1988, 1990) in their wide ranging study of the demersal fishes of the North Atlantic (Figure 6) could find very little evidence to support continuity in horizontal zones. Many species had comparatively small geographical ranges.

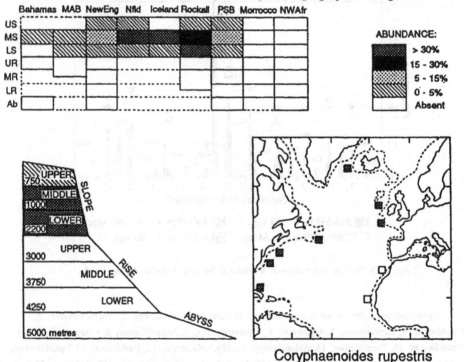

Figure 6. Survey sites for deepwater fishes, upper slope to abyssal depths. The example shown is for *Coryphaenoides rupestris* (from Haedrich and Merrett, 1988)

A knowledge of the vertical distribution of all stages in the life history is essential for the rational exploitation of deep-water species. The equivalent of an inshore nursery area for a shelf species is likely to be a particular depth stratum for a slope-dweller. For example the best catch rates of adult roundnose grenadier in the Rockall Trough would be between 1000 and 1250 m but fishing at these depths might destroy many of the juvenile fish that are also most abundant there. Although juveniles might escape through the meshes of the trawl they will probably suffer a high mortality because of scale loss. Deep-water species have delicate tissues generally and are not well endowed with mucus. Thus they suffer considerable scale loss and skin damage in the trawl. Even though small fish may escape through the meshes, there is likely to be a high mortality among those juvenile escapees of target species which could have important implications for recruitment. Most of the target species are piscivores and a high mortality of potential prey species that are of small adult size could have important implications for the deep-water food webs.

Because there are so many species with overlapping depth ranges it will be impossible to control catches by varying the depth of fishing and fishermen will trawl the depths which will maximise the catch-rates of the target species irrespective of the consequence to other species. For the purposes of assessment it is important to understand the depth range and life history of all the species. Many assessments for deep-water species carried out using data from commercial vessels are of limited value because the whole depth range has not been sampled and is sometimes unknown and the trawls are selective for larger fish.

The vertical distribution of deep-water species in the water column is poorly understood because large midwater trawls have seldom been used in deep-water (Merrett et al., 1986). The reason for the apparently relatively low abundance of juvenile stages compared with the adults of many species such as deep-water sharks and the black scabbard fish (*Aphanopus carbo*) in bottom trawl catches is probably that, in their early life history, they live in midwater and are unavailable to demersal trawls which catch fish which live at only a few meters off the bottom. Diurnal vertical migrations have been described also for some commercially exploited deep-water species, such as roundnose grenadier (Savvatimskii, 1988). The orange roughy (*Hoplostethus atlanticus*) in New Zealand waters can be found at distances of at least 50 m off the bottom, mainly associated with spawning aggregations (Clark and Tracey, 1992).

6. How does the vertical distribution pattern of fish biomass correlate with the trophic input to the oceanic environment?

The vertical distribution of demersal fish biomass on the slope has been shown to peak at around 1000 m in the eastern North Atlantic and thereafter decrease rapidly with depth (Figure 7; Merrett et al, 1991a,b; Haedrich and Merrett, 1992; Gordon and Bergstad, 1992). The food supply for these demersal fishes ultimately originates from surface production, since virtually all trophic input to the ocean is derived from solar energy.

14

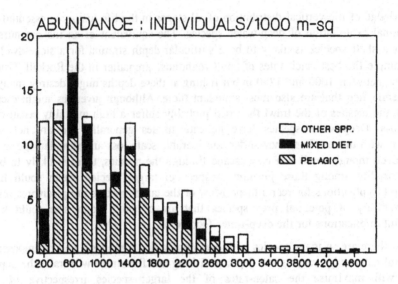

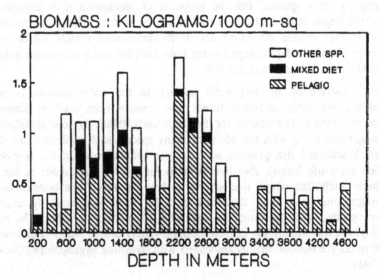

Figure 7. Abundance and biomass of dominant species by depth in Porcupine Seabight. Feeding type, whether essentially pelagic or mixed (demersal/pelagic) diet is also indicated(from Haedrich and Merrett 1992)

Thus vertical profiles of oceanic macroplankton and micronekton (Angel and Baker, 1982; Mauchline, 1990) and benthic biomass (Lampitt et al. 1986) offer a measure of relative food availability to the demersal ichthyofauna and each shows a decrease in biomass with depth. An increase in primary production along the shelf-slope break (e.g. Holligan and Groom, 1986), shelf production and terrigenous input (e.g. Walsh et al. 1981), slope currents and tidal effects (e.g. Huthnance, 1986) and the impingement of pelagic organisms, both horizontal and vertical, around the oceanic rim (Marshall and Merrett, 1977; Gordon and Mauchline, 1990; Mauchline and Gordon, 1991) are likely to be among the factors that allow an increase in the fish biomass to depths of 1000 m or more in this area.

The seabed acts as the ultimate sink for oceanic primary production. Food particles of all sizes sink and accumulate there. Yet, while this offers the diverse deepwater fauna far greater food supplies than are encountered in the overlying midwaters, deep demersal fish biomass has been found to be similar to invertebrate benthic biomass (cf. Lampitt et al. 1986 with Merrett,1987). If the fishes depend on this benthic biomass for food, then only high turnover rates in the benthos could support the demersal fish populations, but there is no evidence of such rates (Gage, 1991). Demersal fish must therefore obtain much of their trophic input from other sources. As mentioned above, the impingement of mesopelagic organisms on the slope provides such a source, and many feeding studies have shown that these organisms contribute largely to the diet of slope demersal fishes. In a detailed study of the diets of the demersal fishes of the Rockall Trough (Mauchline and Gordon, 1991 and references therein) it was shown that between about 800 and 1300 m depth over 80 % of the demersal fish had consumed pelagic prey. Merrett (1986) found reports of 24 different species of meso- and bathypelagic fishes in the stomachs of 34 species of demersal fishes in the Atlantic.

Deepwater continental slope fishes feeding on pelagic organisms from the scattering layers is another example of the way in which the situation in one part of the ocean influences matters elsewhere. A common strategy among scattering layer animals is to migrate into the very surface layers at night to feed, and then to retreat at daylight into the darker deeper layers. This habit means that food chains that could link slope fish species to surface contamination are probably not at all unusual. Just because the continental slope is somewhat remote, there is no reason to consider it a "pure" environment. Contaminants of many sorts, including the heavy metals that accumulate through food webs are just as likely to be found in the deep ocean as on the shallow shelf. Indeed, because the deepwater fishes may feed higher on the food web than do shallow water fishes, the likelihood of finding metals in high concentrations is increased. For example in the relatively unpolluted waters off New Zealand the deep-water shark *Deania calcea* had mercury levels ranging from 1.36 - 2.65 ppm which is greater than permissible limit of 0.5 ppm (Clark and King, 1989).

In higher latitudes food input at the base of the food chain as a consequence of the spring bloom can be very seasonal (Prasad and Haedrich, 1993). Dispersal and storage through the food webs tend to spread this out over the year, with different fishes deriving their energy according to individual feeding and foraging strategies. But the important point is that, no matter where in the deep sea, the basic food comes from the thin and shallow productive layer at the surface, and that the overall abundance of food, again no matter where, tends to decline with the logarithm of the depth, i.e. by orders of magnitude (see above). On an area basis, the continental slope itself is thus incapable of sustainably supporting enormous stocks of fish.

7. What is known about deep-sea fish population structure and breeding biology?

A common feature of many deep-sea fish and invertebrate populations is that they can have markedly bimodal length/frequency distributions, with a peak of juveniles clearly separated from a broader peak of adults. Such a distribution will arise from a stacking of the older year classes of fish resulting from the exponential decrease in growth rate with age. Gage (1991) has used computer models to simulate these distributions in deep-sea invertebrates. In an exploited shelf species this stacking of the older year classes is seldom observed because the older fish are selectively removed by the fishery. Exploitation of a deep-sea population would be expected to reduce the peak of older fish with a consequent decrease in catch per unit effort. The effective elimination of a high proportion of the older fish from a population has several important consequences.

If it is assumed that the carrying capacity of the deep-sea is limited by food availability, then the removal of the older fish of a target species will free energy for others in the ecosystem. (It is assumed that these older fish continue to feed at a similar rate but have diverted energy from somatic growth to reproduction.) While it is possible that the surplus energy might enhance the growth rate of the recruiting cohorts it is more likely, in view of the considerable overlap in food preferences (Mauchline and Gordon, 1985), that other species might be the beneficiaries.

Many exploited deep-sea species mature only at a large size and/or age, for example orange roughy at 20 -25 years (Clark this volume); roundnose grenadier in the Atlantic up to 40-50 cm total length (Atkinson, this volume). High fishing mortality on the reproductively active, normally long-lived adults could seriously reduce the recruitment as has been recognised for the orange roughy fishery of New Zealand by Pankhurst and Conroy (1987). While this is a problem of species breeding annually, as among shelf dwellers, the situation would be compounded in species which have more protracted breeding cycles. There is evidence to suggest that in some deepwater species females may not have the resources to be able to breed every year. In *Coryphaenoides armatus*, a large, highly fecund grenadier of the continental rise and abyssal plain, reproduction may occur only once, at the end of a supposedly long life (Stein 1985). In the orange roughy some 45% of adult females at a major spawning aggregation site off eastern Tasmania were non-reproductive, suggesting that scarcity of food or the cost of joining a spawning aggregation may be the cause of intermittent spawning in this species (Bell et al., 1992). Such temporal limitations may be greater among species of large adult size feeding at high trophic levels. Evidence from the Rockall Trough (Gordon, 1979b) and the Porcupine Seabight (Gordon and Merrett, unpublished data) suggest that those species of smaller adult size and depending mainly upon epibenthic invertebrate fauna for food, are less prone to such limitations in sequential breeding. Because of the general logarithmic decline in food resources, semelparity (spawning once in a lifetime) is a possible reproductive strategy for many slope species, particularly the larger ones.

In long-lived shelf fishes, such as the gadids, females become mature while they are still in the fast growth stage and continue to spawn annually into old age. Our observations on slope dwelling fishes in the Rockall Trough and the Porcupine Sea Bight suggest that in many species females become mature only after they reach adult size when somatic growth has slowed down or ceased. This is illustrated for the roundnose grenadier (*Coryphaenoides rupestris*) where the length at maturity of females as determined by the gonosomatic index (Figure 8a) corresponds to the start of the second peak in the bimodal length distribution (Figure 8b). This would seem to imply that slope-dwellers have a choice imposed upon them; energy is available for either growth or reproduction, but not both. Hence growth essentially ceases at adulthood and resources are channelled into reproduction.

All marine bony fish (teleosts) display essentially one of two fundamental fecundity patterns. A small proportion have a low relative fecundity with large eggs which develop into small precocious juveniles. The majority of species, however, have a high relative fecundity with small eggs (<2 mm diameter, e.g. Duarte and Alcaraz, 1989, Ware, 1975) which develop as small pelagic larvae. Indeed, Duarte and Alcaraz (1989), suggested that colonization of the oceanic environment involved the evolution of such pelagic eggs. These patterns tend to be phylogenetically fixed within major taxa, to constrain diverse adaptive accommodation to the deep-sea environment.

Merrett (1994) investigated the breeding biology of deep demersal fishes of the North Atlantic Basin. He examined the relationship between maximum species size and maximum fecundity, as a measure of the potential reproductive effort per species. Owing to the sparseness of information generally, data from only 19% of the 505 demersal species could be incorporated. These included 5 species of cod-like fishes (Gadiformes - roundnose grenadier, *Coryphaenoides rupestris*; roughhead grenadier, *Macrourus berglax*; hake, *Merluccius merluccius*; blue ling, *Molva dypterygia*; mora, *Mora moro* - Figure 9) and 7 species of miscellaneous groups (angler, *Lophius piscatorius*; silver roughy, *Hoplostethus mediterraneus*; redfish, *Sebastes marinus*; witch, *Glyptocephalus cynoglossus*; American plaice (long rough dab), *Hippoglossoides platessoides*; halibut, *Hippoglossus hippoglossus*; Greenland halibut, *Reinhardtius hippoglossoides* - Fig. 10), which are currently commercially exploited in the North Atlantic. In accordance with the trend outlined above, all produce small eggs which are presumed to be pelagic. These exploited species are all forms of large adult size with maximum fecundities in excess of 32,000 eggs (Fig. 9, 10).

Otherwise, very little is known of the biology and distribution of deep-water fish eggs and larvae and this might be because they remain close to the bottom, a region that has been poorly sampled. Discussion has centred around the grenadier family (Macrouridae) whose species exceed 300 in number and largely dominate in abundance in slope and rise waters (Marshall and Iwamoto, 1973). Free eggs of less

18

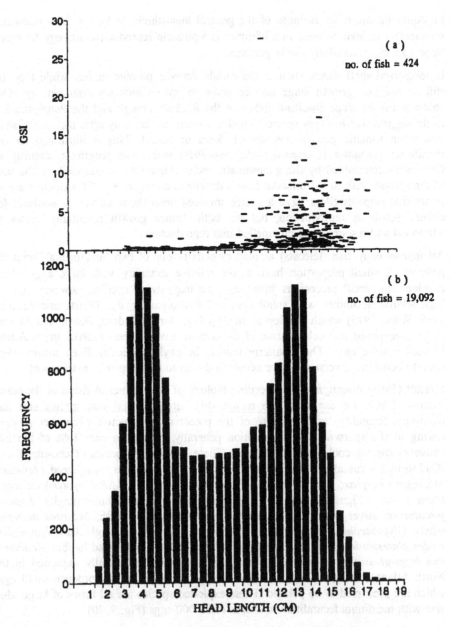

Figure 8. Frequency distributions for *Coryphaenoides rupestris*. a: gonosomatic index. b: length.

19

than 10 species have been reported (Merrett and Barnes, unpublished data) and the larvae of only about double that number have been described. Marshall and Iwamoto's (1973) view of development of macrourid eggs was that their buoyancy took them up into the food-rich waters near the thermocline to hatch. The young grew there and later descended to their adult living depth. In contrast, demersal development of at least some macrourid species has been postulated (Merrett, 1989: Merrett and Barnes, in prep.). Recent studies in the Skaggerak have shown that the eggs and larvae of one species, the roundnose grenadier (*Coryphaenoides rupestris*), appear to remain deep (Bergstad and Gordon, 1994). Similar knowledge in other bony species is equally sparse, although unpublished reports on the orange roughy (*Hoplostethus atlanticus*) have described the initial ascent of eggs to within 200 m of the surface followed by their sinking to about 100-200 m off the bottom where they hatch (M. Clark, pers.comm.).

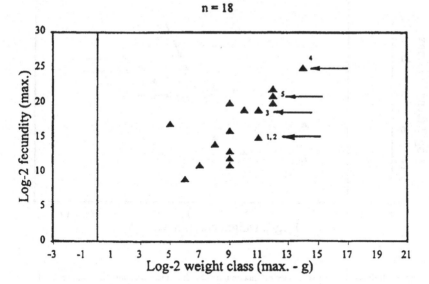

Demersal: Gadiformes

n = 18

Figure 9. Relationship between species' maximum fecundity and maximum size among the demersal oceanic representatives of the order Gadiformes of the North Atlantic Basin (log₂ groupings). (Arrowed species are 1) roundnose grenadier, *Coryphaenoides rupestris*; 2) roughhead grenadier, *Macrourus berglax*; 3) hake, *Merluccius merluccius*; 4) blue ling, *Molva dypterygia*; 5) mora, *Mora moro*.) (After Merrett, 1994)

20

More is known about the reproduction of deep-water sharks most of which belong to the family Squalidae, the spiny sharks. This group are live-bearers, producing large, free living young as top predators in the food web. Yet their fecundity is low (usually only up to about 30 young (Compagno, 1984a; 1984b; 1990)) and, from what is known, their gestation period is long (e.g. Yano and Tanaka, 1988; Capape, 1985). Such a reproductive style renders them particularly vulnerable to excessive exploitation.

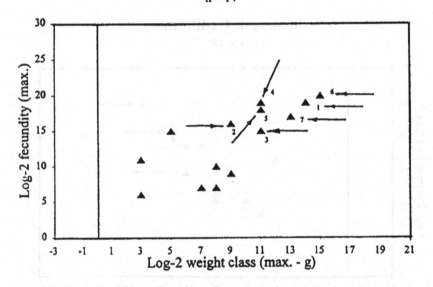

Demersal: miscellaneous species

n = 14

Figure 10. Relationship between species' maximum fecundity and maximum size among the demersal oceanic representatives of the orders Lophiiformes, Beryciformes, Syngnathiformes, Scorpaeniformes and Pleuronectiformes of the North Atlantic Basin (log₂ groupings). (Arrowed species are 1) angler, *Lophius piscatorius;* 2) silver roughy, *Hoplostethus mediterraneus;* 3) redfish, *Sebastes marinus;* 4) witch, *Glyptocephalus cynoglossus;* 5) American plaice (long rough dab), *Hippoglossoides platessoides;* 6) halibut, *Hippoglossus hippoglossus;* 7) Greenland halibut, *Reinhardtius hippoglossoides).* (After Merrett, 1994)

8. Conclusions

Despite the fact that the areas of ocean bottom on the shelf and slope are roughly comparable, the readily accessible portions of the slope are far more restricted. This is due not just to the depth involved, but also to the complex topography and lack of flat areas. The system is a layered one, with strong gradients of many kinds associated with the change in depth, and oceanographic influences from regions far away make themselves felt quite locally.

Because the physical environment is layered, biological phenomena on the slope display vertical zonation patterns as well. Food availability, which declines logarithmically according to depth and which may also be seasonal in nature, is one of the most critical factors. Food availability leads to zonation in the fauna, with different species arranged in differential patterns of abundance and range across the slope. The species most characteristic of the slope are different from those of the shelf, and are far more diverse although individually less abundant.

The distinctiveness of the slope fauna extends to life histories, including especially patterns of reproduction as far as is known. For the great majority of species, the earlier life history stages are as yet unknown, and even the places where they occur in the ocean remain to be discovered. Many questions are still unanswered about deepwater fishes including the small handful (out of the total present) that might have commercial potential.

The adaptations of slope fishes appear to stem mostly from the low food supply. Thus their growth (as confirmed by studies on ageing), population regeneration, and time for recovery from an impact such as heavy fishing mortality will all be very slow. The overall community impact of such disturbance cannot be predicted.

Acknowledgements

Two of the authors (JDMG and NRM) acknowledge the support of EU MAST 2 Contract No. MAS2-CT920033 for some of the data included in this paper.

22

References

Andriyashev, A.P. (1953) Ancient deep-water and secondary deep-water fishes and their importance in a zoogeographical analysis, in Notes on special problems in Ichthyology. Akad. Nauk SSSR Ikhtiol. Kom., 58-64, Moscow.

Angel, M.V. and Baker, A.de C. (1982) Vertical distribution of the standing crop of plankton and micronekton at three stations in the Northeast Atlantic. Biological Oceanography, **2**, 1-30.

Atkinson, D.B. (This volume) The biology and fishery of roundnose grenadier (*Coryphaenoides rupestris* Gunnerus, 1765) in the Northwest Atlantic, in A.G. Hopper (ed.), Deep Water Fisheries of the North Atlantic Oceanic Slope (proceedings of the NATO Advanced Research Workshop, March 1994), Kluwer, Dordrecht, The Netherlands.

Bakken, E., Lahn-Johannesson, J. and Gjøsæter,J. (1975) Bunnfisk på den norske kontinentalskraning. Fiskets Gang, **61**, 557-565.

Bell, J.D., Lyle, J.M., Busman, C.M., Graham, K.J., Newton, G.M., Smith, D.C. (1992) Spatial variation in reproduction and occurrence of non-reproductive adults, in orange roughy, *Hoplostethus atlanticus* Collett (Trachichthyidae), from south-eastern Australia. Journal of Fish Biology, **40**, 107-122.

Bergstad,O.A. and Gordon, J.D.M. (1994) Deep-water ichthyoplankton of the Skagerrak with special reference to *Coryphaenoides rupestris* Gunnerus, 1765 (Pisces, Macrouridae) and *Argentina silus* (Ascanius, 1775) (Pisces, Argentinidae). Sarsia, **79**, 33-43

Bergstad, O.A. and Isaksen, B. (1987) Deep-water resources of the northeast Atlantic: distribution, abundance and exploitation. Fisken og Havet, 1987 (3), 1-56.

Capape, C. (1985) Nouvelle description de *Centrophorus granulosus* (Schneider, 1801) (Pisces, Squalidae). Données sur la biologie de la reproduction et le régime alimentaire des spécimens des côtes tunisiennes. Bulletin de L'Institut National Scientifique et technique d'Oceanographie et de Peche de Salammbo, **12**, 97-141.

Clark, M. (This volume) Experience with management of orange roughy (*Hoplostethus atlanticus*) in New Zealand waters, and the effects of commercial fishing on stocks over the period 1980-1993, in A.G. Hopper (ed.), Deep Water Fisheries of the North Atlantic Oceanic Slope (proceedings of the NATO Advanced Research Workshop, March 1994), Kluwer, Dordrecht, The Netherlands.

Clark, M.R. and King, K.J (1989) Deepwater fish resources off the North Island, New Zealand: results of a trawl survey, May 1985 to June 1986. New Zealand Fisheries Technical Report, No 11, 56pp.

Clark, M.R. and Tracey, D.M. (1992). Trawl survey of orange roughy in southern New Zealand waters, June-July 1991. New Zealand Fisheries Technical Report, No.32, 27pp.

Compagno, L.J.V. (1984a) FAO species catalogue, Vol.4. Sharks of the world. An annotated and illustrated catalogue of shark species known to date. Part 1. Hexanchiformes to Lamniformes. FAO Fisheries, Synopsis, 125, 1-249.

Compagno, L.J.V. (1984b) FAO species catalogue, Vol.4. Sharks of the world. An annotated and illustrated catalogue of shark species known to date. Part 2. Carcharhiniformes. FAO Fisheries Synopsis, 125, 251-655.

Compagno, L.J.V. (1990) Alternative life-history styles of cartilaginous fishes in time and space. Environmental Biology of Fishes, **28**, 33-75.

Duarte, C.M. and Alcaraz, M. (1989) To produce many small or few large eggs: a size-dependent reproductive tactic of fish. Oecologia, **80**, 401-404.

Ellett, D.J, Edwards, A. and Bowers, R. (1986) The hydrography of the Rockall Channel - an overview. Proceedings of the Royal Society of Edinburgh, **88B**, 61-81.

Ellett, D.J. (1993) Transit times to the NE Atlantic of Labrador Sea water signals. International Council for the Exploration of the Seas; Council Meeting, C.M.1993/C:25, 11 pp.

Gage, J.D. (1991) Biological rates in the deep-sea: a perspective from studies on processes in the benthic boundary layer. Reviews in Aquatic Sciences, **5**, 49-100.

Gjøsæter, J. and Kawaguchi, K. (1980) A review of the world resources of mesopelagic fish. FAO Fisheries Technical Paper, (193), 151p.

Gordon, J.D.M. (1979a) Life style and phenology in deep sea anacanthine teleosts. Symposium of the Zoological Society of London, **44** , 327-356.

Gordon, J.D.M., (1979b) Seasonal reproduction in deep sea fish, in E.Naylor and R.G. Hartnoll, (eds.), Cyclic phenomenon in marine plants and animals. Proceedings 13th European Marine Biology Symposium, Pergamon Press, Oxford, pp. 223-229.

Gordon, J.D.M. and Bergstad, O.A. (1992) Species composition of demersal fish in the Rockall Trough, north-eastern Atlantic, as determined by different trawls. Journal of the Marine Biological Association of the United Kingdom, **72**, 213-230.

Gordon, J.D.M. and Mauchline, J. (1990) Depth-related trends in the diet of a deep-sea bottom-living fish assemblage of the Rockall Trough, in Barnes, M. and Gibson, R.N. (eds.), Trophic relationships in the Marine Environment, Proceedings 24th European Marine Biology Symposium 1990, Aberdeen University Press, pp 439-452.

Haedrich, R.L. (This volume) Structure over time of an exploited deep water fish assemblage, in A.G. Hopper (ed.), Deep Water Fisheries of the North Atlantic Oceanic Slope (proceedings of the NATO Advanced Research Workshop, March 1994), Kluwer, Dordrecht, The Netherlands.

Haedrich, R.L. and Krefft, G. (1978) Distribution of bottom fishes in the Denmark Strait and Irminger Sea. Deep-Sea Research, **25**, 705-720.

Haedrich, R.L. and Merrett, N.R. (1988) Summary atlas of deep-living demersal fishes in the North Atlantic Basin. Journal of Natural History, **22**, 1325-1362.

Haedrich, R.L. and Merrett, N.R. (1990) Little evidence for faunal zonation or communities in deep-sea demersal fish faunas. Progress in Oceanography, **24**, 239-250

Haedrich, R.L. and Merrett, N.R. (1992) Production/biomass ratios, size frequencies, and biomass spectra in deep-sea demersal fish, in G.T. Rowe and V. Pariente (eds.), Deep-Sea Food Chains and the Global Carbon Cycle, (proceedings of the NATO Advanced Research Workshop, April 1991), Kluwer, Dordrecht, The Netherlands, pp. 157-182

Haedrich, R.L., Rowe, G.T. and Polloni, P.T. (1975) Zonation and faunal composition of epibenthic populations on the continental slope south of New England. Journal of Marine Research, 33, 191-212.

Hareide, N.-R. (This volume) Comparisons between longlining and trawling for deep water species - selectivity, quality and catchability - a review, in A.G. Hopper (ed.), Deep Water Fisheries of the North Atlantic Oceanic Slope (proceedings of the NATO Advanced Research Workshop, March 1994), Kluwer, Dordrecht, The Netherlands.

Holligan, P.M. and Groom, S.B. (1986) Phytoplankton distributions along the shelf break. Proceedings of the Royal Society of Edinburgh, 88B, 239-263.

Huthnance, J.M. (1986) The Rockall slope current and shelf-edge processes. Proceedings of the Royal Society of Edinburgh, 88B, 83-101.

Jørgensen, O.A. (This volume) A comparison of deep water trawl and long-line research fishery in the Davis Strait, in A.G. Hopper (ed.), Deep Water Fisheries of the North Atlantic Oceanic Slope (proceedings of the NATO Advanced Research Workshop, March 1994), Kluwer, Dordrecht, The Netherlands.

Lampitt, R.S., Billett, D.S.M. and Rice, A.L. 1986. Biomass of the invertebrate megabenthos from 500 to 4100m in the Northeast Atlantic Ocean. Marine Biology, 93, 69-81.

Marshall, N.B. and Iwamoto, T. (1973) Family Macrouridae. Memoir Sears Foundation for Marine Resarch, No 1, Fishes of the Western North Atlantic, Part 6, 496-665.

Marshall, N.B. and Merrett, N.R. (1977) The existence of a benthopelagic fauna in the deep sea. Deep-Sea Research Supplement, 24, 483-497.

Martins, R. and Ferreira, C. (This volume) Line fishing for black scabbardfish (Aphanopus carbo Lowe, 1839) and other deep water species in the eastern mid-Atlantic to the north of Madeira, in A.G. Hopper (ed.), Deep Water Fisheries of the North Atlantic Oceanic Slope (proceedings of the NATO Advanced Research Workshop, March 1994), Kluwer, Dordrecht, The Netherlands.

Mauchline, J. (Editor) (1986) The Oceanography of the Rockall Channel. Proceedings of the Royal Society of Edinburgh, 88B, 356pp.

Mauchline, J. (1990) Aspects of production in a marginal oceanic region, the Rockall Trough, northeastern Atlantic Ocean. Reviews in Aquatic Sciences, 2, 167-183

Mauchline, J. and Gordon, J.D.M., (1985) Trophic diversity in deep-sea fish. Journal of Fish Biology, 26, 527-535.

Mauchline, J. and Gordon, J.D.M., (1991) Oceanic pelagic prey of benthopelagic fish in the benthic boundary layer of a marginal oceanic region. Marine Ecology Progress Series, 74, 109-115.

McCormick, R. (This volume) The Irish experience of deepwater fishing in the NE Atlantic. in A.G. Hopper, (ed.), Deep Water Fisheries of the North Atlantic Oceanic Slope (proceedings of the NATO Advanced Research Workshop, March 1994), Kluwer, Dordrecht, The Netherlands.

Merrett, N.R. (1986) Biogeography and the oceanic rim: a poorly known zone of ichthyofaunal interaction. UNESCO Technical Papers in Marine Science, 49, 201-209.

Merrett, N.R. (1987) A zone of faunal change in assemblages of abyssal demersal fish in the eastern North Atlantic: a response to seasonality in production? Biological Oceanography, 5, 137-151.

Merrett, N.R. (1989) The elusive macrourid alevin and its seeming lack of potential in contributing to intrafamilial systematics of gadiform fishes. Los Angeles Natural History Museum Science Series, (32) 175-185.

Merrett, N.R. (1994) Reproduction in the North Atlantic oceanic ichthyofauna and the relationship between fecundity and species' sizes. Environmental Biology of Fishes

Merrett, N.R., Badcock, J., Ehrich, S. and Hulley, P.A. (1986). Preliminary observations on the near-bottom ichthyofauna of the Rockall Trough: a contemporaneous investigation using commercial-sized midwater and demersal trawls to 100m depth. Proceedings of the Royal Society of Edinburgh, **88B**, 312-314.

Merrett, N.R and Domanski, P.A. (1985) Observations on the ecology of deep-sea bottom-living fishes collected off northwest Africa: 2. Moroccan slope (27^0 -34^0 N) with special reference to *Synaphobranchus kaupi*. Biological Oceanography, **3**, 349-399.

Merrett, N.R and Marshall, N.B. (1985) Observations on the ecology of deep-sea bottom-living fishes collected off northwest Africa (08^0 27^0 N). Progress in Oceanography, **9**, 185-244.

Merrett, N.R., Gordon, J.D.M., Stehmann, M. and Haedrich, R.L. (1991a) Deep demersal fish assemblage structure in the Porcupine Seabight (eastern North Atlantic): Slope sampling by three different trawls compared. Journal of the Marine Biological Association of the U.K., **71**, 329-358.

Merrett, N.R., Haedrich, R.L., Gordon, J.D.M. and Stehmann, M. (1991b) Deep demersal fish assemblage structure in the Porcupine Sea Bight (eastern North Atlantic): results of single warp trawling at lower slope to abyssal soundings. Journal of the Marine Biological Association of the U.K., **71**, 359-373.

Nakamura, I. and Parin, N.V. (1993): FAO species catalogue. Vol. 15. Snake mackerels and cutlassfishe s of the world (Families Gempylidae and Trichiuridae). An annotated and illustrated catalogue of the snake mackerels, snoeks, escolars, gemfishes, sackfishes, domine, oilfish, cutlassfishes, scabbardfishes, hairtails, and frostfishes known to date. FAO Fisheries Synopsis. No. 125, Vol. 15. 1993. 136 p., 200 figs.

Pankhurst, N.W. and Conroy, A.M. (1987). Size-fecundity relationships in the orange roughy, *Hoplostethus atlanticus*. New Zealand Journal Marine Freshwater Research, **21**, 295-300.

Prasad, K.S. and Haedrich, R.L. (1993) Satellite observations of phytoplankton variability on the Grand Banks of Newfoundland during a spring bloom. International Journal of Remote Sensing, **14**, 241-252.

Prasad, K.S. and Haedrich, R.L. (1994) Satellite-derived primary production estimates from the Grand Banks: comparison to other oceanic regimes. Continental Shelf Research.

Reinert, J. (This volume) Deep water resources in Faroese waters to the south, southwest and west of the Faroes: a preliminary account, in A.G. Hopper (ed.), Deep Water Fisheries of the North Atlantic Oceanic Slope (proceedings of the NATO Advanced Research Workshop, March 1994), Kluwer, Dordrecht, The Netherlands.

Rice, A.L., Billett, D.S.M., Thurston, M.H. and Lampitt, R.S. (1991) The Institute of Oceanographic Sciences Biology Programme in the Porcupine Sea Bight: background and general introduction. Journal of the Marine Biological Association of the U.K., **71**, 281-310.

Rowe, G.T. and Haedrich, R.L. (1979) The biota and biological processes on the continental slope, in Doyle and Pilkey (eds.), The Continental Slope. American Association of Petroleum Geologists, Special Publication, No. 27, 49-59, Tulsa.

26

Savvatimskii, P.I. (1988). Diurnal vertical migrations of cod and rock grenadier in the northwest Atlantic. In Behaviour of commercial fishes, Collection of scientific works of VINRO, pp 67-77, 1985. Canadian Translation of Fisheries and Aquatic Sciences no. 5338, 16pp.

Somero, G.N., Siebenaller, J.F. and Hochachka, P.W. (1983). Biochemical and physiological adaptations of deep-sea animals, in Rowe, G.T. (ed.) The Sea 8. Deep-Sea Biology. New York, Wiley, pp 261-330.

Stein, D.L. (1985) Towing large nets by single warp at abyssal depths: methods and biological results. Deep-Sea Research, 32, 183-200.

Sundermann, J. (Ed.), 1986. Landolt-Bornstein Numerical Data and Functional Relationships in Science and Technology. New Series. Group V: Geophysics and Space Research, 3. Oceanography, Subvolume a. Berlin, Springer-Verlag.

Thorsteinsson, H.P. and Valdimarsson, G. (This volume) Experimental utilisation and marketing of by-catches and deep water species in Iceland, in A.G. Hopper (ed.), Deep Water Fisheries of the North Atlantic Oceanic Slope (proceedings of the NATO Advanced Research Workshop, March 1994), Kluwer, Dordrecht, The Netherlands.

Troyanovsky, F.M. and Lisovsky, S.F. (This volume) Russian (USSR) fisheries research in deep waters (below 500 m) in the North Atlantic, in A.G. Hopper (ed.), Deep Water Fisheries of the North Atlantic Oceanic Slope (proceedings of the NATO Advanced Research Workshop, March 1994), Kluwer, Dordrecht, The Netherlands.

Tyler, P.A. (1988) Seasonality in the deep sea. Oceanography and Marine Biology: an Annual Review, 26, 227-258.

Walsh, J.J., Rowe, G.T., Iverson, R.L. and McRoy, C.P (1981) Biological export of shelf carbon is a sink of the global CO_2 cycle. Nature, London. 291,196-201.

Ware, D.M. (1975) Relation between egg size, growth and natural mortality of larval fish. Journal of the Fisheries Research Board of Canada, 32, 2503-2512.

Woodward, A.S. (1898) The antiquity of the deep-sea fish-fauna. Natural Science, 12, 257-260.

Yano, K and Tanaka, S. (1988) Size at maturity, reproductive cycle, fecundity and depth segregation of the deep-sea squaloid sharks, *Centroscymnus owstoni* and *C. coelolepis* in Suruga Bay, Japan. Suisan Gakkaishi, 54, 167-174.

STRUCTURE OVER TIME OF AN EXPLOITED DEEP-WATER FISH ASSEMBLAGE

RICHARD L. HAEDRICH
Department of Biology
Memorial University of Newfoundland
St. John's, Newfoundland, A1B 3X7 Canada

ABSTRACT. *The structure of a deep water fish assemblage on the upper continental slope of Canada's Labrador Sea was studied using a survey time series dataset from the years 1978 to 1991. Greenland halibut (Reinhardtius hippoglossoides) and roundnose grenadier (Coryphaenoides rupestris) were the dominant regular members of the assemblage, and both were commercially exploited. Less abundant but regular members of the upper slope assemblage (Antimora, Centroscyllium, Macrourus, and Nezumia) were not the subject of a fishery, although all occurred as by-catch. Overall biomass of the commercial species declined significantly during the period, and the mean size of Greenland halibut was reduced by half. The biomass of non-target species was less affected, and the mean sizes of less abundant species changed very little, except for broadhead wolffish (Anarhichas denticulatus) which almost doubled in size. There were no consistent temperature changes in the geographic area occupied by the assemblage. Population shifts, mostly from north to south and including dominant species such as Greenland halibut and witch flounder (Glyptocephalus cynoglossus), took place on the shallower adjacent continental shelf but did not occur within the region. The structure of the assemblage on Labrador's upper continental slope is similar to other upper slope assemblages. Focus on faunal regions and overall species ranges seems preferable to arbitrary management zones. Scientific and socio-economic matters relating to fisheries on the upper slope must be addressed in the same time and space frameworks.*

1. Introduction

The great demersal trawl fisheries of Canada's Atlantic continental shelf have been under severe pressure for the past 30 years (Harris, 1993). In 1990, Newfoundland's northern cod stock finally collapsed (Bishop et al., 1993), and today the fishery from Labrador to the U.S. border is closed under a moratorium. The entire demersal fish

A. G. Hopper (ed.), Deep-Water Fisheries of the North Atlantic Oceanic Slope, 27–50.

community appears to have been affected, especially since the mid-1980's. Both the dominant commercially exploited species and a number of far less abundant non-commercial species have all showed declines in biomass (Haedrich et al., 1993). Unusually cold water temperatures, predation by seals, and a changing ocean environment have all have been suggested as the underlying cause for the calamitous decline, but critical examination of all the data available indicates that overfishing must have been the prime agent (Hutchings and Myers, 1994).

With the collapse and closure of the shelf fishery, attention has turned to the deep water stocks on the continental slope (Duthie and Marsden, this volume). Exploitation of the dominant commercial species on the slope; the Greenland halibut (*Reinhardtius hippoglossoides*) and the roundnose grenadier (*Coryphaenoides rupestris*), has been going on for at least a decade, (Atkinson, this volume; Bowering and Brodie, this volume). As these species now come under even more pressure, it is reasonable to expect that the other species which co-occur on the slope will be affected as well, repeating what already has happened on the shelf. What little is known of the ecology of even the common and widespread fish species that inhabit the continental slope suggests that their life histories, age and growth, diets, and population dynamics are quite different from those of shelf species (Haedrich and Merrett, 1992; Gordon et al., this volume). At present, therefore, it would be difficult to predict what the side effects on deep water fish communities might be if dominant members of the community were lost or decimated by modern intensive fisheries. This study is a retrospective monitoring of a 14-year time series of groundfish survey data, and was undertaken to determine what changes could be traced in the demersal fish community of the upper slope as fishing of the commercial species increased during the time period 1978-91.

1.1. DATA OVER TIME

Time series data of any kind are rare for the deep ocean. Part of the reason is due to the cost involved; deep ocean investigations require specialised ships and dedicated time commitments. Furthermore, except in the case of mineral resources which require relatively simple one-time "look-see" sampling, there has been little economic incentive to undertake long-term monitoring of the kind necessary to produce adequate time series. Most deep-sea studies have been motivated by scientific curiosity. Finally, the science of the deep sea has mostly operated, until recently, under the assumption that there are stable environments and unchanging situations over very broad areas (Gage and Tyler, 1991).

Many oceanographers, therefore, have been quite comfortable with the idea that one or a few cruises provided an adequate basis for generalization, especially in regard to the deep ocean and central gyres. This essentially static view underlies much of the biological literature of the deep sea. Examples are those concerned with questions of biodiversity (McGowan and Walker, 1985; Haedrich, 1987; Rex et al., 1993; Gray, 1993) and distribution patterns (Backus et al., 1966; McGowan, 1986). The "Summary Atlas of Deep-living Demersal Fishes in the North Atlantic " (Haedrich and Merrett, 1988) also falls into this category.

But the deep ocean is not static. While evidence that this is so has been accumulating for some time, the most dramatic demonstration has come from the Bathysnap pictures (Lampitt and Burnham, 1983; Rice et al., 1986) made in the Porcupine Seabight. These pictures showed that conditions at the surface, long recognised as being a dynamic place, could be rapidly translated to the deep seabed, and that the fauna there responded at the same pace. Even on time scales as short as a few weeks or months, the conditions do not remain constant.

In sharp contrast to the situation in deep-sea studies, time series have been a keystone of fisheries research since the late 1800's. Fisheries science is fundamentally concerned with the dynamics of natural populations. Data from regular monitoring surveys are used to study those dynamics, to predict the health of stocks, and to set catch quotas. Usually, survey data take the form of species numbers (abundance) and weight (biomass) obtained from net samples taken from a pre-established grid or stratified set of standard stations. The time series derived, however, are mostly confined to the continental shelf regions or, on the high seas, to a few commercially important pelagic species such as tunas and billfishes. Work in the fisheries moreover tends to focus only on the species of interest, taking an autecological (single species) perspective in contrast to the synecological (community) view of the ocean ecologist.

As demersal trawl fisheries have been extended beyond the continental shelves and onto the slope, survey data over time have begun to be assembled. One very important North Atlantic dataset is that for Northwest Atlantic Fisheries Organisation (NAFO) Divisions 2 and 3. This area on Canada's east coast, off Labrador and northeast Newfoundland, is home to several important deep water stocks (Pechenik and Troyanovski, 1970). Canada's Department of Fisheries & Oceans (DFO) has conducted annual demersal trawl surveys there every autumn since 1978. The commercial species of interest have been primarily cod, *Gadus morhua*, redfish, *Sebastes spp.*, Greenland halibut, *Reinhardtius hippoglossoides* (Bowering and Brodie, this volume), and roundnose grenadier, *Coryphaenoides rupestris* (Atkinson, this volume), but data on other species also have been kept .

1.2. THE DEEP GROUP

Villagarcía (1994) has used the DFO time series data in a study of the demersal fish community in NAFO Divisions 2J+3KL. She identified four fish assemblages that, except in latter years, occupied distinct regions within the area in a rather stable and persistent manner (Fig. 1). Since about 1987, however, the distribution of the assemblages has changed as the major commercial fishery, based on the northern cod stock, declined and collapsed catastrophically in the 1990's (Haedrich et al., 1993). One of Villagarcía's (1994) assemblages is a Deep Group on the upper continental slope off Labrador and northeast Newfoundland (Fig. 2). This group forms the basis for this paper.

The Deep Group is comprised of a mix of shallow and deep-sea demersal species, ranging from the dominant Greenland halibut and roundnose grenadier through to typical upper slope species such as black dogfish,(*Centroscyllium fabricii)*, roughhead

30

grenadier,(*Macrourus berglax)*, marlinspike, *(Nezumia bairdii)*, blue hake,(*Antimora rostrata)* and witch flounder, *(Glyptocephalus cynoglossus)*, to the deep-sea spiny eels *(Polyacanthonotus rissoanus* and *Notacanthus chemnitzii)*. The deepwater redfish, *Sebastes mentella*, which is mainly pelagic, was very important in some years. Although deep water fisheries had been established in the region before 1978, it was from that time to the present , that fishing seems to have had a significant impact on the stocks (Bowering and Brodie, this volume). The study of the Deep Group assemblage offers an opportunity to investigate possible impacts on non-target fish species and on the demersal fish com-

munity in general in an increasingly exploited upper continental slope eco-system.

2. Methods

Since 1978, DFO has conducted regular stratified-random groundfish surveys each autumn in NAFO Divisions 2J and 3K. Stratification of the sampling stations is by latitude, longitude and depth. A standard otter trawl tow of 30 minutes duration is made at each station; the mean depth of tow and bottom temperature are recorded. All fish taken are identified, counted and weighed, and, for selected commercial species, other more detailed biological information is also recorded. The raw catch data for these surveys from 1978 to 1991 were made available to researchers at Memorial University by DFO's Science Branch, Northwest Atlantic Fisheries Centre, St. John's.

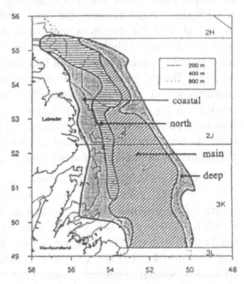

Figure 1. Approximate areas occupied by fish faunal assemblages on the Newfoundland-Labrador shelf (Villagarcia,1994)

Fish assemblages were identified based on the relative biomass (log transformed) of the individual species present in each sample (Haedrich et al., 1993; Villagarcía, 1994).The TWINSPAN program (Hill, 1979) was the primary analytical tool, and final results were checked by cluster analyses and by mapping the resultant groups. For further details, Gomes (1993) and Villagarcía (1994) should be consulted. Stations assigned to the Deep Group (Fig. 2) ranged from 230 to over 1330 m depth, with a mean depth of 565 m. In the Deep Group, the year 1978 contained far fewer stations than did subsequent years (6 *vs* a mean number of stations in 1979-1991 = 32, range 14-55), and therefore was omitted from most analyses, including cluster and multi-dimensional scaling (MDS).

All the data for the demersal species were used. Clearly pelagic taxa were excluded, for example the fairly common mesopelagic *Chauliodus sloani* and *Stomias boa*.

31

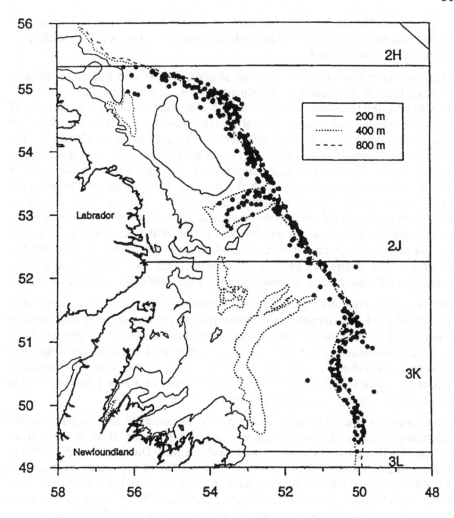

Figure 2. Location of all stations comprising the Deep Group, 1978 - 91. Based on Villagarcia 1994. NAFO Divisions 2J and 3K lie completely within the area

Pelagic species that are an important part of demersal food webs were retained (for example capelin, *(Mallotus)*, Arctic cod, *(Boreogadus)*, and deep-sea blacksmelt, *(Bathylagus)*, were retained.

Since most identifications were made on shipboard, some care had to be taken in accepting them without question, particularly as regards rare or unusual species. For the most part, the taxonomy of deep water fishes is rather well known for this part of the North Atlantic, and regional general reference texts are available (e.g. Scott and Scott, 1988). Nevertheless the taxonomy for some groups remains problematical. For example, the Newfoundland species of Zoarcidae *(Lycodes and Lycenchelys)* are in need of critical modern taxonomic assessment (J. Nielsen, in litt.); identification is difficult and confusion can arise with the more northerly species in this generally Arctic group. There is also the problem that different technicians may have been involved in making the identifications on the research survey vessel in different years. Some quality control on the raw data has been exercised and , for example, extremely doubtful records have been removed, some others have been combined. Bathylagidae, *Bathylagus*, and *Bathylagus euryops*, for example, were all combined into a single taxon since only one species is known from the area (McKelvie and Haedrich, 1985).

Emphasis in fisheries studies tends to be on biomass, since the primary interest is in commercial yield. For comparative purposes with other fishery-oriented work, data are reported in this study on an annual basis as catch-per-unit-effort (CPUE), i.e. kg of catch in a 30-minute survey tow. Ecological studies tend to emphasize numbers of individuals (abundance), and this has been done in the case of community comparisons over time within the Deep Group. Within each survey year, catch data (numbers and weights) were summed by species.

Cluster analyses and MDS analyses were used to determine the similarity of fish community structure in any one year with that in other years. Input to the analyses was the abundance data by year and by latitude (north = Division 2J, above 52°30N, south = 3K, below 52°30N). The measure of overlap in the cluster analyses was Bray-Curtis dissimilarity (where 0 = identical and 1 = no overlap) and the clustering strategy was UPGMA (unweighted pair-group method, arithmetic average, i.e. nearest neighbour). Calculations were carried out using programs in the NTSYS-pc package (Rohlf, 1988). The mean size within a year was calculated as the total weight taken of a species divided by the total number taken (to give mean weight in gm/individual for each species).

3. Results

Sixty taxa of fishes were recorded in the Deep Group (Appendix I). The greatest number of species reported in any one year was 43 (1984) and the least was 26 (1990; there were 22 species reported in 1978 when only 6 stations were sampled).

The assemblage structure was strongly dominated, both in numbers and in weight, by a relatively few species all of which were commercially exploited. Redfish *(Sebastes mentella)* alone comprised over 69.5% by numbers and 49% by weight of the entire recorded catch. A small group of the next ranking species - roundnose grenadier

(*Coryphaenoides rupestris*), Greenland halibut (*Reinhardtius hippoglossoides*), and cod (*Gadus morhua*) together accounted for 21.5% of the numbers and 33.6% of the weight. The 15 top ranking species by numbers and top 10 by weight are listed in Table 1. It should be noted that cod was only abundant in a few of the sample years.

TABLE 1. Top ranking species by numbers in the Deep Group. The numbers and weights (kg) indicated are the total catch for all the survey years 1978_1991 and all stations, n=422

SPECIES	RANK BY No.	TOTAL No.s	RANK BY Wt	TOTAL Wt
Sebastes mentella	1	157394	1	68016.7
Coryphaenoides rupestris	2	31280	4	10833.8
Reinhardtius hippoglossoides	3	11977	2	24848.9
Gadus morhua	4	5372	3	10955.0
Macrourus berglax	5	4239	6	3295.3
Glyptocephalus cynoglossus	6	2139	10	1611.4
Sebastes marinus	7	2016	8	1832.9
Antimora rostrata	8	1929	12	647.1
Centroscyllium fabricii	9	1699	7	2785.1
Nezumia bairdii	10	1393	19	231.1
Anarhichas denticulatus	11	1218	5	7962.5
Hippoglossoides platessoides	12	876	18	330.9
Anarhichas lupus	13	559	16	353.1
Synaphobranchus kaupi	14	500		83..3
Bathytoctes sp.	15	459	9	1659.1

Faunal composition (relative numbers) was compared year to year and north (NAFO Division 2J) to south (NAFO Division 3K) by cluster analysis and MDS. The results for the Deep Group indicated that although temporal changes in abundance could be detected, spatial changes in abundance at the scale of the divisions were not obvious. For this reason, comparisons are reported from year to year but generally not north to south. The analyses based on years between 1979 and 19991 showed three major clusters at about the 55% level of dissimilarity that were chronologically associated; one cluster included the years 1979-85, a second included 1986-88, and the third included 1989-91 (Fig. 3). MDS revealed a similar pattern; the years 1979-85 had negative loadings on the first two factors, 1989-91 were positive, and 1986-88 fell in between (Fig. 4).

34

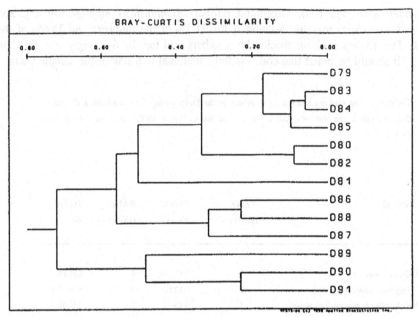

Figure 3. Cluster analysis of Deep Group assemblage by year. The index of overlap is Bray-Curtis dissimilarity and the clustering strategy is UPGMA

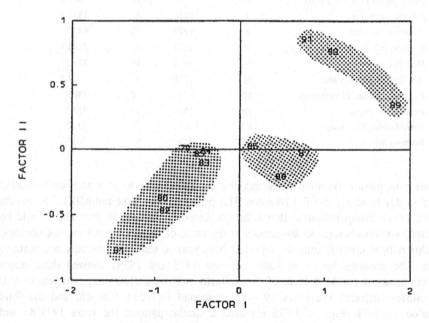

Figure 4. Multi-dimensional scaling (MDS) of Deep Group assemblage by year. The shading indicates the three groups mentioned in the results

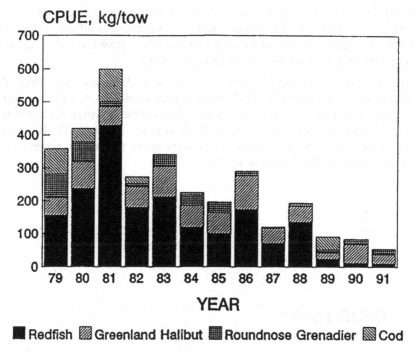

Figure 5. Deep Group catch-per -unit-effort (kg./tow) in surveys from 1979 to 1991- dominant species including redfish

Catches by weight of the dominant species, which are also those commercially exploited, declined rather steadily from 1979 to 1991 (Fig. 5), a pattern that was clear whether or not the generally overwhelmingly dominant redfish was included (Fig. 6). Catches of the less abundant, middle-ranking species such as black dogfish (*Centroscyllium fabricii*), roughhead grenadier (*Macrourus berglax*), marlinspike (*Nezumia bairdii*), blue hake (*Antimora rostrata*) and witch flounder (*Glyptocephalus cynoglossus*) showed no consistent trends throughout the period (Fig. 7). The first four of these are widely-distributed typical upper slope species and are not fished commercially. The witch is thought to spend its juvenile life on the upper slope and then much of its adult life on the deeper parts of the shelf (Powles and Koehler, 1970, but see also Walsh, 1987, for a critique of this view), and is a valuable commercial species taken now in relatively small but stable amounts in deep water of the Labrador-East Newfoundland region (Bowering and Brodie, 1991).

The mean sizes (weight/individual) of most species showed little variation over time (Fig. 8). Greenland halibut was a dramatic exception; its average size declined from 3 kg in 1978 to 1 kg in 1991. In the same period, witch flounder went from an average

36

size of 0.7 kg in 1978 to about 1 kg in 1982-83 and then declined steadily to 0.5 kg in 1991 (Fig. 9). One of the largest species, broadhead wolffish (*Anarhichas denticulatus*), actually showed an increase in average size from less than 6 kg in the early years of the time series to over 10 kg in the 1990's.

Bottom temperatures and depths were recorded on each survey tow. These data provide an index of the extent to which environmental conditions in the area occupied by the Deep Group varied over time. Table 2 shows these statistics. As might have been expected for this upper slope region (Gordon et al., this volume), the range in values is rather small and the average temperature at the time of the surveys in the autumn was rather similar from year to year.

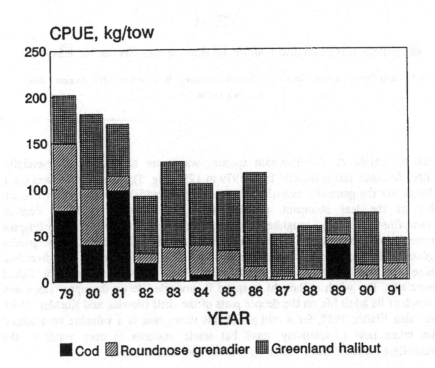

Figure 6. Deep Group catch-per-unit -effort (kg./tow) in surveys from 1979 to 1991- dominant species excluding redfish

4. Discussion

4.1. CHANGES IN CATCH RATES

The fauna comprising the Deep Group, with about 60 species in 25 families, is similar to that reported from other upper slope regions. It is low in diversity and there is a high degree of dominance by a very few species, all of which are exploited by commercial fishing to some degree. The species found, not surprisingly, show strong affinities with the fauna reported from recent surveys off Greenland (Rätz, 1991; Nielsen and Bertelsen, 1992). Rätz (1991) records 57 demersal species. Based on far fewer trawl samples than were available for the Deep Group (33 *vs.* 422), Snelgrove and Haedrich (1985) recorded 39 species from the Newfoundland slope (Carson Canyon) as a whole.

There were 72 species taken in the Middle Atlantic Bight (Hudson Canyon, 55 stations) with the same deep-sea gear used in Carson Canyon. On the Grand Banks, there were 30 species whose biomass exceeded 0.1% of the total weight in the catch (Gomes et al., 1992; therefore rarer species were omitted), similarly about 31 species in shelf assemblages off Nova Scotia (Mahon et al., 1984; Mahon and Smith, 1987), and about 40 species in the Middle Atlantic Bight (Gabriel, 1992). The fauna of the Skagerrak deep region, with 29 species recorded, is also of low diversity and roundnose grenadier comprises over half the biomass (Bergstad, 1990), This fauna has been shown to depend on immigration from outside the area, i.e. somewhat expatriate and thus likely to have somewhat unusual characteristics.

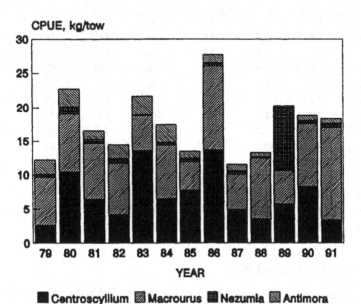

Figure 7. Deep Group catch-per-unit-effort (kg./tow) in surveys from 1979 to 1991- less abundant upper slope species. Note the range of CPUE relative to Figs 5 and 6.

Changes over time in the catch rates of fishes in the Deep Group were most apparent in the dominant species. In this low diversity assemblage, the dominant forms are also the ones that were commercially exploited. Declines in catch rate were clear in the demersal species - Greenland halibut, cod and roundnose grenadier - but were just as apparent in the more pelagic redfish (Fig. 5). The less abundant species on the upper slope, but nonetheless ranking species and not yet the subject of directed fisheries, remained at about the same catch level throughout the time series (Fig. 7). This situation is different from that on the continental shelf, where relative declines in the less abundant, non-commercial species match or even exceed those in the dominant, commercial forms (Haedrich et al., 1993). Even so, the catch rates for the upper slope species were not flat, but seem to show a periodicity of sorts. There is no ready explanation for this, but in this context it is worth noting a study of cycles in the California Dungeness crab fishery. There, it was suggested that intense fishing pressure could produce a destabilized population that was characterised by strong oscillations in density (Berryman, 1991). Such cycles can be considered the "harbingers of chaos" according to Berryman (1991 citing May, 1974).

Table 2. Labrador upper slope environment: average, minimum, and maximum temperatures (T, °C), and minimum and maximum depths (Z, m) sampled. Based on stations assigned to the Deep Group; n = number of stations in each year.

YEAR	Avg T.	Min T.	Max T.	Min Z	Max Z	n
1978	3.9	3.3	4.5	282	392	6
1979	3.3	1.7	4.0	233	1334	34
1980	2.7	1.5	3.8	253	838	42
1981	3.6	2.8	4.4	252	812	42
1982	3.4	0.5	4.8	231	946	55
1983	4.2	3.8	5.0	350	816	30
1984	2.8	0.7	4.9	268	935	36
1985	3.6	1.2	4.5	313	932	50
1986	3.5	3.5	3.9	314	960	21
1987	3.3	3.5	4.1	328	895	26
1988	3.7	3.3	3.9	343	950	21
1989	3.3	1.7	3.7	278	798	14
1990	3.7	3.3	4.1	386	970	20
1991	3.6	1.7	5.1	288	990	25

4.2. CAUSES OF THE DECLINES

Because both commercial and non-commercial species declined on the shelf, it was at least plausible that a large-scale environmental change was the cause (all such arguments, e.g. deYoung and Rose (1994), have been based on a postulate of colder temperatures) . But on the upper slope, the response as far as the fish community was concerned was apparent in some species but not in others (i.e. selective not global), and the temperature data themselves showed small ranges and no consistent pattern of change (Table 2). The steady decline in survey catch rates of commercial species in the Deep Group indicates that these species are already over-exploited in the deep water off Labrador. The commercial catch data for roundnose grenadier, available over a far longer time period, show that even by the time the scientific surveys began in 1978 the stocks were sorely diminished (Atkinson, this volume; Ommer, this volume).

The evidence now seems indisputable that overfishing has been the major agent in the demise of Newfoundland's great commercial fishery on the continental shelf (Hutchings and Myers, 1994). It is therefore quite plausible, as Rätz (1992) has argued for the fish community off West Greenland, that by-catch mortality contributed to the observed parallel decline in many non-commercial species on the shelf. The geographic ranges of these shelf species are similar to that of the target commercial species. Because the shelf species were commonly associated, trawls which took the dominant commercial species caught and killed numbers of the less abundant non-commercial species at the same time.

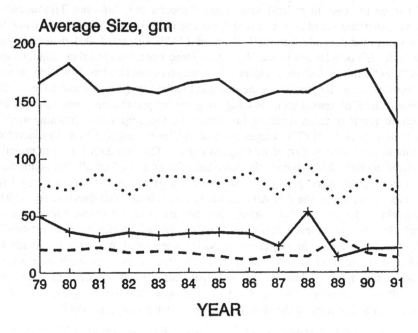

Figure 8. Mean sizes (gm./individual) of selected typical upper slope species in the Deep Group, 1979 to 1991.

On the upper slope, the distribution patterns of ranking shelf and slope species may overlap within certain depth ranges but most likely do not broadly coincide. The survey depths for the most part, and probably the commercial fishing depths too only reached a relatively small portion of the full depth range of the deep water species. Spanish trawlers fish to 1700m in the survey area (Junquera and Zamarro, 1994). Nonetheless, as far as by-catch mortality goes, deep water species tend to be more lightly built and thus more fragile than shallow-living species, and probably would die due to scale loss even if they escaped through the meshes of a net (Gordon et al., this volume). Furthermore, because of the likelihood that these are long-lived, slow-growing and slow-reproducing species (Gordon et al., this volume), the 14-year time period of the series must be considered quite short relative to the lifespan of the deep water fish involved. It is worth considering the example of the cod in which the basic stock size was declining steadily for many years (Hutchings and Myers, 1994) before changes could be tracked in the standard survey catches.

4.3. PREDICTING IMPACTS ON THE COMMUNITY

It is difficult to predict even what changes might be expected in the deep water species populations as local predation pressure (in the form of commercial fishing) on them increases. Is the local population a part of the spawning stock or is it, as is thought to be the case in at least some pelagic species (e.g. *Sebastes*; Troyanovsky, 1992), an expatriate population removed from and unlikely to return to areas suitable for spawning and juvenile survival? Bergstad (1990) suggested that demersal and mesopelagic fish populations in the Norwegian Deep were maintained by immigration via the Atlantic inflow, but more recent work (Bergstad and Gordon, 1994) has shown that a number of species, including the dominant roundnose grenadier and silver smelt (*Argentina silus*), do spawn there. Fishing on spawning populations could certainly be expected to result in catch declines, but where are the deep water spawning areas? Wenner and Musick (1977) suggested that *Antimora* undertook a considerable migration to the northerly part of its range to spawn. The data do not exist to provide a definitive answer, although juveniles have been found in the Rockall/ Porcupine Sea Bight area (Merrett, pers. comm.) The roundnose grenadier has been suggested to migrate for spawning on the mid-Atlantic Ridge (Dushchenko and Savvatimsky, 1992; Troyanovsky and Lisovsky, this volume), but data are lacking here too. Knowledge of the early life history of even this already over-exploited species remains extremely sketchy because the juvenile stages are so poorly represented in any samples made to date (Bergstad and Isaksen, 1987; Merrett, 1989). But, again, it is worth considering the cod. Only recently has it emerged that this species spawns all over the Newfoundland continental shelf and not, as has been believed for years, in a few special places at the edge of the shelf far offshore (Hutchings et al., 1993).

Gomes et al. (1994) identified population shifts during the period 1978 to 1991 in various species offshore on the Labrador Shelf. The catch rates (biomass) for cod and witch flounder declined in the north but increased in the south. Catch rates of broadhead wolffish (*Anarhichas denticulatus*) declined in the north, but stayed at about the same level in the south. North-south comparisons in the Deep Group, comparing catches in Division 2J to those in 3K, revealed no such changes. Whatever it was that affected catch rates on the shelf did not translate onto the upper slope. On the other hand, the dramatic reduction in size of commercial species such as cod (which went from a mean size of 1.8 kg in 1978 to 0.9 kg in 1991), Greenland halibut, and even witch flounder (Fig. 9) was observed in both places, as was the increase in size of the broadhead wolffish. Even though catch rates appear to have behaved independently, population structure as reflected in the size of species showed a coherent pattern on the shelf and on the slope. Bowering (1987) and Bowering and Brodie (1991) pointed out how sizes and other related life history characteristics have changed in flatfish species, including Greenland halibut and witch, in the region. Clearly catch rates alone do not reveal the full story.

The clusters identified in Figure 2 suggest an evolution over time within the assemblage, with the years 1979-85 being rather more similar than the years 1986-88 and the years 1989-91 (the catch in 1981 had a high dominance of redfish (Fig. 5), which accounts for the fact that it stands somewhat apart). This same sort of grouping was identified by Gomes et al. (1994) who, using their fixed frame of reference approach, called the period 1978-86 *First Signs* (of coming problems), the period 1987-89 as the *Critical Period* (where community changes hinted at during 1985-86 did not reverse), and the years 1990-91 as *Collapse*. The suggestion, based on cluster and MDS analyses, of a similar tri-partite division over the years in the Deep Group data provide support for Gomes et al.'s (1994) concept, which was derived primarily from a study of changes in biomass in the shelf community.

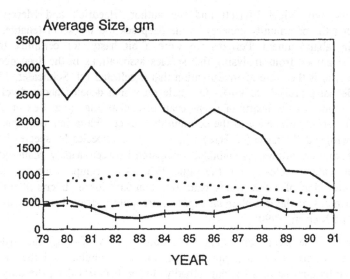

Figure 9. Mean sizes (gm./individual) of commercially exploited species in the Deep Group, 1979 to 1991

4.4.COMMUNITIES AND FRAMES OF REFERENCE

The perspective of the present study is a synecological one, i.e. what has been called in fisheries a "multi-species" approach (Mercer, 1982). Basic in this approach was an objective definition of a community (some may prefer the term "assemblage") and the spatial area occupied by that community. The result is a frame of reference that seems to contain more information and provide a greater depth of understanding than does the normal fisheries frame of reference, in this case single species (cod, Greenland halibut) catch statistics in the various NAFO Divisions (2J, 3K, etc.). At the autecological end of the spectrum, however, the individual species/individual station charts over a wide area shown by Atkinson (this volume) and Bowering and Brodie (this volume) are very interesting in providing time series that reveal population changes in an informative and spatially explicit manner. These, too, make little reference to the NAFO Divisions.

The NAFO Divisions have an historical basis and are established as the fundamental unit for statistical analysis and resource management. But they are arbitrary, some like 2J and 3K extending far offshore to waters exceeding 3000 m in depth (Fig. 2), quite outside the range of the commercial stocks they were established to delineate. The regions defined by the faunal analyses (cluster and MDS) have quite a different basis. They are based on the biological system and use the regularity and persistence of associations among species to define the geographical areas within which biological and statistical analyses should proceed.

4.5. COMMUNITIES AND SCALE

Several years ago, Nigel Merrett and the author (Haedrich and Merrett, 1990) suggested that the community concept in the deep sea did not seem tenable, and we suggested its abandonment. Perhaps we were a bit hasty, for certainly there are insights to be gained from analysing fish species associations in the deep ocean. The important matter is the scale of investigation that is selected (cf. Schneider, 1994). On a basin-wide basis (1000's of km) - the scale on which debates about diversity and other matters dear to the hearts of deep-ocean ecologists rage (e.g. Rex et al., 1993, Gray, 1993) - strict adherence to the community concept while ignoring spatial scale issues has befogged the issue (cf. Haedrich, 1987). On a species by species basis, and where the taxonomy allows, assemblages compared between widely scattered locales at the basin scale are clearly not the same. When such comparisons are made, the presence quite often of a few widespread and dominant forms diverts attention from the fact that many of the less abundant species are quite restricted in their distribution (Haedrich and Merrett, 1990).

On the smaller mesoscale of banks and eddies (100's of km), the criterion of comparability seems to be met. Species lists are very similar, and the distribution patterns of individual species overlap broadly. In the fisheries data, the only datasets where faunal composition over time are available, stability and persistence over at least a few generations are characteristic of assemblages identified on those scales (Overholtz and Tyler, 1985; Gabriel, 1992; Gomes, 1993). These data provide valuable material for a community to be analysed and understood within an holistic ecological framework. Alverson 1993, citing Alverson and Larkin, in press, expects that management within such a framework must be the way of the future.

5. Conclusions

The identification of a biologically-defined region on the upper slope was a sensible and important first step in considering the dynamics of the fish community and its individual components (see Gomes, 1993). From this perspective, insights were derived that otherwise probably would not have emerged. Wherever deep water fisheries are planned, or before those now being prosecuted proceed very far, it would seem wise to define the natural biological regions in the area, and to use these as the basis for monitoring and management. At the same time, it would also seem wise to combine this approach with the wide area but spatially explicit time series approach of Atkinson (this volume) and Bowering and Brodie (this volume) as a means to understanding the overall dynamics of target species. Even single composite maps, as illustrated very nicely in the work off Iceland by Magnússon and Magnússon (this volume), can be very informative when expressed in this way. Such a combination of approaches would seem to fit well into the developing field of landscape ecology (Wiens et al., 1993).

But, as this NATO Workshop has shown, scientific knowledge is only a part of the picture. Fisheries are indeed about people (Ommer, this volume) and their economic viability (Thorsteinsson and Valdimarsson, this volume) is as important as biological sustainability of the stocks. Nonetheless, from an ecological point of view, these two matters should operate within the same framework of time. Scale mis-match, where economic forecasts (for example) operate at the scale of years but the biology operates at the scale of decades or centuries, leads among other things to reduced options, perpetual surprise, more fragile natural systems, and more dependent and distrustful citizens (Holling, 1993). To exploit slow-growing, late-maturing deep water fishes in short-term, ill-conceived government programs to take up the social slack of a fishing moratorium on the shelf, as in the deep water gillnetting described by Duthie and Marsden (this volume), is surely not the path to follow.

Acknowledgements - This work was supported by an NSERC/DFO Subvention, and is a part of a long-term NSERC-supported program dealing with oceanic distribution patterns. Thanks are due to Nigel Merrett, John Horne, David Methven and Jeff Hutchings for their comments on the manuscript. The author is grateful to Roy Ficken for his excellent photographic help with the figures and to Lynn Anne Bussey for her cheerful assistance with the data processing.

44

Appendix I Fish species recorded in the Deep Group during the years 1978-1991. Number is the total number of individuals of that species taken and weight is the total weight in kg.

Canadian (English) name	Latin name	Number	Weight
Baird's smoothhead	*Alepocephalus bairdii*	3	2.6
Broadhead wolffish	*Anarhichas denticulatus*	1218	7962.5
Atlantic wolffish	*Anarhichas lupus*	559	353.1
Spotted wolffish	*Anarhichas minor*	107	382.5
Blue hake	*Antimora rostrata*	1929	647.1
Atlantic argentine	*Argentina silus*	13	7.9
	Artediellus sp.	25	0.7
Snowflake hookear sculpin	*Artediellus uncinatus*	4	0.2
Alligatorfish	*Aspidophoroides monopterygius*	14	0.3
Goitre blacksmelt	*Bathylagus euryops*	144	6.4
Spinytail skate	*Bathyraja spinicauda*	109	1432.3
	Bathytroctes sp.	459	1659.1
Arctic cod	*Boreogadus saida*	30	0.4
Cusk	*Brosme brosme*	6	25.9
Black dogfish	*Centroscyllium fabricii*	1699	2785.1
Rock grenadier	*Coryphaenoides rupestris*	31280	10833.8
Polar sculpin	*Cottunculus microps*	91	17.4
	Cottunculus sp.	67	67.9
Pallid sculpin	*Cottunculus thomsoni*	16	7.1
Wrymouth	*Cryptacanthodes maculatus*	4	0.9
Lumpfish	*Cyclopterus lumpus*	41	125.3
Atlantic cod	*Gadus morhua*	5372	10955.0
Greenland cod	*Gadus ogac*	1	0.2
	Gaidropsaras sp.	35	15.3
Witch flounder	*Glyptocephalus cynoglossus*	2139	1611.4
Plaice	*Hippoglossoides platessoides*	876	330.9
Atlantic halibut	*Hippoglossus hippoglossus*	32	405.9
Atlantic poacher	*Leptagonus decagonus*	5	0.2
Monkfish	*Lophius americanus*	2	2.3
Snakeblenny	*Lumpenus lampretaeformis*	1	0.0
	Lycenchelys sp.	7	1.5
Esmark's eelpout	*Lycodes esmarki*	60	55.0
Newfoundland eelpout	*Lycodes lavalei*	1	0.5
Arctic eelpout	*Lycodes reticulatus*	15	3.6

Canadian (English) name	Latin name	Number	Weight
Checker eelpout	*Lycodes vahlii*	368	88.5
Roughhead grenadier	*Macrourus berglax*	4239	3295.3
Capelin	*Mallotus villosus*	270	5.4
Blue whiting	*Micromesistius poutassou*	4	0.7
Arctic sculpin	*Myoxocephalus scorpioides*	4	0.7
Atlantic hagfish	*Myxine glutinosa*	2	0.2
Marlinspike	*Nezumia bairdii*	1393	231.1
Spiny eel	*Notacanthus chemnitzii*	243	355.9
White barracudina	*Notolepis rissoi krøyeri*	21	1.2
Longfin hake	*Phycis chesteri*	11	2.2
Shortspine tapirfish	*Polyacanthonotus rissoanus*	24	33.2
Round skate	*Raja fyllae*	30	14.1
Shorttail skate	*Raja jenseni*	9	16.4
Thorny skate	*Raja radiata*	289	350.9
Smooth skate	*Raja senta*	89	52.9
Greenland halibut	*Reinhardtius hippoglossoides*	11977	24848.9
Golden redfish	*Sebastes marinus*	2016	1832.9
Deepwater redfish	*Sebastes mentella*	157394	68016.7
Snubnose eel	*Simenchelys parasiticus*	20	4.9
Slatjaw cutthroat eel	*Synaphobranchus kaupi*	500	83.3
Roughnose grenadier	*Trachyrhincus murrayi*	80	20.1
Moustache sculpin	*Triglops murrayi*	1	0.1
	Triglops sp.	7	0.3
White hake	*Urophycis tenuis*	1	0.5
Bluntsnout smoothhead	*Xenodermichthys copei*	2	0.1

46

References

Alverson, D.L. (1993) 'The management challenge', in: K. Storey (ed.), The Newfoundland Groundfish Fisheries: Defining the Reality, Conference Proceedings, Institute of Social and Economic Research, Memorial University, St. John's, pp. 78-93.

Alverson, D.L. and Larkin, P.A. (in press) Fisheries: Fisheries Science and Management - Century 21, American Fisheries Society.

Atkinson, D.B. (this volume) 'The biology and fishery of roundnose grenadier (Coryphaenoides rupestris Gunnerus, 1765) in the Northwest Atlantic', in: A.G. Hopper (ed.), Deep Water Fisheries of the North Atlantic Oceanic Slope, Kluwer Academic Publishers, Dordrecht.

Backus, R.H., Craddock, J.E., Haedrich, R.L. and Robison, B.H. (1977) Atlantic Mesopelagic Zoogeography, Fishes of the Western North Atlantic, Mem. 1, Part 7. Sears Foundation for Marine Research, New Haven, pp. 266-287.

Bergstad, O.A. (1990) 'Ecology of the fishes of the Norwegian Deep: Distribution and species assemblages', Netherlands Journal of Sea Research, 25(1/2), 237-266.

Bergstad, O.A. and Gordon, J.D.M. (1994) 'Deep-water ichthyoplankton of the Skagerrak with special reference to Coryphaenoides rupestris Gunnerus, 1765 (Pisces, Macrouridae) and Argentina silus (Ascanius, 1775) (Pisces, Argentinidae)', Sarsia, 78, 000-000.

Bergstad, O.A. and Isaksen, B. (1987) 'Deep-water resources of the Northeast Atlantic: Distribution, abundance and exploitation', Fisken og Havet, 1987(3), 1-56.

Berryman, A.A. (1991) 'Can economic forces cause ecological chaos? The case of the Northern California Dungeness crab fishery', Oikos, 62(1), 106-109.

Bishop, C.A., Murphy, E.F., Davis, M.B., Baird, J.W. and Rose, G.A. (1993) 'An assessment of the cod stock in NAFO Divisions 2J+3KL', NAFO SCR Document 93/86, 1-50.

Bowering, W.R. (1987) 'Distribution of witch flounder (Glyptocephalus cynoglossus) in the southern Labrador and eastern Newfoundland area and changes in certain biological parameters after 20 years of exploitation', Fishery Bulletin U.S., 85, 611-629.

Bowering, W.R. and Brodie, W.B. (1991) 'Distribution of commercial flatfishes in the Newfoundland-Labrador region of the Canadian Northwest Atlantic and changes in certain biological parameters since exploitation', Netherlands Journal of Sea Research, 27, 407-422.

Bowering, W.R. and Brodie, W.B. (this volume) 'Greenland halibut (Reinhardtius hippoglossoides): A review of the dynamics of its distribution and fisheries off Eastern Canada and Greenland', in: A.G. Hopper (ed.), Deep Water Fisheries of the North Atlantic Oceanic Slope, Kluwer Academic Publishers, Dordrecht.

deYoung, B. and Rose, G.A. (1993) 'On recruitment and distribution of Atlantic cod (Gadus morhua) off Newfoundland', Canadian Journal of Fisheries and Aquatic Science, 50, 2729-2741.

Dushchenko, V.V. and Savvatimsky, P.I. (1987) 'The intraspecific structure of the roundnose grenadier, Coryphaenoides rupestris Gunnerus, in the North Atlantic: variability in local groups and conditions for their forming', Voprosy Ikhtiologii, 5, 784-793.

Duthie, A. and Marsden, A. (this volume) 'Canadian experience: Gillnetting in the Atlantic Ocean', *in*: A.G. Hopper (ed.), Deep Water Fisheries of the North Atlantic Oceanic Slope, Kluwer Academic Publishers, Dordrecht.

Gabriel, W.L. (1992) 'Persistence of demersal fish assemblages between Cape Hatteras and Nova Scotia, northwest Atlantic', Journal of Northwest Atlantic Fisheries Science, 14, 29-46.

Gage, J.D. and Tyler, P.A. (1991) Deep-Sea Biology: A natural history of organisms at the deep-sea floor, Cambridge University Press, Cambridge.

Gomes, M. do C. (1993) Predictions Under Uncertainty: Fish Assemblages and Food Webs on the Grand Bank of Newfoundland, ISER Books, St. John's.

Gomes, M.C., Haedrich, R.L. and Rice, J.C. (1992) 'Biogeography of groundfish assemblages on the Grand Bank', Journal of Northwest Atlantic Fisheries Science, 14, 13-27.

Gomes, M.C., Haedrich, R.L. and Villagarcía, M.G. (1994) 'Spatial and temporal changes in the groundfish assemblages on the Northeast Newfoundland/Labrador Shelf, Northwest Atlantic, 1978-1991'.

Gordon, J.D.M., Merrett, N.R. and Haedrich, R.L. (this volume) 'Environmental and biological aspects of slope-dwelling fishes', *in*: A.G. Hopper (ed.), Deep Water Fisheries of the North Atlantic Oceanic Slope, Kluwer Academic Publishers, Dordrecht.

Gray, J. (1993) 'Coastal diversity: Is it as high as that of the deep sea?', Proceedings of the 28th European Marine Biology Symposium, Hersonissos, Crete, September 1993. Abstract.

Haedrich, R.L. (1987) 'The species number-area relationship in the deep sea', Marine Ecology-Progress Series, 24, 303-306.

Haedrich, R.L. and Krefft, G. (1978) 'Distribution of bottom fishes in the Denmark Strait and Irminger Sea', Deep-Sea Research, 25, 705-720.

Haedrich, R.L. and Merrett, N.R. (1988) 'Summary atlas of deep-living demersal fishes in the North Atlantic', Journal of Natural History, 22, 1325-1362.

Haedrich, R.L. and Merrett, N.R. (1990) 'Little evidence for faunal zonation or communities in deep sea demersal fish faunas', Progress in Oceanography, 24, 239-250.

Haedrich, R.L. and Merrett, N.R. (1992) 'Production/biomass ratios, size frequencies, and biomass spectra in deep-sea demersal fish', *in*: G.T. Rowe and V. Pariente (eds.), Deep-Sea Food Chains and the Global Carbon Cycle, Kluwer Academic Publishers, Dordrecht, pp. 157-182.

Haedrich, R.L., Villagarcía, M.G. and Gomes, M.C. (1993) 'Spatial and temporal changes in the fish community of the Newfoundland-Labrador Shelf', Proceedings of the 28th European Marine Biology Symposium, Hersonissos, Crete, September 1993. Abstract.

Harris, L. (1993) 'Seeking equilibrium: an historical glance at aspects of the Newfoundland fisheries', *in*: K. Storey (ed.), The Newfoundland Groundfish Fisheries: Defining the Reality, Conference Proceedings, Institute of Social and Economic Research, Memorial University, St. John's, pp. 1-8.

Hill, M. (1979) TWINSPAN - A FORTRAN Program for arranging multivariate data in an ordered two-way table by classification of the individuals and attributes, Ecology and Systematics, Cornell University, Ithaca.

Holling, C.S. (1993) 'Investing in research for sustainability', Ecological Applications, 3(4), 552-555.

Hutchings, J.A. and Myers, R.A. (1994) 'What can be learned from the collapse of a renewable resource? Atlantic cod, Gadus morhua, of Newfoundland and Labrador', Canadian Journal of Fisheries and Aquatic Science, in press.

Hutchings, J.A., Myers, R.A. and Lilly, G.R. (1993) 'Geographic variation in the spawning of Atlantic cod, Gadus morhua, in the northwest Atlantic', Canadian Journal of Fisheries and Aquatic Science, 50, 2457-2467.

Junquera, S. and Zamarro, J. (1994) 'Sexual maturity and spawning of Greenland halibut (Reinhardtius hippoglossoides) from Flemish Pass area', NAFO Scientific Council Studies, no. 20, 47-52.

Lampitt, R.S. and Burnham, M.P. (1983) 'A free-fall time-lapse camera and current meter system 'Bathysnap' with notes on the foraging behaviour of a bathyal decapod shrimp', Deep-Sea Research, 30A, 1009-1017.

Magnússon, J.V. and Magnússon, J. (this volume) 'On deep sea fishes of the Icelandic Slope and Reykjanes Ridge. Distribution, relative abundance and biology', in: A.G. Hopper (ed.), Deep Water Fisheries of the North Atlantic Oceanic Slope, Kluwer Academic Publishers, Dordrecht.

Mahon, R. and Smith, R.W. (1989) 'Demersal fish assemblages on the Scotian Shelf, Northwest Atlantic: spatial distribution and persistence', Canadian Journal of Fisheries and Aquatic Science, 46(suppl.1), 134-152.

Mahon, R., Smith, R.W., Bernstein, B.B. and Scott, J.S. (1984) 'Spatial and temporal patterns of groundfish distribution on the Scotian Shelf and in the Bay of Fundy, 1970-81', Canadian Technical Report, Fisheries and Aquatic Sciences, No. 1300, 1-164.

May, R.M. (1974) 'Biological populations with non-overlapping generations: stable points, stable cycles, and chaos', Science, 186, 645-647.

McGowan, J.A. (1986) 'The biogeography of pelagic ecosystems', in: S. van der Spoel and A. Pierrot-Bults (eds.), Pelagic Biogeography, Unesco technical Papers in Marine Science 49, pp. 191-199.

McGowan, J.A. and Walker, P.W. (1985) 'Dominance and diversity maintenance in an oceanic ecosystem', Ecological Monographs, 55(1), 103-118.

McKelvie, D.S. and Haedrich, R.L. (1985) 'Mesopelagic fishes (excluding Myctophidae) from research cruises off Newfoundland and Labrador', NAFO Scientific Council Studies, 8, 67-75.

Mercer, M. (ed.) (1982) Multispecies approaches to fisheries management advice, Canadian Special Publications in Fisheries and Aquatic Science, 59, Ottawa.

Merrett, N.R. (1989) 'The elusive macrourid alevin and its seeming lack of potential in contributing to intrafamilial systematics', in: D.M. Cohen (ed.), Papers on the systematics of gadiform fishes, Los Angeles County Natural History Museum Science Series, No. 32, Los Angeles, pp. 175-185.

Nielsen, J.G. and Bertelsen, E. (1992) Fisk i grønlandske farvande, Atuakkiorfik, Nuuk.

Ommer, R.E. (this volume) 'Deep water fisheries, policy/management issues, and the sustainability of fishing communities', in: A.G. Hopper (ed.), Deep Water Fisheries of the North Atlantic Oceanic Slope, Kluwer Academic Publishers, Dordrecht.

Overholtz, W.J. and Tyler, A.V. (1985) 'Long-term responses of the demersal fish assemblages of Georges Bank', Fisheries Bulletin U.S., 83, 507-520.

Pechenik, L.N. and Troyanovski, F.M. (1970) Trawling resources of the North-Atlantic continental slope, Israel Program for Scientific Translations, 1971, no. 5977, pp. 1-66.

Powles, P.M. and Koehler, A.C. (1970) 'Depth distribution of various stages of witch flounder (*Glyptocephalus cynoglossus*) off Nova Scotia and in the Gulf of St. Lawrence', Journal of the Fisheries Research Board of Canada, 27, 2053-2062.

Rätz, H.-J. (1991) 'Notes of the structures and changes in the ichthyofauna off West Greenland', NAFO SCR Document 91/36, Serial no. N1916, 1-16.

Rätz, H.-J. (1992) 'Decrease in fish biomass off West Greenland (Subdivision 1B-1F) continued', NAFO SCR Document 92/40, Serial no. N2088, 1-8.

Rex, M.A., Stuart, C.T., Hessler, R.R., Allen, J.A., Sanders, H.L. and Wilson, G.D.F. (1993) 'Global-scale latitudinal patterns of species diversity in the deep-sea benthos', Nature, 365, 636-639.

Rice, A.L., Billett, D.S.M., Fry, J., John, A.W.G., Lampitt, R.S., Mantoura, R.F.C. and Morris, R.J. (1986) 'Seasonal deposition of phytodetritus to the deep-sea floor', Proceedings of the Royal Society of Edinburgh (B), 88, 265-279.

Rohlf, F.J. (1988) NTSYS-pc, Numerical Taxonomy and Multivariate Analysis System, Version 1.50, Exeter Publishing, Setauket.

Schneider, D.C. (1994) 'Scale-dependent patterns and species interactions in marine nekton', in: P.S. Giller, A.G. Hildrew and D. Raffaelli (eds.), Aquatic Ecology: Scale, Pattern and Process, Blackwells, London.

Scott, W. and Scott, M. (1988) 'Atlantic fishes of Canada', Canadian Bulletin of Fisheries and Aquatic Science, 219, 1-731.

Snelgrove, P.V.R. and Haedrich, R.L. (1985) 'Structure of the deep demersal fish fauna off Newfoundland', Marine Ecology - Progress Series, 27, 99-107.

Thorsteinsson, H.P. and Valdimarsson, G. (this volume) 'Experimental utilisation and marketing of by-catches and deep water species in Iceland', in: A.G. Hopper (ed.), Deep Water Fisheries of the North Atlantic Oceanic Slope, Kluwer Academic Publishers, Dordrecht.

Troyanovsky, F.M. (1992) 'Observations on non-maturing redfish (*Sebastes mentella* Travin) in the Northwest Atlantic', Journal of Northwest Atlantic Fishery Science, 14, 145-148.

Troyanovsky, F.M. and Lisovsky, S.F. (this volume) 'Russian (USSR) fisheries research in deep waters (below 500 m) in the North Atlantic', in: A.G. Hopper (ed.), Deep Water Fisheries of the North Atlantic Oceanic Slope, Kluwer Academic Publishers, Dordrecht.

Tyler, A.V., Gabriel, W.L. and Overholtz, W.J. (1982) 'Adaptive management based on structure of fish assemblages of northern continental shelves', in: M. Mercer (ed.), Multispecies approaches to fisheries management advice, Canadian Special Publications in Fisheries and Aquatic Science, 59, Ottawa, pp. 149-156.

Villagarcía, M.G. (1994) 'Structure and Distribution of Demersal Fish Assemblages on the northeast Newfoundland/Labrador Shelf', M.Sc. Thesis, Department of Biology, Memorial University of Newfoundland, St. John's.

Walsh, S.J. (1987) 'Habitat partitioning by size in witch flounder, *Glyptocephalus cynoglossus*: a reevaluation with additional data and adjustments for gear selectivity', Fishery Bulletin U.S., 85, 147-153.

Wenner, C.A. and Musick, J.A. (1977) 'Biology of the morid fish *Antimora rostrata* in the western North Atlantic', Journal of the Fisheries Research Board of Canada, 34, 2362-2368.

Wiens, J.A., Stenseth, N.C., Van Horne, B., and Ims, R.A. (1993) 'Ecological mechanisms and landscape ecology', Oikos, 66, 369-380.

THE BIOLOGY AND FISHERY OF ROUNDNOSE GRENADIER (*Coryphaenoides rupestris* Gunnerus, 1765) IN THE NORTH WEST ATLANTIC

D. B. ATKINSON

Science Branch,
Department of Fisheries and Oceans,
PO Box 5667,
St. John's, Newfoundland,
Canada A1C5X1

ABSTRACT

Exploratory fishing by the Soviet Union in the north west Atlantic during the early 1960s revealed substantial concentrations of roundnose grenadier (Coryphaenoides rupestris Gunnerus 1765) inhabiting slope areas in depths greater than about 500 m. Because of this, a commercial fishery began in 1967, and has continued to the present, albeit at relatively low levels in recent years. The biology and population dynamics of this species have been difficult to study because of their deep distribution. As a result, although many hypotheses have been put forward, much of the research has been generally inconclusive. Their relative unimportance as a commercial species in the north west Atlantic has also contributed to our general lack of understanding of the species. Nonetheless, over the years considerable information has been gathered, and this report summarises the biological investigations to date, as well as describing the history of the fishery and its management.

A. G. Hopper (ed.), Deep-Water Fisheries of the North Atlantic Oceanic Slope, 51–111.
© *1995 Kluwer Academic Publishers.*

52

1. Introduction

Investigations of the macrourid group of marine fishes began during the second half of the 1700s when Gunnerus first described the roundnose grenadier (*Coryphae-noides rupestris* Gunnerus 1765). Since then, over 300 species of macrourid fish (Family *Macrouridae*) have been identified throughout the oceans of the world. Of these, more than 90% occupy the continental slopes in depths ranging from about 200 to 2000 m (Marshall 1965). Whether found in slope or abyssal depths, nearly all macrourids have a well developed swim bladder. In general, they are similar to gadids and related groups.

Overall, about 80 species occur in the Atlantic Ocean, and approximately 30 species are known to dwell in the western North Atlantic including the Caribbean and Gulf of Mexico (Marshall 1965). Only two species, the roundnose grenadier and roughhead grenadier (*Macrourus berglax* Lacépède, 1802), are found on both sides of the Atlantic, possibly because of the continuous extent of suitable depths between the two areas.

At least seven species of macrourids have been found off the east coast of Canada (Liem and Scott 1966). The most abundant of these, the roundnose grenadier, is widely distributed along the continental slope (Parsons 1976). Exploratory fishing during the 1960s located substantial concentrations and because of this, commercial exploitation of the resource was initiated in 1967. Catches in the north west Atlantic averaged between about 20,000 to 30,000 tonnes annually prior to 1977, but after extension of national jurisdiction to 200 miles by Canada and Denmark (on behalf of Greenland), dropped significantly to average less than 10,000 tonnes a year. At present, roundnose grenadier in the north west Atlantic is fished under quota regulation only within the zones of Canada and Greenland.

The roundnose (also called rock, black, bluntnose, blunt-snouted) grenadier is a demersal fish with an elongate body (Liem and Scott 1966). It is rounded in front, strongly compressed posteriorly, and tapers gradually from the origin of the first dorsal fin to the tail. A blunt nose extends over the snout and beyond the mouth which is situated on the underside of the head with viliform teeth in both jaws. The eyes are large and egg-shaped. The largest recorded specimens were about 130 cm in total length (Sahrhage 1986).

Perhaps because of their deep distribution coupled with their relatively low commercial value, comparatively little research has been conducted on roundnose grenadier in the north west Atlantic. Scientists from the Soviet Union (now Russia) have regularly reported on roundnose grenadier, primarily because of their ongoing fisheries. These reports include discussions of the biology and distribution of this species (e.g. Pechenik and Troyanovsky 1969, 1970, Chumakov and Savvatimsky 1990, Savvatimsky 1969, 1972, 1993). Parsons (1976) described the distribution and relative abundance off the east coast of Canada based on the results of Canadian research surveys conducted in the area from 1958 to 1973. Sahrhage (1986) summarised the available literature for this species in the North Atlantic. Atkinson and Power (1987b) provided an update of information from the Canadian database. Annual assessments of the resource, first carried out by the International Commission of the Northwest Atlantic Fisheries (ICNAF) continue to be conducted annually by the Northwest Atlantic Fisheries Organization (NAFO).

Much of the information available on the biology and fishery for roundnose grenadier in the north west Atlantic is unavailable in the primary English literature, but ˜rather can be found in non-English publications, or other lesser known manuscripts. It is the purpose of this document to review this literature base, and in combination with results from more recent data analyses present a comprehensive review of our current state of knowledge of this marine resource as it exists in the north west Atlantic (Figure 1).

2. Distribution

2.1 GEOGRAPHIC RANGE

Roundnose grenadier are found along slopes of continental shelves in both the north west and north east Atlantic in depths of about 200 to 2000+ m, as well as in the area of the Mid-Atlantic Ridge (Figure 2). Suitable depths extend between the two sides of the Atlantic via the Faroe-Iceland-Greenland Ridge, around Greenland, and between Greenland and Labrador which allows for movement between the areas (Marshall 1965). In the north west Atlantic, the species can be found from Cape Hatteras in the south to between Baffin Island and Greenland in the north.

The earliest records in the north west Atlantic are from the southern Scotian Shelf and Georges Bank area (Figure 1). Bigelow and Schroeder (1953) reported that roundnose grenadier had been taken often enough on the continental slope of the Gulf of Maine and off New England to be considered common in depths greater than 635 m. Schroeder (1955) found roundnose grenadier in depths >365 m from the outer edge of Scotian Shelf south to Cape Charles. The abundance was greatest in the Scotian Shelf area although he did not consider them to be present in sufficient amounts to be of any commercial interest. Interestingly, in 1993, a Canadian fishing company found a commercially acceptable concentration of roundnose grenadier in deep waters off the Scotian Shelf (pers. comm. M. O'Conner of National Sea Products). Markle and Musick (1974) found the species along a transect at 900 m extending from Hudson Canyon to Norfolk Canyon. They occurred in almost all fishing sets north of about 38° 50'N but in only 3% of them south of that location. Markle et al. (1988) found roundnose grenadier in depths of 552 to 1085 m along the outer edge of Scotian Shelf although the numbers caught were not high.

Detailed explorations of areas off Newfoundland and northward including Davis Strait began in 1957 (Travin and Pechenik 1962). The main investigations were for cod, haddock and redfish, but incidental catches of other species were made, although these were not described in any detail. In 1968, the Soviet Union first reported that a 'special' fishery for roundnose grenadier had taken place in Division 3K in 1967 (Konstantinov and Noskov 1968). This resulted from earlier explorations in deep water, and reports of these investigations, which began around 1962, soon followed.

54

Savvatimsky (1969) provided details of Soviet and German Democratic Republic (GDR) investigations to the areas off Newfoundland, Labrador and Baffin Island dating back to 1962. They were originally targeted at determining the distribution of redfish and Greenland halibut, but significant quantities of roundnose grenadier were found in depths greater than 650 m. Although the relative proportion of the three species fluctuated in depths of 500-700 m, at greater depths roundnose grenadier consistently dominated. Two main areas of concentration were identified, one in Division 3K and the other in divisions 2G and 2H (Figure 3).

Pechenik and Troyanovsky (1969, 1970) noted that roundnose grenadier had shown up as incidental catches in the fisheries for cod and redfish. They reported that from 1965 to 1968 investigations had been carried out from the north-eastern slope of the Grand Banks to the Davis Strait in 500 to 1400 m. During the first two years, they caught fish in 500 to 650 m but the concentrations were quite mobile and did not provide stable catches. Catches improved sufficiently in 1967 and 1968 to enable a commercial fishery to begin first in Division 3K in 1967, then in Subarea 0 in 1968. These investigations were carried out during the second half of the year when the waters were ice free. Their analysis suggested more extensive distribution of major concentrations than indicated by Savvatimsky (1969) (Figure 4).

The first comprehensive analysis of Canadian research vessel data was carried out by Parsons (1976). Catches were greatest off Funk Island Bank (Division 3K), Hamilton Bank (Division 2J) and the northern third of Labrador (divisions 2G and 2H). More recent Canadian data (Atkinson and Power 1987b) suggest that the largest concentrations reside off West Greenland (Subarea 1) - an area not included to a large extent in Parson's (1976) database - and in divisions 2G, 2H and 3K with somewhat smaller catches in Division 2J. Chumakov and Savvatimsky (1987) also found the largest catches to be in divisions 2G, 2H and 3K. Brodie et al. (1992) found the highest density of roundnose grenadier to be in Division 2G during a line transect survey of divisions 2G,2H,2J,3K,3L and 3M, but the highest density of small fish was found in Division 3K. Overall, Canadian data agree with the general distribution pattern described by Pechenik and Troyanovsky (1970).

Marshall and Iwamoto (1973) reported that roundnose grenadier distribution extends to 66° N in both the north west and north east Atlantic, although Eliassen (1983) found the species further north in the coastal depressions and fjords of Norway. Forest et al. (1978) caught specimens as far north as about 65° N during fishing in Davis Strait, but Berth et al. (1979) did not catch the species north of about 63° N in Subarea 0 although they also fished to 66° N. The results of a Canadian survey in Davis Strait (Atkinson and Bowering 1987), to about 70° N, confirm a northern limit of about 66° N. More recent surveys (Yamada, et al. 1988, Yatso and Jørgensen 1989a, Jørgensen and Akimoto 1990, 1991, Yano and Jørgensen 1992, and Satani et al. 1993) to subareas 0+1 consistently recorded roundnose grenadier only as far north as about 65° N.

Savvatimsky (1972) and Atkinson and Power (1987b) concluded that relatively few roundnose grenadier were present south of Division 3K. Canadian data revealed only small catches in deeper water around the edge of Flemish Cap, Grand Bank and St. Pierre Bank (Parsons 1976). Konstantinov (1980) reported relatively small catches of roundnose grenadier amounting to less than one tonne per hour trawling in depths of 1000 to 2000 m north-east of Flemish Cap. Baird and Bowering (1986) concluded, based on the commercial fishery, that there were probably few roundnose grenadier in Division 3L. A Russian survey in 1992 included 9 sets in the Flemish Pass area. Grenadier catches made up to 15-20% of catch rates of 20-180 kg. per hour (Savvatimsky 1993).

2.2 DEPTH RANGE

Parsons (1976), Forest et al. (1978), Berth and Vaske (1980) and Savvatimsky (1986) all found that the largest catches were from depths greater than 500 m. Because all sampled fish were either immature or maturing, Pechenik and Troyanovsky (1969, 1970), Grigorev (1972) and Savvatimsky (1972, 1982) concluded that distribution must extend to, and spawning take place in water deeper than 1400 m. Parsons (1976) also concluded that the distribution extended deeper than waters currently fished either commercially or by research vessels. Roundnose grenadier have been caught as deep as 1750 to 2600 m in Division 3K and as deep as 1000 to 1200 m in Division 3M (Konstantinov 1980). In 1991, Savvatimsky reported that the 1990 trawl survey conducted by the Soviets did not detect any concentrations of roundnose grenadier in depths less than 1200 to 1300 m. Possible reasons for this were not given.

Konstantinov (1980) noted that this species forms commercial concentrations in shallower waters in the north (600 to 800 m in Subarea 0) than in the south (1000 to 1300 m in Division 3K) and suggested a geographic cline in depth distribution. This possibility has not been explored further.

2.3 TEMPERATURE RANGE

Parsons (1976) reported that the fish were generally found in water temperatures of 3.5 to 4.5deg. C. Temperatures where roundnose grenadier were found in the greatest concentration in Subarea 0 in 1977 ranged from 3.0 to 4.0deg. C (Minet et al. 1978). In Davis Strait, Burmakin (1978) found that the fish formed concentrations at the warm boundary (3 to 5deg. C) of fronts, and that the grenadier accumulate under the centre of warmer water when this lies above the bottom. He concluded that catches of roundnose grenadier fluctuate in relation to temperature, increasing when warmer and decreasing when temperatures are cooler. Chumakov and Savvatimsky (1983, 1984, 1987, 1990) and Savvatimsky (1986, 1988) found the highest catch rates in Subarea 2 in 800 to 1100 m in 1969 to 1971 but somewhat deeper, 900 to 1100 m, during 1972 to 1980. They suggested that roundnose grenadier form the most dense concentrations as the core of Atlantic warm water (>4deg. C) approaches the continental slope at about 800-1000 m, and hypothesised that the cooling of the water masses in the north west Atlantic in the 1975 to 1983 period resulted in a dispersal of roundnose grenadier along with movement into deeper, warmer waters since they contended that the species forms the densest concentrations when temperatures are above 4.0deg. C. Therefore, during the period of general cooling beginning in the 1970s and extending into the 1980s, roundnose grenadier gradually moved into deeper waters. This argument was also put forward by Ernst (1984). Atkinson and Power (1987b) found the largest concentrations associated with the warmest bottom temperatures encountered; 3.5 to 3.9deg. C. Jørgensen and Akimoto (1990) found that in Subarea 1 in depths of 1000 to1500 m where almost all of the biomass of roundnose grenadier was found between 1988 and 1990, the bottom temperatures averaged about 3.4deg. C.

In the north west Atlantic, there is good correlation between depth and bottom temperature. In the depths occupied by roundnose grenadier, bottom temperatures do not fluctuate seasonally, and only fluctuate by a few tenths of a degree between years. It therefore seems unlikely that major changes in distribution of these fish occur because of these minor temperature fluctuations. Nonetheless, the movement of roundnose grenadier between deeper and shallower water in relation to temperature has been a recurring theme in much of the Soviet literature since the 1960s, where it has been discussed in relation to variations in the fishery. This is discussed in more detail later.

Interestingly, in the north east Atlantic roundnose grenadier are usually found in warmer water, and Eliassen (1983) has suggested that roundnose grenadier adapt to quite different temperatures in different regions of the continental slopes of the North Atlantic. Around Iceland, they were most abundant in 600 to 1000 m (Magnússon 1977, 1987) at bottom temperatures of 3.2 to 6.8deg. C, and along the continental slopes off the Hebrides and Rockall the largest catches were taken in 600 to 900 m at temperatures ranging from 7.6 to 11.5deg. C (Bridger 1978, Ehrich 1983). None were caught in temperatures <1.5deg. C. In the coastal waters and fjords of Norway (north of about 66° N), the species was taken in 350 to 707 m of water (Eliassen 1983). In this area, bottom temperatures were quite warm, being about 6.0 to 6.7deg. C. Further south, in the Biscay area, the highest concentrations were found deeper, at about 1000 m (Ehrich 1983). He attributed this difference in the distribution with depth to a preferred temperature range (6.0 to 7.5deg. C) which occurs deeper in the more southern area. Kotthaus and Krefft (1967) found this species in depths of 601 to 650 m and temperatures of 3.0 to 4.0deg. C on the Iceland-Faroe Ridge. This temperature range is more similar to that found in the north west Atlantic.

The reasons why roundnose grenadier are found in different temperature regimes in the north west and north east Atlantic is unknown. It may be that they are tolerant of, and can adapt to, a wide range of temperatures from about 3 to >10deg. C, and some other factor exerts greater control over their distribution than purely temperature. In the area of preferred habitat, they tolerate the ambient temperature which just happens to be different in the north west and north east Atlantic.

2.4 DIEL VERTICAL MOVEMENTS

Based on sounder records, Pechenik and Troyanovsky (1969, 1970) concluded that the fish make vertical migrations off bottom at certain times of the day to heights of about 120 m. Savvatimsky (1969) also noted that besides near bottom, sounder records revealed concentrations to 300 to 500 m up in the water column. In contrast, Jørgensen and Akimoto (1991) found that with the exception of one tow, catches in the pelagic zone were small during a survey in Subarea 1. They concluded that the fish were moving up in the water column to feed.

Pechenik and Troyanovsky (1969, 1970) described a diurnal migration which they also related to feeding. They concluded that roundnose grenadier do not follow their prey up into the water column at night, but remain close to the bottom resulting in higher catch rates during that time. As their food descends closer to the bottom, the fish move up somewhat to feed resulting in lower catch rates during the day. Savvatimsky (1985) also concluded that these fish exhibit diel movements off the bottom, but in contrast to the conclusions above, reported that catches were generally smaller at night in both the north west and north east Atlantic.

Savvatimsky (1985) also found diel movement to be more pronounced for smaller fish. Although Merrett (1978) found both larvae and juveniles up in the water column in the north east Atlantic with a tendency for smaller individuals to be higher, he found no evidence of diel movement of larvae or juveniles. It should be noted however, that "smaller" fish described by Savvtimsky (1985) were larger than the juveniles Merrett (1978) refers to.

In the north east Atlantic Haedrich (1974) reported a modal distance off bottom of 480 m in the Denmark Strait. Eliassen (1983) reported catches up to 100 m off the bottom along the coast and in the fjords of Norway, but the largest pelagic concentrations were found only about 25 m off bottom. In the same area, Bergstad and Isaksen (1987) found single acoustic echoes to 300 m from the bottom.

Korotkov and Yakimovets (1985) found evidence of diel movement in the Mid-Atlantic Ridge area with fish being detected as high up as 30 to 140 m. Zaferman (1991) found dense aggregations of roundnose grenadier at a height of 10-20 m in the area of the North Atlantic Ridge and Reykjanes Ridge in water depths of 700-1300 m. They appeared to prefer areas of weak currents in troughs and gorges. The overall quantity of fish in the pelagic zone was less than near the bottom. They concluded that the fish moved up and down in the water column in relation to the tides, moving up with rising tides and down during low tide.

Available data do indicate that roundnose grenadier are found up in the water column as well as near the bottom. The evidence remains conflicting however with regard to the nature of these movements; both their timing and purpose. Further studies are required to clarify the issue.

2.5 SEASONAL AND INTER-ANNUAL MOVEMENTS

Differences in catches at different times of the year suggested that the fish moved up and down the slope seasonally (Pechenik and Troyanovsky 1969, 1970, Po-drazhanskaya 1971, Grigorev 1972, Zilanov 1976, and Savvatimsky 1972, 1982). They appeared to begin to migrate into shallower water around the end of summer. The concentrations were stable in the second half of the year off north-east Newfoundland, Labrador and in Davis Strait, then the fish moved back to deeper water in winter. Soviet investigations, based on stomach content data, attributed these migrations to be in pursuit of their prey. Off west Scotland, roundnose grenadier also exhibit seasonal migrations, moving to deeper water during the winter (Bridger 1978).

There have also been a number of Soviet reports describing the inter-annual movement of roundnose grenadier between deeper and shallower water in relation to temperature. As noted previously, this is discussed in more detail later.

2.6 DISTRIBUTION BY SIZE

Savvatimsky (1969) and Pechenik and Troyanovsky (1970) concluded that the largest roundnose grenadier live off Iceland and are, on average, about 20 cm (total length) longer than those off north-eastern Newfoundland (Division 3K). In the north west Atlantic, they, and Podrazhanskaya (1971) found a gradual decline in size of these fish from north to south. Parsons (1976) also found the smallest fish in the south, as did Atkinson and Power (1987b) who reported a north-south decline in fish size in subareas 2 and 3. They did, however, find smaller fish in subareas 0 and 1. Berth et al (1979) made similar observations. In contrast, Savvatimsky (1993) found a gradual increase in mean fish size from divisions 0B to 3K, although the survey extended to deeper water, where larger fish were found, in the more southern areas. Thus his findings may well have been an artifact of the depths surveyed, although Jørgensen and Akimoto (1991) found that fish size increased from north to south in Subarea 1.

Savvatimsky (1969) stated that the length of the fish may increase gradually with depth but never decreases. His data from more recent years (Savvatimsky, 1989, 1990, 1991, 1993) support this conclusion, as do the findings of Brodie et al. (1992). Jørgensen and Akimoto (1991) found that in Subarea 1, fish size increased with depth, and that fish size also increased from north to south. Parsons (1976) did not find any relationship between average weight of fish caught and depth. He did note that the largest fish were occasionally found in relatively shallow water (e.g. Division 3K), similar to the results of the 1991 Soviet survey reported by Savvatimsky (1992). Atkinson and Power (1987b) found considerable variability with fish size and depth although in the more northern areas (subareas 0 and 1) there was a distinct increasing trend in size with depth. Further south, in water of less than 500 m, large fish were often encountered although present in relatively few numbers. In subareas 0 and 1, small fish were mainly present in depths <800 m but were caught down to about 1000 m. In Division 2H, these small fish were only present between 600 and 800 m, while in divisions 2J and 3K they were caught from 600 to 1000 m. They were caught in all depths fished in Division 3L (500 to 800 m).

Magnússon (1987) reported that catches of small fish (<20 cm total length, or 50 mm anal fin length (AFL)) were rare in depths of <800 m but were frequent in depths >1000 m off the south-west of Iceland, north of Reykjanes Ridge. Gordon (1978) found small fish (<25 cm total length at about 1000 m off the west of Scotland while Bridger (1978) reported that most juvenile fish (<25 cm total length) were caught in depths >1050 m in the same area. Ehrich (1983) also found that the proportion of juveniles increased with increasing depth in the area, but that the length of females did not change with depth. The overall decrease in size with depth was related to the increasing proportion of males with depth.

Overall, the evidence concerning size distribution is somewhat conflicting and confusing. It is possible that seasonal movements between deeper and shallower water, and the timing of the various studies contribute to this confusion, and more detailed examination of the databases would be necessary to clarify this. In general however, it does appear that there is probably a cline in size distribution in the north west Atlantic both in a north-south direction within at least Subarea 2, and also with depth. Larger fish are generally further north and in deeper water. The finding of very small fish in deep waters off Iceland and in the north east Atlantic may be related to reproduction occurring in very deep water.

3. Biology

3.1 SEX RATIO, SPAWNING AND STOCK STRUCTURE

Savvatimsky (1969) reported that in the north west Atlantic the sex ratio was quite constant from 1965 to 1968 along the whole of the continental slope regardless of season or depth, being about 65% males. Nevertheless he reported that the fishery in a small area of Division 3K from 1967 to 1969 did have an apparent effect on the sex ratio which changed from about 61% to 69% males over this period (Savvatimsky 1972). Atkinson and Power (1987b) also found that the sex ratio was fairly constant over the entire area during 1974 to 1986 at about 60% males. Geistdoerfer (1979a) found that overall, 69% of the roundnose grenadier caught were males during experimental fishing on the continental slope of Newfoundland and Labrador in 1975. More recently, Savvatimsky (1982) concluded that the sex ratio is subject to seasonal as well as year-to-year fluctuations. In the depths sampled (to 1500 m), the proportion of females increased from January to April, then decreased again around September-October, perhaps reflective of spawning migrations between deeper and shallower waters.

Changes in the sex ratio have also been reported for other areas. Danke (1987) found the sex ratio to vary between about 3% and 58% females during different hauls in the Mid-Atlantic Ridge area. In the north east Atlantic the sex ratio has also been shown to vary with location. Whereas Savvatimsky (1969) reported that only about 30% of the roundnose grenadier caught off Iceland were males, Magnússon (1978) found about 64% males off East Greenland, a similar ratio to that in the north west Atlantic. No small fish were found in the area. In some areas off Iceland the percentage females was as high as 91% (Magnússon 1977), but the ratio of females to males differed between areas ·and years (cf. Magnússon 1979). Gordon (1978) found that off west Scotland, males and females tended to be separated by depth. Ehrich (1983) found an increasing proportion of males with depth in the same area.

Grigorev and Serebryakov (1981) have briefly described the eggs of roundnose grenadier and their development, however the spawning time of roundnose grenadier is uncertain. Phleger (1971), in discussing macrourids generally, considered that there is a specific spawning season, and Andriyashev (1954) first suggested that roundnose grenadier spawn in late autumn. Savvatimsky (1969) found spawning males and post-spawning females during October in 1968. In contrast, Marshall (1965) concluded that spawning occurs in both spring and autumn, and Geistdoerfer (1979) found that spawning takes place in spring based on information from both the north east and north west Atlantic. Pechenik and Troyanovsky (1970) considered that the spawning period is drawn out, and Grigorev (1972) concluded that spawning takes place throughout the year.

Investigations around Iceland led Magnússon (1977, 1987) to conclude that although the period of most intense spawning is during January-April, there was evidence that some spawning takes place throughout the year and that the main spawning time may be somewhat different in different areas. Her data were insufficient to determine the depth at which spawning occurs but there was some indication that it may take place off the bottom since few mature individuals were taken in bottom trawls. Gordon (1978) also concluded that spawning may take place year round off west Scotland, but was most intense during September-October. Alekseeva and Alekseev (1984) considered that spawning takes place over an extended period since mature individuals were found in the area of Reykjanes Ridge from March to October, with the maximum during June to September. There is evidence of a late autumn and winter spawning period in the waters of Norway (Bergstad and Isaksen 1987).

When taken together, the current data support the conclusions of Magnússon (1977, 1987) that spawning takes place to some degree throughout the year but that there are periods of more intense spawning during particular periods. These periods appear to vary between areas. Further study of this is necessary, and may be informative in helping to clarify stock structure.

Debate concerning stock structure has existed for many years, and continues today. The difference in the sex ratios between Iceland and the north west Atlantic, along with the fact that larger and mature fish are found off Iceland, led Zakharov and Mokanu (1970), Podrazhanskaya (1971) and Alekseeva and Alekseev (1984) to suggest that one stock exists in the North Atlantic, and that the spawning area is off Iceland. They hypothesised that the eggs and larvae are carried passively to West Greenland and off Baffin Island and eventually to the continental slope of Canada. Larger, maturing fish (40-50 cm) migrate back to Iceland to spawn. Pechenik and Troyanovsky (1970) however, concluded that because of the shape of these fish, roundnose grenadier do not make long migrations. The single stock hypothesis was also dismissed by Grigorev (1972) and Savvatimsky (1972, 1982) who also argued that they were probably not capable of the long migrations back from Canadian waters to Iceland. Dushchenko (1985) and Dushchenko and Savvatimsky (1987) also concluded that there is no evidence of any large scale migrations of mature or maturing fish.

Dushchenko (1983, 1988) concluded that based on variable egg and larval drift, roundnose grenadier exhibit a number of hierarchically linked population structures. Dushchenko and Savvatimsky (1987) considered that the main spawning areas are in the area of the Mid-Atlantic Ridge and in the north east Atlantic. The period of passive dispersal in the currents of the eastern and western North Atlantic may last 1-2 years until mass settlement begins when the fish are 10-15 cm long. They believe that the overall population is represented by a series of local groups between which exchanges take place at early developmental stages. They also hypothesised that cooler water in the north west Atlantic prevents maturation, and roundnose grenadier in this area represent a dead-end population, being dependent upon drift of eggs and larvae from other areas for recruitment. However, Grigorev (1972), noted that juveniles (about 8 cm) were found along all of the Canadian coast as well as along West Greenland and even in south west Icelandic waters and suggested that this would be unlikely if there was localised spawning off Iceland only. Savvatimsky (1972, 1982) reported sexually mature individuals in depths of 1400 to 1500 m in the north west Atlantic as far south as Scotian Shelf with the percentage of mature individuals increasing with depth, and he also found specimens as small as 7 cm. He stated that these supported the hypothesis of spawning in deep water in the area. Savvatimsky (1972) and Grigorev (1972) both proposed that spawning takes place in the north west Atlantic in waters deeper than those generally fished (>1000 m). Geistdoerfer (1979a) agreed with the observations of Grigorev and Savvatimsky above, and confirmed the presence of sexually mature individuals in the north west Atlantic. Atkinson and Bowering (1987) found mature and maturing males during a survey in Davis Strait in August of 1986.

Savvatimsky (1972, 1982) concluded that the difference in the weight-at-length of the fish off Iceland and those in the north west Atlantic was indicative of different stocks. Atkinson (1989) also concluded, based on length-weight differences, that separate stocks exist in the North Atlantic. He considered that differences in weight-length relationships may be useful in discussions of stock separation, and the results of his analysis tentatively suggested separate populations in the north east and north west Atlantic, with a third group existing along the north Mid-Atlantic Ridge and up into subareas 0 and 1. Savvatimsky and Atkinson (1993) found, however that differences in length-weight relationships can be artefacts of sampling and advised caution in interpretation of this type of data. Nonetheless, sexually mature fish as well as small juveniles have been found in the north west Atlantic as noted above. Also, Danke (1986) found mature and maturing individuals in the Mid-Atlantic Ridge area, and concluded that spawning takes place in that area. In the north east

Atlantic, Merrett (1978) found both larvae and juveniles over bottom depths of 2700 to 2800 m at 60° N 20° W, and considered them to be of local origin. Dushchenko (1980) concluded that separate populations exist on Bill Baileys Bank and Hatton Plateau. Although very few spawning individuals were caught, Magnusson (1979) concluded that spawning does take place around Iceland. The presence of small specimens (<25 cm) in the west, south and south-west areas (Magnusson 1980) was interpreted as further confirmation that these areas serve as nursery grounds and spawning occurs in the area. Pechenik and Troyanovsky (1970) also reported that spawning does take place off Iceland in depths greater than 1000 m from about July to February.

Savvatimsky (1982), Dushchenko (1985) and Dushchenko and Savvatimsky (1987) also stated that the differences in parasites in fish from different areas indicated isolation of groups. Logvinenko et al. (1983) concluded, on the basis of genetics, that roundnose grenadier of the North Atlantic are divided into a number of reproductively isolated groupings and that the eastern, central and western areas are clearly separable.

Savvatimsky and Shibanov (1987) have summarised the hypotheses concerning the inter-relationships of roundnose grenadier in the North Atlantic. They pointed out that two schools of thought existed; one supporting the single stock hypothesis, and the other in favour of the "dead-end" idea for roundnose grenadier in the north west Atlantic. Besides the hypothesis of spawning off Iceland, they described a slightly different idea put forward by Zubchenko who suggested that roundnose grenadier in the north west Atlantic migrate via Flemish Cap to Reykjanes Ridge to spawn. The authors concluded that the dispersion of eggs and larvae over a wide area is indicative of a single population. It appears however, that these authors ignored much of the information available concerning the presence of mature and maturing individuals as well as juveniles in many different parts of the North Atlantic. There are no data to support the idea of any return migration from the northwest Atlantic, nor any data to support the idea of significant numbers of large maturing individuals in the Flemish Cap area.

Overall, they correctly concluded that more extensive studies are needed to resolve the question of stock structure, and that international co-operation is necessary so as to include as extensive an area as possible for study. Until such work is carried out, definitive conclusions regarding stock structure will not be possible. At present, it seems obvious that spawning does take place in many different areas of the North Atlantic. Whether this is significant in all of the areas such that local stocks can be supported or not remains unclear. It is obvious however, that resolution of the issue with regard to the resource is important in relation to the development and sustaining of any commercial fishery.

3.2 AGE, GROWTH AND MATURITY

Over the years, debate has existed concerning the best way to measure the length of roundnose grenadier. Because of tail breakage and regeneration, many believed that an alternative measurement to total length is required. Jensen (1976) provided early information on the relationship between total length and pre-anal fin length. Based on work by Atkinson (1981b) NAFO adopted anal-fin length (tip of snout to base of first anal fin ray) to the nearest 0.5 cm as the standard for measuring roundnose and roughhead grenadiers. Savvatimsky (1981, 1984) argued that because pre-anal length was different for males and females, and because the relationship changed with increasing length, partial lengths should not be used, and total length is more "accurate." In drawing these conclusions however, he failed to consider that total length measurements by Soviet scientists had routinely been done in 3 cm groupings, and this negated any potential gains in accuracy using total length. Other Soviet scientists (Bajdalinov et al. 1986) concluded that because of the high incidence of tail breakage (>50%) a transition to the use of partial lengths was necessary. Magnusson (1987) also uses pre-anal length routinely.

The debate came to an end in 1987 when Savvatimsky (1987b) published information on the conversion of pre-anal length to total length in different parts of the North Atlantic, and concluded that "Mass measurements of roundnose grenadier should be carried out by ante-anal distance......" Nonetheless, the different measurement standards in use by different laboratories over the years have made comparisons of results of different studies of age and growth more difficult to compare.

Roundnose grenadier are considered to be fairly long lived and slow growing (Savvatimsky 1969). Using scales, he found that samples from Iceland, Davis Strait and north east Newfoundland in the 60-80 cm range were between ages 10 and 15 (Savvatimsky 1971a, 1971b, 1975). Fish of lengths about 95 cm caught off Iceland were estimated to be about 25 years of age (Figure 5). These results are similar to those reported by Kosswig (1980) who also used scales (Figure 5). Bridger (1978) used otoliths to age fish taken off west Scotland. He concluded that the fish lived to age 40 years or more, and that these fish were between 90 and 100 cm in length (Figure 5). He also concluded that natural mortality is about 0.2. This seems somewhat high for fish estimated to live beyond 40 years. German Democratic Republic scientists have also reported that based on otoliths the growth rate is slower (cf. Atkinson et al. 1982). Savvatimsky (1971b) was not successful in ageing roundnose grenadiers using otoliths, but later work (Savvatimsky et al. 1977) suggested that no differences existed when ageing by scales or otoliths. Gordon (1978) also aged specimens using otoliths, and found no significant differences in growth rate from that reported by Savvatimsky (1971). A number of different researchers have found that males are smaller at age than females (Savvatimsky 1969, 1971a, 1971b, Kosswig 1986, Magnusson 1986, 1987). Little to no work has been carried out related to age validation, and the differences in estimated age-at-length between different aging experts, and scales versus otoliths remains to be resolved.

Various authors have described the weight-length relationship of roundnose grenadier in different regions of the North Atlantic. Savvatimsky (1969) published the first relationships for males and females, and these indicated that females are heavier at length than males. He found differences in the relationships between north east Newfoundland and Iceland, with fish off Iceland being almost twice as heavy at length. He later (1971) revised his relationships for the north west Atlantic. Borrmann (1976) provided information from subareas 0 and 1 and Division 2H separately, but Atkinson (1980b) found some discrepancies based on age/length data submitted to NAFO, and using sample weights for collected frequencies concluded that while Borrmann and Savvatimsky had derived similar relationships for Subarea 2, Borrmann's relationship for subareas 0 and 1 was probably incorrect. Atkinson recommended using a combination of the relationships of Savvatimsky (1971) and Borrmann (1976) for roundnose grenadier in the north west Atlantic until further work was done. Atkinson (1989) carried out a comprehensive analysis of weight-length relationships in various areas of the North Atlantic (Figure 6) combining information from previous studies using both total length measurements and partial length measurements. As noted above, he concluded that the data suggested three possible populations. Most recently, Savvatimsky and Atkinson (1993) found that results of different studies on samples from Subarea 2 and Division 3K indicated general agreement while results of different studies on samples from subareas 0 and 1 were more variable. It was concluded that at least some of the variability may be due to differences in condition between seasons and years, and because of this the authors advised caution in interpretation of results as indicators of different stocks.

Zakharov and Mokanu (1970) suggested that sexual maturity begins at lengths of 40 to 50 cm, and Gordon (1978) also found that the first length of sexual maturity was about 50 cm. Eliassen (1986) and Bergstad and Isaksen (1987) reported the age of first maturity to be 8 and 10 years respectively for males and females off Norway. It is not known how the ageing was carried out, but based on scale readings from fish sampled off West Greenland by Kosswig (1980) these would represent fish from 40-50 cm, whereas using otoliths would result in somewhat smaller lengths of between 35 and 45 cm (Bridger 1978). Magnússon (1987) also found differences between the sexes. For males, 50% maturity occurred at an anal fin length of about 15 cm, whereas it was about 17.8 cm for females. Based on Atkinson's (1981) calculations, these would be approximately 70 and 83 cm total length respectively.

3.3 FOOD AND FEEDING

Marshall (1965) noted that the general physical features of the mouths of grenadiers is such that they probably feed on plankton and micronekton, but the position of the mouth, obliquely directed downwards, would facilitate bottom feeding. Further, by using their projecting snout, they may actively pursue benthic food. Geistdoerfer (1975) produced a comprehensive analysis of the food and feeding of *Macrouridae* in general, although little detail related to roundnose grenadier was given. McLellan (1977) has also described feeding strategies of Macrourids, and found that the stomach contents are generally well correlated with differences in head shape.

Savvatimsky (1969) commented that the snout of roundnose grenadier is less elongated than in other members of the family, and when coupled with their fine teeth and fairly wide mouths would lead one to conclude that they are primarily plankton feeders. He, as well as Pechenik and Troyanovsky (1969), reported that *Pandalus sp.* predominated in sampled stomachs from subareas 2 and 3, while *euphausiids, Themisto sp.*, and *Calanus sp.* were also present. Houston (1983) and Houston and Haedrich (1986) sampled specimens taken from the Carson Canyon area of the Grand Banks off Newfoundland. They found that pelagic and benthopelagic specimens constituted >70% of the total number of food items, with *calanoid copepods* numerically dominant (69%). Other prey included, in descending order, *amphipods, cumaceans, euphausiids, polychaetes, isopods* and *ostracods.* Podrazhanskaya (1971) also reported the incidence of small fish and squid in the stomachs. Since sand, mud and stones were also occasionally found, he believed that besides feeding pelagically, they must also forage on the bottom for food. Based on parasite loads, Zubchenko (1981) also concluded that in the north west Atlantic roundnose grenadier feed on bottom organisms as well as pelagically. Geistdoerfer (1979b) indicated that contribution of benthos to the diet of roundnose grenadier was negligible in the north east Atlantic, with the predominant items in the stomach consisting of *Natantia, copepods, euphausiids, cephalopods, amphipods* and small fishes in declining proportion.

Gushchin and Podrazhanskaya (1984) compared the feeding of roundnose grenadier at various locations throughout the North Atlantic. They found that the diets consisted of a very wide range of species, and that the stomach contents varied from location to location. In the north west Atlantic, shrimp, *hyperiids, copepods* and *euphausiids* were important although the relative proportion differed from area to area, and between fish sizes. In the areas of Reykjanes Ridge and the Mid-Atlantic Ridge on the other hand, they fed predominantly on fish such as *myctophids,* but also consumed *cephalopods* and shrimp. They concluded that roundnose grenadier fill the niche of a non-specialised pelagic predator, and that differences in feeding in the different areas is reflective of differences in prey species. The trophic relationships were described for the north west Atlantic and Reykjanes Ridge/Mid-Atlantic Ridge areas (Figure 7).

Podrazhanskaya (1971) found that stomach fullness indices increased during September to December and supportive evidence of increased feeding during this period came from fatness indices. As noted earlier, Savvatimsky (1971a, 1975) and others also believed that during the second half of the year when feeding is most intense, the fish move up the slope of the shelf into shallower water. This makes them more susceptible to commercial fishing during that time of year.

That roundnose grenadier exhibit diel movements has also been described earlier. As was noted, this movement has been linked by some to feeding and the pursuit of prey (cf. Savvatimsky 1969, 1985, Pechenik and Troyanovsky 1969).

Roundnose grenadier are also prey to other species (Figure 7). Chumakov and

Podrazhanskaya (1983) recorded that roundnose grenadier (as well as redfish) were the most important prey in the northern areas where the largest Greenland halibut are found. Yatso and Jørgensen (1989b) also found that roundnose grenadier constituted an important part of the diet of Greenland halibut in the Davis Strait area.

3.5 PARASITES

It was noted above that Savvatimsky (1982) concluded that there were different isolated groups of roundnose grenadier based on differences in parasite infestation. The information available to him indicated that 74% of the samples taken off Labrador were infected with trematodes, whereas no trematodes were found in samples taken off Norway. Zubchenko (1981) looked at the parasitic fauna of roundnose grenadier in the north west Atlantic. He found a total of 18 different species of parasites. Of these, 8 characterize the parasitic fauna of roundnose grenadier. These are: *Myxidium profundum, Diclidophora macruri, Glomericirrus macrouri*, and *Gonocera macrouri* (all specific to grenadiers) and *Myxidium melanocetum, M. melanostigmum, Auerbachia pulchra* and *Philobythos atlanticus* (all specific to deepwater fish only). There were differences in the incidence of different parasites between Davis Strait, Labrador and the northern Grand Bank, but 50% of the parasites were found in all three areas. The differences appeared to be related to fish size and age. In addition to many of the above parasites, Zubchenko (1984) found incidence of other types such as *Anasakis sp.* in samples from Outer Baileys and Bill Baileys Banks and Reykjanes Ridge.

4. The Fishery

4.1 BEGINNINGS

The conclusion from all of the early investigations during the early to mid 1960s was that roundnose grenadier were the most promising marine resource prospect for a new commercial fishery in deeper waters of the north west Atlantic, and the best catches could be obtained in the second half of the year when the fish moved into shallower waters to feed. Although some catches were taken in earlier years by scouting vessels, Konstantinov and Noskov (1968) first reported to ICNAF that a 'special' fishery for roundnose grenadier had taken place in Division 3K in 1967 from October to December in 600-800 m. This was the first reported use of the commercial fleet to exploit this resource, and a total catch of 15,902 tonnes was reported. Even at this early stage, fishermen reported that the behaviour of roundnose grenadier appeared to be related to Greenland halibut (Savvatimsky 1969). When the incidence of Greenland halibut increased in an area, catches of roundnose grenadier declined. This has been a recurring theme throughout the history of the fishery.

Savvatimsky (1969) also described the importance of this species as a food source, and paid particular attention to the fat content of the liver. In the north west Atlantic, the highest content was found to occur in October at about 7%, and this was related to increased feeding during the second half of the year. The relative weight of the liver increased with increasing fish size so it was considered advantageous to catch as large fish as possible. The meat was also found to be satisfactory to the consumer.

These both also improved the attraction for this new fishery.

Savvatimsky (1969) pointed out that this new fishery was also advantageous to the Soviet fleet since the fish movements up the slope to shallower water in the second half of the year made them most amenable to fishing at a time when other fisheries were reduced because of movement of other target species inshore (Figure 8).

Pechenik and Troyanovsky (1970) described the bottom relief and type in the north west Atlantic as an aid to fishermen. They also described the meteorological conditions with information on weather, tides, currents and ice. Early work described the method of fishing particular to these deep waters. The main gear in use to catch roundnose grenadier was (and remains) the otter or bottom trawl, although since concentrations are often found up in the water column, midwater trawls have also been employed. Particular difficulties were associated with fishing in these relatively deep (for the period) waters, and early on trawl winches and rigging used for shallower operations were adapted. Special floats and foot gear were required, and it was necessary to reduce towing speeds to about 2.0 to 2.5 knots in order to allow the net to make good contact with the bottom but not tear. Because of the time required to "shoot" and retrieve the gear, Savvatimsky (1969) pointed to the importance of accurate navigational and sounder equipment. Pechenik and Troyanovsky (1970) described the evolution of specialised deepwater gear, and also provided information on how to best "shoot" in the deepwater areas of the north west Atlantic so as to avoid fouling of the gear.

Soviet scientists were not the only ones interested in this fishery during this period. Nodzynski and Zukowski (1971) also described the species and early fishery in the north west Atlantic from the Polish perspective. They too indicated the important food value of this fish based on the results of comprehensive technological and chemical analyses.

4.2 FISHING AREAS AND GEAR

Since its beginnings, the fishery has been prosecuted primarily using bottom trawls, although there have been some midwater catches recorded through the years. Although some fisheries were prosecuted by vessels of 500 to 999 GRT in the early period, most catches have been taken by large factory freezer trawlers of >2000 GRT. Codend mesh size commonly used is 60 mm, and this species is currently exempt from the 130 mm minimum mesh size requirement for other species in subareas 0, 2 and 3 inside the Canadian Zone. The fishing areas were fairly well defined based on the early exploratory work (Figure 9) and there was little change in these over time until recently when the largest catches have been recorded south of Division 3K and outside Canada's 200 mile limit taken as bycatch in the rapidly expanding Greenland halibut fishery in divisions 3L, 3M and 3N.

Early fishing was carried out in about 600 to 800 m, but was extended to deeper waters in conjunction with technology improvements such as larger and more powerful winches, and more efficient deepwater fishing gear. At present the fishery is generally prosecuted down to about 1500 m, although there are a few records of sets being made as deep as 1800 to 2000 m.

4.3 CATCH HISTORY

Roundnose grenadier catches did not appear in the Statistical Bulletins of ICNAF prior to 1967, although varying amounts were undoubtedly caught before this time and included with "other groundfish." The USSR began reporting catches from subareas 2 and 3 beginning in 1967, and some catches were also recorded for non-member countries in the same year (Table 1, Figure 10).

In subareas 0 and 1, catches gradually increased, and peaked at 12,318 tonnes in 1974 (Table 1, Figure 11). After extension of jurisdiction by Canada and Denmark (on behalf of Greenland), catches declined substantially due to reduced allocations to the nations which had traditionally prosecuted the fishery. Throughout the 1980s and until the present, catches have represented only by-catch in the offshore Greenland halibut fishery using bottom trawls. Effort in this Greenland halibut fishery has increased in recent years, particularly in Subarea 0 (Bowering 1994).

During the period of relatively high catches in subareas 0 and 1, the USSR took the majority until about 1977 after which their catches declined, and those of the Federal Republic of Germany (FRG) increased (Table 2a). There is some anecdotal information that suggests that the high catches reported by FRG were actually cod, but mis-reported as roundnose grenadier. Until 1977, most of the catches were taken during the second half of the year (Table 3a). The FRG catches, on the other hand were reported to be from the first half, and from further south than the earlier catches.

The most important fishing areas have been further south in subareas 2 and 3. Catches almost doubled between 1967 and 1968, then declined substantially in 1969 (Table 1, Figure 10) probably because of reduced effort. The highest catch in the time series was 75,445 tonnes taken in 1971. This high amount was due to an abnormally high reported catch of 54,179 tonnes from Division 2G, whereas in most years, the greatest proportion of the total catch was taken in Division 3K. After 1971, catches again declined and averaged around 20,000 to 25,000 tonnes until 1979. Quotas imposed during this period were not restrictive as they were never achieved. After 1978, the catches declined significantly as a result of more stringent controls placed on the fishery inside Canada's 200 mile limit by Canadian authorities. This is described in more detail in the next section.

Beginning in 1990, the proportion of catches taken from the remainder of Subarea 3 south of Division 3K began to increase. These catches were primarily by-catches in the expanding Greenland halibut fishery outside Canada's 200 mile limit in divisions 3L 3M and 3N. It is possible that the total reported landings actually represent a mixture of roundnose grenadier and roughhead grenadier since some statistical bodies (including Canada) routinely record log entries of "grenadier" as roundnose.

The Soviet Union accounted for over 90% of the total reported landings annually prior to 1980 (Table 2b). Since then, their share from inside Canada's 200 mile limit

TABLE 1. Nominal catches of roundnose grenadier in the north west Atlantic 1967 to 1994 by area

Year	SA 0	SA 1	Total SA 0+1	TAC SA 0+1	2G	2H	2J	3K	Other SA 3	Total SA 2+3	TAC SA 2-3	Grand Total
1967	1,129	6	1,135			868	217	16,009	210	17,304		18,439
1968	5,996	284	6,280		2,536	4,089	479	23,553	606	31,263		37,543
1969	2,642	68	2,710		387		264	11,682		12,333		15,043
1970	545	5,980	6,525				468	22,267	129	22,864		29,389
1971	4,172	4,132	8,304		54,179	2,738	81	18,392	55	75,445		83,749
1972	5,783	2,311	8,094		2,161	655	293	21,122	155	24,386		32,480
1973	1,054	3,830	4,884		5,880	232	632	10,855	165	17,564		22,448
1974	2,661	9,657	12,318		3,220	2,007	333	22,816	40	28,416		40,734
1975	204	4,749	4,953	10,000	6,489	3,536	1,754	15,388	258	27,425	32,000	32,378
1976	2,610	5,893	8,503	14,000	3,841	1,460	1,381	13,636	275	20,593	32,000	29,096
1977	721	2,214	2,935	8,000	2,597	525	206	11,935	123	15,386	32,000	18,321
1978		5,839	5,839	8,000	3,112	1,412	913	15,250	15	20,702	35,000	26,541
1979	106	6,815	6,921	8,000	1,035	3,090	438	3,200	18	7,781	35,000	14,702
1980	32	1,752	1,784	8,000	279	493	728	451	104	2,053	30,000	3,837
1981	87	392	479	8,000	967	1,693	463	3,920	42	7,085	27,000	7,564
1982	43	48	91	8,000	719	734	182	2,709		4,344	27,000	4,435
1983	46	22	68	8,000	140	1,390	36	1,916	87	3,569	11,000	3,637
1984	25	25	50	8,000	107	269	3	3,362	112	3,873	11,000	3,923
1985	16	39	55	8,000		80	13	4,642	213	4,948	11,000	5,003
1986	1	85	86	8,000	60	117	58	7,222	32	7,427	11,000	7,513
1987		377	377	8,000		254	213	6,682	1,069	8,298	11,000	8,675
1988	120	398	518	8,000	329	226	9	4,658	1,071	6,293	11,000	6,811
1989	28	54	82	8,000	32	202	47	4,361	314	4,956	11,000	5,038
1990*		156	156	8,000	86	52	2	606	3,284	4,030	11,000	4,186
1991*		155	155	8,000	41	129	46	125	4,086	4,427	11,000	4,582
1992*	89	18	107	8,000	34		18	274	7,161	7,487	11,000	7,594
1993*	0	0	0	8,000					3,839	3,839	11,000	
1994			0	8,000							4000**	

* Provisional
** Canadian Zone only

has been more variable, and the former German Democratic Republic (GDR) caught a greater proportion due to a shift in relative allocations. Catches by the European Union (EU) began to increase in 1990 as a result of increased fishing effort to deeper waters in Subarea 3, south of Division 3K and outside the Canadian 200 mile limit directed at Greenland halibut as noted above.

During the earliest years of the fishery in this area, effort and catches were highest during the second half of the year (Table 3b), and throughout most of the 1980s, almost no catch was reported during the first six months of each year. This was timed to the movement of the fish to shallower water in pursuit of prey during this period, as well as a re-direction of effort from other species as described previously. Beginning around 1990 however, catches began to be more evenly spread across all months since as described earlier, catches were primarily as bycatch in the Greenland halibut fishery outside 200 miles which occurs over the 12 months.

5. Resource Management and Stock Trends

Gulland (1968) was the first to propose a potential exploitation level for roundnose grenadier. He suggested that it may be possible to harvest as much as 10,000 tonnes annually from each of Subarea 1, Subarea 2 and Subarea 3. It was noted at the time that these were very preliminary estimates, and were really not much more than guesses. In 1971 the Standing Committee on Research and Statistics (STACRES) of ICNAF noted that based on ageing information that suggested roundnose grenadier is a relatively slow growing and long lived species, the potential productivity of fisheries may be limited similar to the situation with redfish (Anon. 1971).

In 1972, the STACRES Working Group on Co-ordinated Groundfish Surveys reported that data from research surveys could be used to evaluate major changes in the status of fish stocks (Anon. 1972). It was requested that all countries conducting surveys in the north west Atlantic prepare information on their catches per haul for evaluation by the Assessment Subcommittee. One of the species for which data was requested was roundnose grenadier.

During a meeting held in January, 1973, STACRES commented upon quotas established by the Commission for a number of stocks and species for 1973 (Anon. 1973). It noted that insufficient data were available to carry out any assessment of the roundnose grenadier resource and again requested that any available information be brought forward for discussion.

Up until that point in time, roundnose grenadier remained an unregulated species. Things began to change in 1974 however, when the first ever assessment of this species was presented to ICNAF. Pinhorn (1974) estimated long term sustainable yield for subareas 2 an 3 using data available from Savvatimsky (1971). His analysis indicated a maximum sustainable yield (MSY) between 24,000-37,000 tonnes, and he concluded that it would seem prudent to limit removals to no more than about 30,000 tonnes. Although a portion of the stock was thought to exist deeper than depths currently fished, the fished portion was probably fully exploited, and technology didn't exist to fish deeper. STACRES agreed with this and in January,

TABLE 2a. Nominal catches of roundnose grenadier by country from Sub-areas 0 and 1

Year	Den(G)	GDR	FRG	Poland	Russia	Norway	Japan	Canada
1972[2]	-	-	-	147	2164	-	-	-
1973	11	1835	-	-	3036	-	-	-
1974	5	2804	-	-	9509	-	-	-
1975	6	186	33	-	4728	-	-	-
1976	1	181	147	-	8174	-	-	-
1977	10	61	519	-	2345	-	-	-
1978	32	-	5807	-	-	-	-	-
1979	21	-	6794	-	106	-	-	-
1980	-	-	1752	-	32	-	-	-
1981	39	-	353	-	87	-	-	-
1982	37	-	11	-	43	-	-	-
1983	22	-	-	-	46	-	-	-
1984	25	-	-	-	25	-	-	-
1985	36	14	-	-	2	-	3	-
1986	81	-	-	-	1	-	4	-
1987	58	-	-	-	-	-	319	-
1988	138	-	-	-	120	-	260	-
1989	8	-	-	-	1	27	46	-
1990[1]	156	-	-	-	-	-	-	-
1991[1]	155	-	-	-	-	-	-	-
1992[1]	18	-	-	-	75	-	-	14

[1] Provisional   [2] Sub-area 1 only   Den(G) =Greenland

TABLE 2b. Nominal catches of roundnose grenadier by country from Sub-areas 2 and 3

Year	Rom	GDR	FRG	Poland	Russia	Faroe	EU	Japan	Cuba	Nor	Can(N)	Can(MQ)
1972	-	239	-	123	24024	-	-	-	-	-	-	-
1973	-	684	-	294	16586	-	-	-	-	-	-	-
1974	-	1766	199	181	26270	-	-	-	-	-	-	-
1975	-	2705	-	1499	23221	-	-	-	-	-	-	-
1976	-	497	1	101	19994	-	-	-	-	-	-	-
1977	7	613	174	15	14577	-	-	-	-	-	-	-
1978	108	1801	973	60	17760	-	-	-	-	-	-	-
1979	-	480	-	100	7201	-	-	-	-	-	-	-
1980	-	898	32	36	1087	-	-	-	-	-	-	-
1981	-	1407	-	18	5660	-	-	-	-	-	-	-
1982	-	1640	-	15	2689	-	-	-	-	-	-	-
1983	-	2586	-	50	933	-	-	-	-	-	-	-
1984	-	3650	23	51	147	-	-	2	-	-	-	-
1985	-	3740	178	12	1018	-	-	-	-	-	-	-
1986	-	4571	13	17	2801	-	3	13	-	-	-	9
1987	-	4469	-	1	2725	9	1001	79	4	-	-	10
1988	-	3380	8	17	1890	-	911	85	-	-	1	1
1989	-	2352	-	17	2230	-	290	45	-	1	1	19
1990[1]	-	1	-	-	538	-	3211	125	-	-	121	34
1991[1]	-	-	2	-	-	-	4053	-	-	-	220	152
1992[1]	-	-	-	-	30	3	7005	-	-	-	437	12

[1] Provisional   Rom = Romania   Russia inc. former USSR   Nor = Norway

72

TABLE 3a. Nominal catch by month of roundnose grenadier in Sub-areas 0 and 1

year	JAN	FEB	MAR	APR	MAY	JUNE	JULY	AUG	SEPT	OCT	NOV	DEC	U/K
1972[1]	8	-	-	-	-	12	1789	353	138	3	8	-	-
1973	-	-	-	-	4	-	47	494	528	1098	289	2424	-
1974	85	-	-	1	-	1	390	1306	182	528	2289	7527	-
1975	46	158	35	43	-	111	307	672	439	109	1171	1862	-
1976	475	7	1	197	-	-	-	206	632	1793	3276	1917	-
1977	464	94	20	14	2	5	58	1094	1089	38	18	39	-
1978	139	130	723	2554	1942	343	4	2	1	-	-	-	1
1979	605	759	348	626	1658	1122	123	118	1	185	545	831	-
1980	686	385	-	-	-	-	-	418	117	118	23	6	31
1981	1	4	-	-	-	-	-	-	170	245	17	8	34
1982	1	3	9	6	4	11	1	3	-	14	25	7	7
1983	-	3	6	5	1	-	-	-	7	5	21	14	6
1984	-	2	6	8	1	1	-	14	14	2	-	2	-
1985	1	6	8	6	3	1	-	-	5	2	19	4	-
1986	3	3	8	44	11	2	4	1	2	2	2	3	1
1987	-	-	-	-	-	-	-	48	180	-	87	4	58
1988	6	11	6	8	48	26	2	180	43	163	20	5	-
1989	2	3	-	-	11	23	2	-	5	265	8	2	-
1990[2]	-	-	-	-	-	-	-	-	-	-	-	-	156
1991[2]	-	-	-	-	-	-	-	-	-	-	-	-	155
1992[2]	-	-	-	-	-	-	-	-	-	-	-	-	107

TABLE 3b. Nominal catch by month of roundnose grenadier in Sub-areas 2 and 3

year	JAN	FEB	MAR	APR	MAY	JUNE	JULY	AUG	SEPT	OCT	NOV	DEC	U/K
1972	173	-	2	15	2748	8125	5762	3481	465	2643	921	51	-
1973	466	60	37	123	51	277	5202	3663	3514	1785	1453	933	-
1974	205	22	187	5	2	520	2479	1459	2214	4976	9050	7927	-
1975	784	1388	400	807	47	1596	812	6516	7498	3301	2332	1944	-
1976	843	1225	1	605	290	106	257	1856	1170	3961	4530	5749	-
1977	44	8	12	45	13	6	1776	5698	3411	1973	1681	719	-
1978	264	467	13	45	7	405	6416	3963	1812	3964	1484	1862	-
1979	103	32	44	6	136	683	1169	1612	1691	611	745	949	-
1980	3	4	48	13	2	-	-	130	376	794	577	106	-
1981	40	14	1	2	4	1	168	1636	1391	759	1751	1318	-
1982	4	-	3	5	3	4	559	563	410	698	1465	630	-
1983	3	18	4	-	3	1	1	74	1292	861	866	446	-
1984	31	13	6	19	-	5	-	45	460	3018	123	153	-
1985	44	7	1	96	73	-	54	873	1869	1361	537	33	-
1986	9	5	-	-	-	-	117	2818	2093	1555	494	336	-
1987	71	111	45	96	75	5	22	2732	1633	1561	1319	628	-
1988	415	33	38	-	8	87	841	837	690	1485	1608	251	-
1989	76	23	25	23	39	54	579	1497	704	902	946	88	-
1990[1]	108	322	598	1171	488	152	139	393	77	116	212	133	121
1991[1]	84	325	515	835	346	425	251	331	288	811	185	32	-
1992[1]	67	74	185	433	260	960	185	16	362	264	162	476	4043

[1] Provisional   [2] Subarea 1 only   U/K = unknown

1974 put the 30,000 tonne value forward as a recommended TAC for 1974 (Anon. 1974). The Commission responded by establishing a precautionary Total Allowable Catch (TAC) of 32,000 tonnes for 1974, although the reason for setting the actual TAC 2,000 tonnes higher than recommended is obscure.

Later in 1974 STACRES recommended continuation of the 32,000 tonnes TAC level for 1975 (Anon. 1974). It also commented that a developing fishery was also taking place in Subarea 1 and the Baffin Island area (This area west of Subarea 1 was not considered part of the ICNAF area at that time. It was later considered as Statistical Area 0, and became Subarea 0 with the formation of NAFO in 1978.), and Denmark had proposed that the fishery in Subarea 1 be regulated. There was no evidence of stock boundaries for roundnose grenadier in the north west Atlantic, and STACRES noted that it was possible that the species forms one single large stock throughout the entire area. Even if that was true however, they were not expected to rapidly migrate from one area to another, and therefore the two areas could be treated separately for practical purposes. STACRES did not have any data available upon which to base an assessment, but advised the Commission that if they wished to impose a precautionary TAC, it could be imposed for Subarea 1 plus the Baffin Island area and be set at about 6,000 tonnes, close to the average catches of recent years (Anon. 1974). The Commission responded by setting the TAC for these areas in 1975 at 10,000 tonnes. Again the rationale for the increase from the recommended TAC is unclear.

The separation of assessments and TACs between the northern subareas 0 and 1 and southern subareas 2 and 3 has remained through to the present. There has been little to no subsequent discussion of the appropriateness of this separation, although available data indicate that it is not inappropriate, particularly in light of the fact that most of the fishery in the southern area has traditionally occurred in Division 3K. In 1986, the Fisheries Commission of NAFO posed a number of questions to the Scientific Council concerning roundnose grenadier in subareas 2 and 3 (Anon. 1986). These included requests for information on evidence for the existence of different stocks within the area, the proportion of the biomass in the NAFO Regulatory Area (that area of the north west Atlantic outside the various national zones) as opposed to within the Canadian Zone, the proportion of the stock in Division 3L in the Regulatory Area, and catches associated with $F_{0.1}$ and F_{max} in Division 3L. The Scientific Council was unable to provide answers to these questions except to say that existing evidence did not indicate significant quantities of roundnose grenadier south of Division 3K. It was also pointed out that additional research would probably not change this perspective.

In 1975, in response to an increased catch in Subarea 1 and Statistical Area 0 (to 12,000 tonnes), STACRES recommended an increase in the TAC to 12,000 tonnes for 1976 (Anon. 1975). The reason for this is unclear unless it was not desirable to place any restrictions on existing fishing effort. For subareas 2 and 3, it was recommended that the TAC remain at 32,000 tonnes. While the recommendation for the more southern area was accepted, the Commission set the TAC for the northern area at 13,500 tonnes for 1976.

During the mid 1970s, there was considerable discussion within the Commission of ICNAF concerning effort limitations as a means of controlling fisheries, and effort controls were actually put in place for some fisheries. The Commission had excluded roundnose grenadier from the regulations for 1976, but in September 1975, STACRES indicated that it did not agree with this (Anon. 1975). It noted that from a biological perspective, roundnose grenadier were an integral part of the groundfish ecosystem and they should be included in any overall regulation of this system. From a practical perspective, exclusion of roundnose grenadier effort from the controls would result in less of an overall reduction than would be otherwise achieved. STACRES was unable to comment on this matter in 1976 because it was too early to evaluate possible impacts (Anon. 1976). No further comments on this initiative were made by STACRES, and the issue of specific effort controls died with extension of national jurisdictions in 1977.

Borrmann (1976) prepared the first estimates of sustainable yield based on separate cohort analyses for the two stock areas. She determined that for Statistical Area 0 and Subarea 1, the $F_{0.1}$ sustainable yield was between 7,400 and 9,800 tonnes depending on whether natural mortality (M) was assumed to be 0.1 or 0.2 respectively. For subareas 2 and 3, the comparable values were between 31,800 and 40,800 tonnes. Based on these calculations, STACRES recommended that the TACs for 1977 be set at 8,000 and 35,000 tonnes respectively (Anon. 1976). These recommendations were supported by the scientific advisors to Panel 1 and Panel 2 respectively (Anon. 1976).

The management regime for the two stocks of roundnose grenadier changed dramatically at the beginning of 1977. With the declaration of 200 mile economic zones by Canada and Denmark (on behalf of Greenland), the southern stock (at least the traditional fishing areas) came within the jurisdiction of Canada. Although Canada continued to ask for scientific advice from ICNAF (later replaced by NAFO), it became a national prerogative to set TACs. The northern stock fell under the jurisdiction of Canada and Denmark (on behalf of Greenland) and these two nations established final TACs and allocations bilaterally although scientific advice was (and still is) requested through ICNAF then NAFO.

The 1977 TAC levels were carried over to 1978 due to a lack of any additional information to suggest any change (Anon. 1977). Savvatimsky (1977) concluded from his analysis of size, age and sex composition of catches from 1967-1976, that the fishery had not negatively affected the stocks.

In 1978, Borrmann updated her analyses based on a more complete data set. She found that in Statistical Area 0 and Subarea 1, fishing mortalities had approximated $F_{0.1}$. Sustainable yields at $F_{0.1}$ and F_{max} were estimated to range between 6,900 to 7,500 tonnes and 7,000 to 8,700 tonnes respectively at Ms of 0.1 and 0.2. Comparable calculations for the stock in subareas 2 and 3 gave estimates of 26,100 to 30,700 tonnes and 27,700 to 37,700 tonnes. At the same meeting, Parsons et al. (1978) presented the first general production analysis of these stocks using commercial catch and effort data which had been reported to ICNAF. Their analyses indicated an MSY of about 8,000 tonnes for the northern stock, and 32,000 tonnes for the southern one. Based on these two analyses, STACRES concluded that the TAC levels of 8,000 tonnes and 35,000 tonnes for Statistical Area 0 and Subarea 1, and subareas 2 and 3 respectively should remain in place for 1979 (Anon. 1978). These were accepted by Canada and Denmark (on behalf of Greenland).

It is interesting that during this period, TAC levels were established and then maintained based alternately on MSY (approximate F_{max}) and $F_{0.1}$ (approximate 2/3 MSY effort) calculations. There are no indications in the ICNAF Redbooks of any discussions of the implications of these apparent management strategy changes nor on the most appropriate reference exploitation for roundnose grenadier. There were ongoing general discussions during this period of the relative merits of these different target exploitation rates, but mainly in conjunction with exploitation of the more traditional species such as cod and redfish.

Another general production analysis was prepared in 1979 (Atkinson 1979). For Statistical Area 0 and Subarea 1, the MSY was again estimated to be 8,000 tonnes and the yield at 2/3 MSY effort to be 6,700 tonnes. In subareas 2 and 3, comparable calculations were 31,000 tonnes and 27,500 tonnes respectively. Based on these calculations, it was recommended that the TAC for the northern area remain at 8,000 tonnes for 1980, but that the TAC for the southern stock be reduced to 30,000 tonnes (Anon. 1979). Besides the general production analysis, this later decision was made based on the observation that catch rates had been declining during the second half of the 1970s. These recommendations were accepted and put in place for 1980.

During 1979 there was a gradual transition of assessment and management responsibilities from ICNAF to NAFO. This was necessary because with the declaration of 200 mile limits in the north west Atlantic, the original mandate of ICNAF was no longer applicable, and a new body with an updated mandate was necessary. The last round of assessments carried out by ICNAF were those done in 1979. In 1980 an new body, the Scientific Council of NAFO, took over this responsibility. As described earlier, management responsibilities for the traditional fishing areas for roundnose grenadier in subareas 0 and 1 and 2 and 3 were moved from the international forum to respectively Canada and Denmark (on behalf of Greenland) bilaterally, and Canada alone, although both parties have continued to request advice through NAFO up to the present.

Beginning in 1981, analyses of the catch and effort data were done using a multiplicative model (Gavaris 1980). The application of this model was considered superior to using simple annual averages since it allowed for inter-calibration and standardisation of catch and effort between different country-gear-tonnage classes, divisions, and months so that an overall standardised catch-per-unit-effort could be obtained for each year (Figure 12).

General production analyses continued to be attempted annually, but increasing difficulties were encountered with time. In NAFO subareas 0 and 1, the catches were low and amounted to by-catch only, resulting in a loss of the directed effort time series. As a result, the advice remained at 8,000 tonnes through the 1980s and to the present (1994) for this stock since there was no additional acceptable information available upon which to base any change. In 1986, Canada conducted a survey to the area and obtained a biomass estimate of about 110,000 tonnes (Atkinson and Power 1987a). An 8,000 tonne TAC was only 7% of this; well within a reasonable exploitation level. More recent estimates of biomass from this area have been much lower (about 40,000 tonnes in 1987-1989, and only about 5,000 to 8,000 tonnes more recently) (Yamada et al. 1988, Yatso and Jørgensen. 1989, Jørgensen and Akimoto 1990, 1991, Yanu and Jørgensen 1992, and Satani et al. 1993) but the TACs have not been lowered because of uncertainties surrounding interpretation of the survey results. They have been conducted at different times of the year, and have covered different depth ranges. In 1991, two surveys were conducted with very different results (Yanu and Jørgensen 1992) possibly indicating a seasonal component to survey results related to movement of the fish up and down the slopes between deeper and shallower waters. In summary, the low estimates from recent surveys have not been accepted as indicative of stock size, and there is no reason to suspect that the stock in this area is any lower than it was at the time of cessation of the directed fisheries. Leaving the TACs at 8,000 tonnes will allow for possible fisheries to develop if, for some reason, new interest is generated.

Based on data from subareas 2 and 3, the correlations between catch-per-unit-effort and effort (used in general production analysis) began to deteriorate with the inclusion of more years' data (Atkinson 1980a, 1981a). Nonetheless, based on the 1980 analysis, the TAC for 1982 was lowered to 27,000 tonnes. Regression results were not statistically significant when examined in 1982 and subsequent years (Atkinson 1982, 1983, 1984, 1985). Another Cohort analysis was also attempted in 1981 but could not be successfully calibrated using the catch-per-unit-effort data. The Scientific Council considered that the catch rate data were probably unreliable as indicators of stock status (Anon. 1981). It was thought that this may be related to the way the statistical data were grouped as non-directed catches could be included and impact on estimated catch rates. There was therefore no change in the advised TAC level for 1982.

In 1982, a virtual population analysis (VPA) based on data only from Division 3K was reviewed (Savvatimsky and Shafran 1981). The results indicated that assuming M=0.2, the biomass in the division had declined fairly steadily from about 164,000 tonnes in the 1960s to only 92,000 tonnes in 1978 (Figure 13). Fishing mortalities on the fully recruited ages had ranged between about 0.2 and 0.8, and had been about 0.45 in recent years compared to $F_{0.1}$ and F_{max} estimates of 0.3 and 0.65 respectively. The best estimate of yield at both $F_{0.1}$ and F_{max} was put at about 5,500 tonnes based on indications of declining recruitment in recent years (Figure 13) (Anon. 1982). Coupling this with an average catch of about 5,750 tonnes for Subarea

2 (excluding the high catch in 1971) resulted in an estimated long term yield of only 11,000 tonnes for the two subareas combined. The Scientific Council, while recognising that catch rates may not be indicative of stock status, was concerned that these rates had declined since the early 1970s, with the 1980 point being the lowest observed. As a result of these deliberations, beginning in 1983, annual TACs were set at the much lower level of 11,000 tonnes.

It became mandatory beginning in 1977 for non-Canadian vessels fishing in Canadian waters to carry observers on board. As a result, a comprehensive dataset was gradually accumulated containing, among other things, set by set catch and effort information. If the assumption of the NAFO Scientific Council concerning the catch and effort data was correct, then analyses based on the observer data should correct the problem since any analysis could be restricted to those sets meeting specified criteria related to the catch of roundnose grenadier. Analysis of both the NAFO and observer datasets was first carried out in 1986 (Atkinson and Power 1986). The two resultant series showed similar trends in catch rates for the 1980 to 1984 period (Figure 12). Regressions of catch-per-unit-effort on effort were not significant using the observer data (Atkinson and Power 1987a) and it was concluded that these data are not appropriate for evaluating stock status. The previous supposition of the Scientific Council concerning the catch and effort data was apparently not correct, and other factors were obviously important in affecting catch rates. Since then, general production analyses have not been attempted, although Atkinson and Power (1988) have re-examined both catch rate series but concluded that there was insufficient contrast in the data over time for them to be workable in general production models.

Overall, catch rates were lower after the extension of jurisdiction (Figure 12). Soviet scientists noted back in the late 1960s that catches of roundnose grenadier were related to the amount of Greenland halibut in the area (Savvatimsky 1969). After extended jurisdiction, Canada restricted the by-catch of Greenland halibut taken in the roundnose grenadier fisheries by the USSR and GDR to only 10% although there were no scientific studies done to evaluate the practicality of this level. With the apparent gradual increase in the size of the Greenland halibut stock during the early 1980s it became, according to Soviet and GDR scientists, increasingly more difficult to direct for roundnose grenadier in the traditional areas. They attributed the low catch rates and catches to the regulated necessity of fishing marginal areas of distribution and concentration so as to avoid Greenland halibut (Chumakov and Savvatimsky 1983, 1984, Ernst 1984). They also indicated that by-catches were greatest in the north, and decreased as one moved south. Bowering (1983) analysed information contained in the Canadian observer and research vessel databases and also concluded that by-catches were greater further north. In most instances, the proportion of Greenland halibut compared to roundnose grenadier was greater than 10%. It was concluded (Anon. 1983) that a more realistic level of by-catch would be in excess of 20% with an increasing allowance from south to north. This was re-iterated in 1984, 1985 and 1986 (Anon. 1984, 1985, 1986) when the Scientific

Council advised that a 20% limit for Division 3K and a 30% limit for Subarea 2 would be more realistic than the current 10% restriction. Kulka (1985), based on further analyses of the Canadian observer data concluded that by-catches not only decreased from north to south, but that they also declined with increasing depth.

In 1986 and 1987 the GDR was permitted a 30% by-catch when fishing in depths greater than 800 m. Preliminary analyses of the data from these two years from the GDR and Soviet fisheries indicated variable results (Anon. 1988) which did not support the theory that by-catch limitations were restricting catches of roundnose grenadier. The by-catch increase to 30% in depths >800 m was extended to the Soviet Union for 1988. Detailed analysis of the catch and effort data collected from the roundnose grenadier fishery indicated that historically, catch rates were not higher when the by-catch of other species was higher, nor when the by-catch restrictions were changed from 10% to 30% (Atkinson and Power 1989). On the contrary, catch rates appeared to decline. From a scientific perspective, the discussions ended at this point although the by-catch level of 30% when fishing in depths >800 m has remained in place ever since.

Soviet scientists also attributed the lower catches through the 1980s to the fact that while the fishery was being prosecuted to maximum depths of only about 1500 m, the fish were known to be distributed much deeper than this. Chumakov and Postolaky (1979) made reference to a correlation between bottom temperatures and the proportion of roundnose grenadiers and Greenland halibut in the Davis Strait area. Catches of Greenland halibut were found to increase in years of lower temperatures. Burmakin (1978) also made reference to differences in distribution of the two species depending on bottom temperature. Because of the general cooling trend in the north west Atlantic during the 1980s, the roundnose grenadier were considered to be moving to the deeper water and becoming less and less available to the fisheries (Borovkov et al. 1989, Chumakov and Savvatimsky 1983, 1984, 1990, Ernst 1984, Savvatimsky 1987a, 1988).

Savvatimsky (1989) related the increase in catches in 1987 and 1988 to a temperature increase in the main branch of the Labrador current. He predicted that if this warming trend continued, catches would also continue to increase. Standardised catch rates derived for catch and effort data collected by Canadian observers also showed some increase during these years (Figure 12).

Catches and catch rates had both declined again substantially by 1990 (Figure 10, 12). The previous assertions concerning links between water temperatures and roundnose grenadier catches were brought into question by Savvatimsky himself (1991) who reported that the 1990 survey did not find concentrations of roundnose grenadier in depths less than 1000-1200 m. He reviewed earlier literature relating decreased catches to cooling temperatures, and concluded that it remains unclear to what extent this cooling may have played a role. He could find no relationships between the declines in the catches and water temperatures. He even went so far as to say that "little is known about the influence of water temperature on roundnose grenadier behaviour.", and contrary to his earlier interpretation of events, concluded that the reasons for the declines in catches (and by inference catch rates) during the 1980s had not been elucidated yet. He nonetheless felt that the fish had migrated to deeper waters. As a footnote to this, temperature data available during the cooling period of the 1980s indicate that bottom temperatures at the depths under consideration fluctuated only by tenths of degrees. Although one cannot be certain, it seems unlikely that changes on this small scale would precipitate changes in distribution of roundnose grenadier of the magnitude required to explain the fishery trends over the same period.

In 1990, a concerted effort was made by a multi-national group of scientists to re-examine in detail the existing information based on a combined dataset of Soviet Union and GDR (Savvatimsky et al. 1990). Different approaches to virtual population analysis were attempted assuming M=0.15 and calibrating with different catch rate series. Although all estimated parameters were significant based on one of the analyses, and the results from both were quite similar, neither was considered satisfactory. In many years the estimated exploitable biomass was close to or less than the reported catch, and this was thought to reflect the fact that the fish were distributed beyond the maximum depths fished. It was re-iterated that the use of information from trawl surveys was inappropriate because these surveys do not cover the entire depth range of distribution. The analyses presented led the Scientific Council of NAFO to conclude that in its present form, virtual population analysis is an inappropriate tool because the calibration indices do not come from the entire area of distribution (Anon. 1990). Keeping the TAC at a precautionary level of 11,000 tonnes was advised although average catches had been well below this in recent years, because most of the catch had been taken in Division 3K, while the TAC was applicable to all of subareas 2 and 3.

6. The Present

Catches and catch rates in the traditional fishing areas of subareas 2 and 3 have continued to decline during the 1990s (Figure 10, 12), and Russia did not conduct any commercial fishery in 1992 although they had an allocation of 4,100 tonnes. Soviet (now Russian) surveys have continued to show declines in the traditional areas and depths. In 1991, the survey found the highest concentration at about 59° 50' N 1100-1360 m (Borovkov et al. 1992). In the traditional fishing area in Division 3K, catches were small; about 50% of levels found in 1990. Research catch rates declined further in 1991 (Borovkov et al. 1993).

A detailed analysis of the commercial catch and effort data obtained from the Canadian observer program was carried out in 1993 (Atkinson et al. 1993). After a relatively stable period to 1987, catch rates declined steadily. Mapping of the catch rate data from the second half of the 1980s illustrated the same declines beginning around 1988 (Figure 14). Because of statements that the fish had been gradually moving deeper, the observer data were examined by 250 m depth interval. The results suggested that catch rates did not decline in the 500-750 m range, but declined over all other depth ranges simultaneously. This would suggest that if the fish had moved deeper, they must have done it on a large scale in a short period. This was considered unlikely. Canadian autumn survey data, although only extending to 1000 m, were also examined. Results also show declines in the traditional fishing area of Division 3K in recent years (Figure 15). The TAC level of 11,000 tonnes was established partially based on the estimated status of the resource in Division 3K, and catch rates were stable until 1987 when catches from this area averaged about 6,000 tonnes. Because catch rates had declined by about 50% since then and the Scientific Council of NAFO was very concerned about this, it therefore recommended a new lower precautionary TAC level of only 3,000 tonnes (Anon. 1993). It was emphasised that this would apply to the traditional fishing areas within the Canadian zone only. This was subsequently accepted by Canadian authorities and put in place for the Canadian Zone for 1994.

At this point in time, one can only speculate as to why catch rates have declined in the traditional fishing areas of subareas 2 and 3 in recent years. The observed declines are consistent with those observed for many other groundfish species in the same area (Atkinson 1993). It has been hypothesised that changing environmental conditions most commonly recorded as cooling water temperatures and the increasing volume of the Cold Intermediate Layer (CIL - considered that portion of the Labrador current water mass with temperatures <0.0deg. C) may have played a role since the trends are similar for all species whether commercially exploited or not. It is believed that some species may have moved further south and to the east. For example, Greenland halibut is believed to have migrated (Bowering 1994) resulting in a significant increase in effort and catches in divisions 3L,3M,3N and 3O south of the traditional fishing areas. This new fishery has gradually evolved in a southerly direction, perhaps because of continued southerly movement by the fish. In the developing Greenland halibut fishery outside the Canadian zone, by-catches of roundnose grenadier have been low on a percentage basis, but nonetheless represent the highest catches ever reported from areas south of Division 3K. Although this may be most related to the tremendous increase in effort in the area, given the substantial exploratory work conducted by the USSR many years ago, it might be expected that higher catches may have been achieved in earlier years had the fish been there. Thus it is possible that roundnose grenadier may also have to some extent moved south and east from the traditional areas of distribution. Regrettably, it is not possible to investigate this with the existing data, and any hypotheses must remain only speculation.

7. The Future

In 1993, NAFO once again indicated that the necessary data required to carry out more detailed assessments of roundnose grenadier in the north west Atlantic are not available (Anon. 1993). Extensive survey information from the deepwater areas both inside and outside the Canadian zone are required. It is not expected that these data will be forthcoming as a result of any research directed at roundnose grenadier because of the relatively little interest in this species, even in the face of declines in more traditional groundfish resources. On the other hand, there is a great deal of interest in the Greenland halibut resource. It is possible that in the process of investigating Greenland halibut, data useful in forming a better understanding of roundnose grenadier resource in the north west Atlantic will be obtained.

8. Conclusions

Roundnose grenadier were first described in the 18th century, and extensive investigations in the north west Atlantic were carried out beginning in the early 1960s, but a surprising amount remains to be learned about the species in this area. The geographic distribution has been well described for many years, and in particular the location of the highest concentrations are well known because of the commercial fishery. Although the species is distributed continuously from Davis Strait (at about 66° N) to off Cape Hatteras, the best fishing has traditionally occurred in Division 3K centred around 50° N in 750 to 1500 m. That this represents an area of preferred habitat may be assumed, but this is not known for sure nor the possible reasons understood. Previously it had been hypothesised that changes in distribution also occurred in relation to temperature, but the most recent work now suggests that this is not the case. There does appear to be seasonal migration between deeper and shallower water related to feeding, but the consensus at present is that larger scale migrations probably do not occur, at least not within any yearly time scale.

Although small juveniles as well as some mature females have been found in the north west Atlantic, the numbers have been low. This implies that a) spawning may take place at depths greater than those normally fished or otherwise sampled, and/or b) productivity is extremely low, or c) as has been hypothesised, the population in the north west Atlantic is dependent on egg and larval drift from other areas such as Iceland. Although there has been some speculation that larger maturing adults return to Iceland and the Mid-Atlantic Ridge to spawn, many reports describe roundnose grenadier as being incapable of long migrations because of their body shape, and therefore the north west Atlantic population is "dead end." The species occupies different temperatures in the north east and north west Atlantic, and it has been further proposed that maturation does not occur at the lower temperatures found in the north west Atlantic. At this point in time however, all of this is speculative.

There has never been a full and comprehensive examination of all of the data from different studies in any consistent manner, and this is necessary in order to help clarify the numerous remaining questions. This examination would require a collaborative international effort so that all available data sets from the North Atlantic could be included. Such a study would provide useful and important information to our understanding of roundnose grenadier in all of the North Atlantic. This type of information is vitally important for management of any fishery taking place now or in the future. It is of course critical to have a good understanding of stock structure although as yet our ideas about this for roundnose grenadier are vague at best. It is also important to understand the reproductive strategies and processes in order to best evaluate strategies for long term sustainability in the fisheries.

In the north west Atlantic, quota management for the roundnose grenadier has been based on the same criteria as for many of the shallower water species even though all signs were of generally lower productivity. This in itself would result in a gradual decline of the resource. The history of the management of this species in the north west Atlantic can provide us with other important clues about appropriate approaches to the management of deepwater resources. It is obvious, in hind sight, that roundnose grenadier south of Davis Strait were over-exploited during the early years of the fishery, and that science was slow in responding to the limited indicators that suggested this had occurred.

The fishery began after 'commercial' concentrations had been located, but without extensive analysis of the size of the resource nor its population dynamics. In the absence of catch or effort controls, the fishery expanded at a pace dictated by man's desire for the fish rather than based on any *a priori* consideration of what level of removals the resource might be able to absorb. The first lesson should be that in the future, any deepwater fisheries must be developed cautiously, and only in parallel with scientific research.

Shortly after the fishery began, and when assessments were first carried out, the calculated MSYs were in the range of catches of the time, so there was no apparent need for concern. It appears however, that the early assessments were overly optimistic in predicting what the resource could withstand in the way of fishing pressure. It is often the case that only catch and effort data are available from newly developed fisheries upon which to base an assessment. It is common to use these data in production models to calculate long term "equilibrium" yields at their commonly calculated reference points of MSY and 2/3 effort at MSY. It is now well known that this type of model can give overly optimistic predictions about the resources and their sustainability, particularly during a period of stock decline (e.g., during intial fishery development) (Hilborn and Walters 1992). Data collected during the beginnings of a fishery, when the virgin stock is being fished down, do not reflect long term prospects for the stock in question with regard to sustainability. For a slow growing species with low inherent productivity, this over-estimation will be further exacerbated Therefore, 'caution' is advised in the application of "equilibrium" production models as they will paint an overly optimistic picture of long term sustainable yields.

Although updated data analyses done during the late 1970s and early 1980s suggested lower MSYs than had earlier calculations, there was an ongoing reluctance to lower the recommended quotas as far as some analyses suggested they should be. In addition, declines in indices of abundance such as catch rates were interpreted, for various reasons, to not be reflective of trends in the resource. Rather, reasons for not changing recommended catch levels were advocated even though catches were well below the established TACs. It is now apparent that more attention should have been paid to these indices, and that again as a measure of caution, they should have been taken at face value. Although arguments that the distribution of the resource may extend beyond the bounds of the fishery have some validity, and make interpretation of indices difficult, these arguments should not dictate policy. Rather, for deepwater resources, strong trends in indices in the fishing areas quite possibly reflect trends in the total resource, and should therefore not be ignored.

Assuming that overfishing occurred prior to 1977, the sensitivity of this deepwater species to over-exploitation can be seen in the fact that after over 15 years there is no sign of recovery; instead the decline appears to be continuing even in the absence of significant fishing pressure in traditional areas. Because of the remaining uncertainlty of the stock structure and reproductive success in the north west Atlantic, it is not clear at present whether recovery will ever occur.

Perhaps the greatest lesson to be learned is that without adequate scientific knowledge and understanding of the resource coupled with careful management, these deepwater resources may be easily depleted, and once they are fished 'out,' they may well be gone for at least the foreseeable future. It is our obligation and duty to not let this happen.

Acknowledgements

The author would like to thank the library staff of the Department of Fisheries and Oceans, St. John's, Newfoundland for their valuable assistance in obtaining much of the referenced material, and arranging for translations. The staff of Science Branch are to be thanked for careful collection of the available Canadian data, both commerical and research. D. Kulka, H. Clarke and an anonymous reviewer all provided valuable comments on earlier drafts of the manuscript, as did participants of the Workshop.

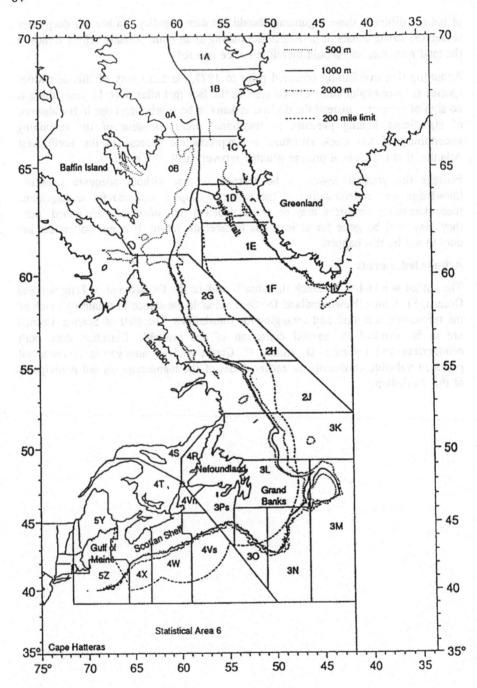

Figure 1: Map of the Northwest Atlantic showing NAFO divisions.

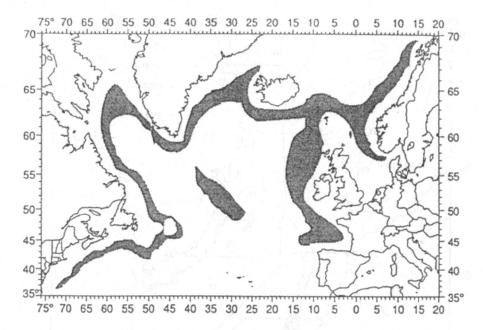

Figure 2: Distribution of roundnose grenadier (*Coryphaenoides rupestris* Gunn.) in the North Atlantic (adapted from Sahrhage 1986).

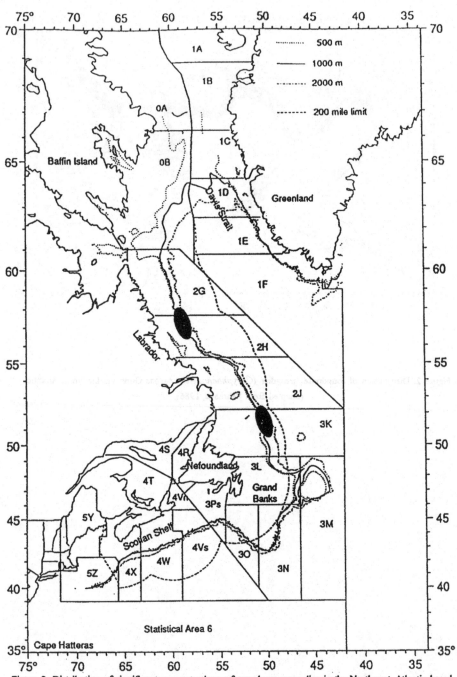

Figure 3: Distribution of significant concentrations of roundnose grenadier in the Northwest Atlantic based on USSR explorations, 1961-1965 (adapted from Savvatimsky 1969).

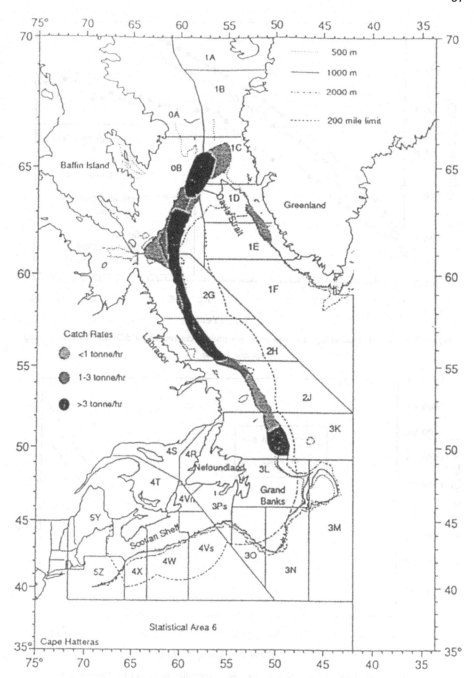

Figure 4: Distribution and density of roundnose grenadier in the Northwest Atlantic (adapted from
Pechenik and Troyanovsky 1970).

88

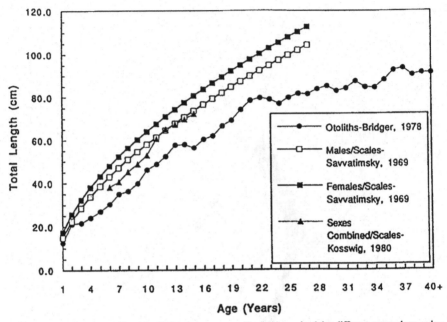

Figure 5: Age-length relationships for roundnose grenadier as determined by different researchers using scales and otoliths.

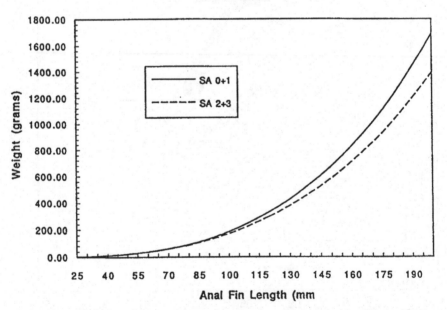

Figure 6: Weight-length relationships for roundnose grenadier as derived by Atkinson (1989).

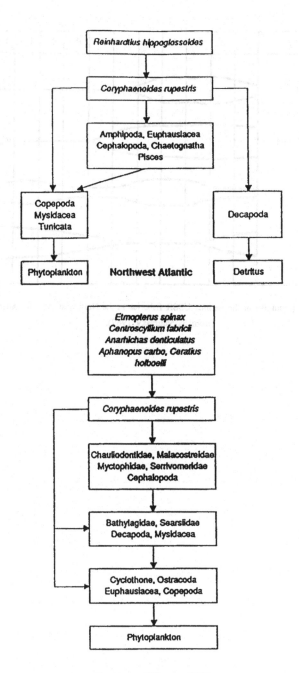

Northwest Atlantic

Reykjanes Ridge and Mid-Atlantic Ridge

Figure 7: Trophic relationships of roundnose grenadier in the Northwest and Northeast Atlantic (from Gushchin and Podrazhanaskaya 1984)

90

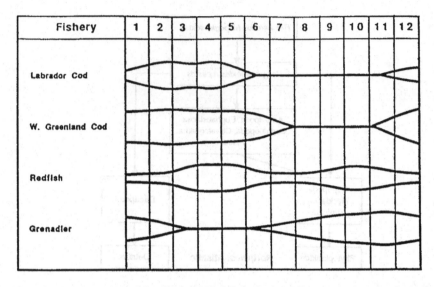

Figure 8: Timing (monthly) of the various USSR fisheries in the Northwest Atlantic during the 1960s (from Savvatimsky 1969).

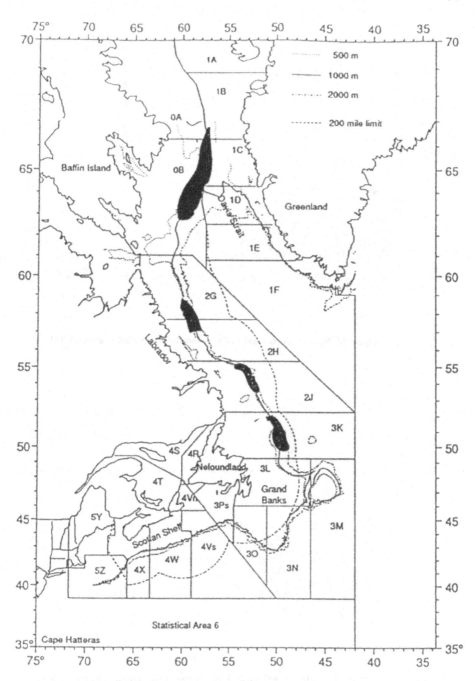

Figure 9: Traditional commercial fishing areas for roundnose grenadier in the Northwest Atlantic.

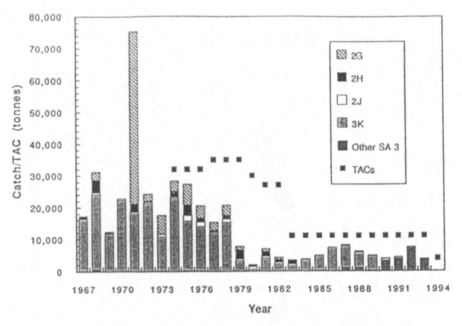

Figure 10: Nominal catches and TACs for roundnose grenadier in subareas 2+3.

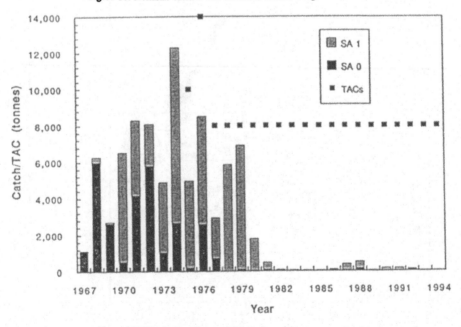

Figure 11: Nominal catches and TACs for roundnose grenadier in subareas 0+1.

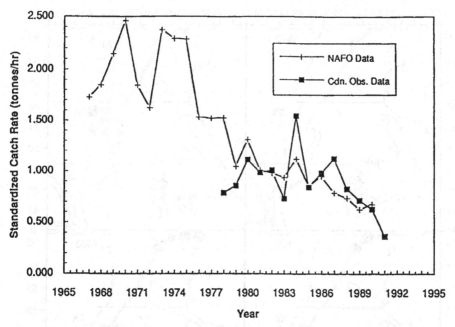

Figure 12: Standardised catch-per-unit-effort (tonnes/hour) for roundnose grenadier in subareas 2+3 as derived from catch and effort data available from NAFO, and the Canadian Observer Program.

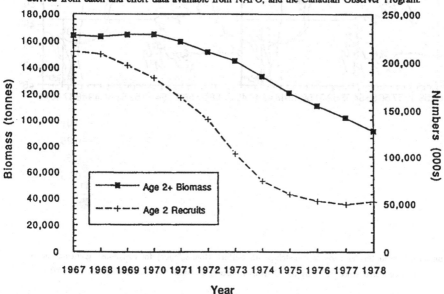

Figure 13: Trends in age 2+ biomass and age 2 recruitment for roundnose grenadier in Division 3K based on the virtual population analysis conducted by Savvatimsky (1981).

94

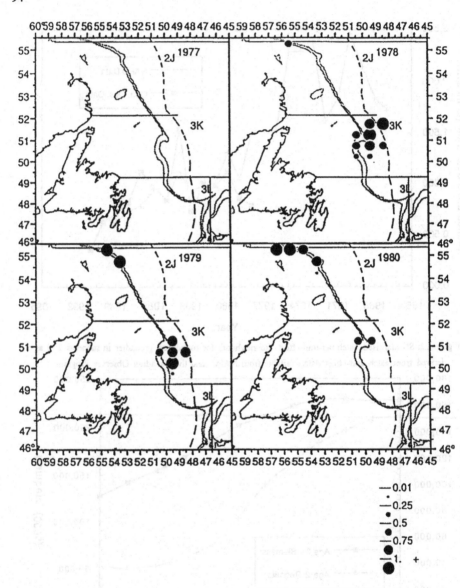

Figure 14: Distribution and trends in catch-per-unit-effort (tonnes/hour) for roundnose grenadier based on data collected since 1977 by the Canadian observers within the Canadian Zone. 1977-80

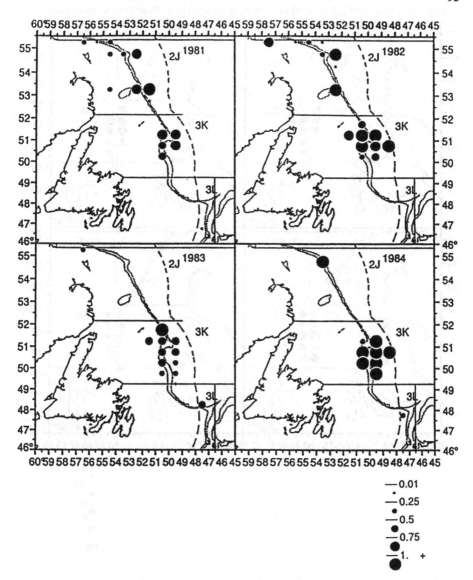

Figure 14: Distribution and trends in catch-per-unit-effort (tonnes/hour) for roundnose grenadier based on data collected since 1977 by the Canadian observers within the Canadian Zone. 1981-84

96

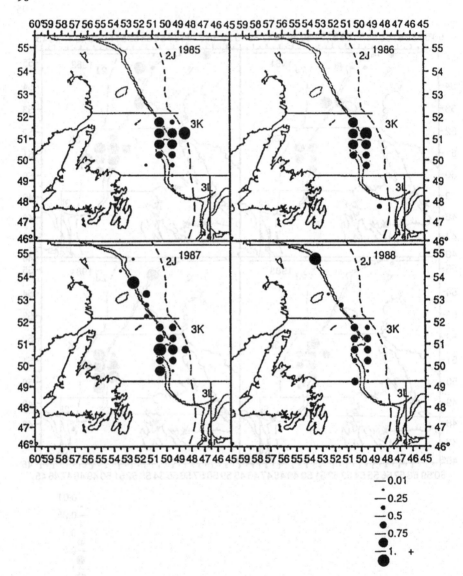

Figure 14: Distribution and trends in catch-per-unit-effort (tonnes/hour) for roundnose grenadier based on data collected since 1977 by the Canadian observers within the Canadian Zone. 1985-88

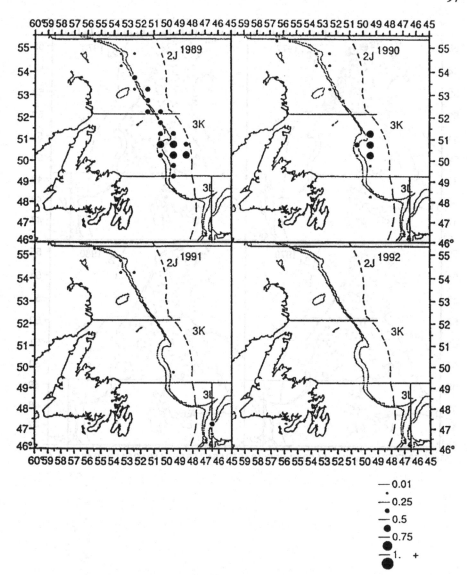

Figure 14: Distribution and trends in catch-per-unit-effort (tonnes/hour) for roundnose grenadier based on data collected since 1977 by the Canadian observers within the Canadian Zone. 1989-92

98

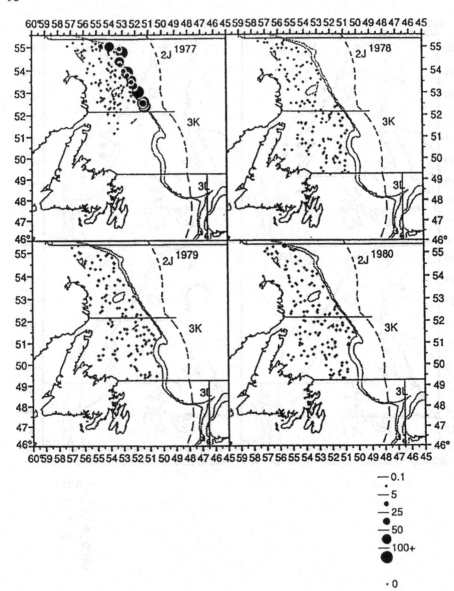

Figure 15: Distribution and trends in catches (kg/30 min. tow) of roundnose grenadier in divisions 2J and 3K as determined from annual Canadian fall stratified random surveys to 1000 m. 1977-80

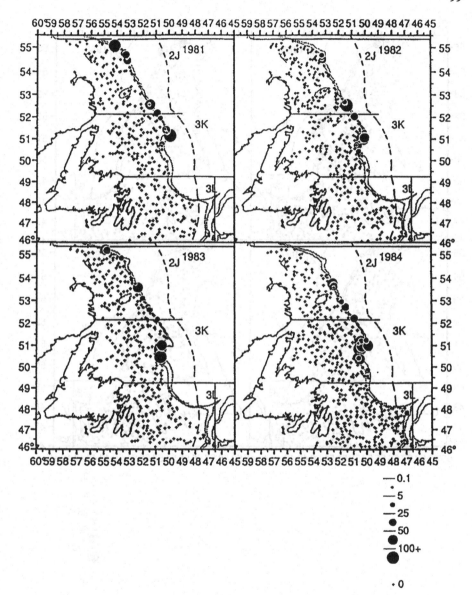

Figure 15: Distribution and trends in catches (kg/30 min. tow) of roundnose grenadier in divisions 2J and 3K as determined from annual Canadian fall stratified random surveys to 1000 m. 1981-84

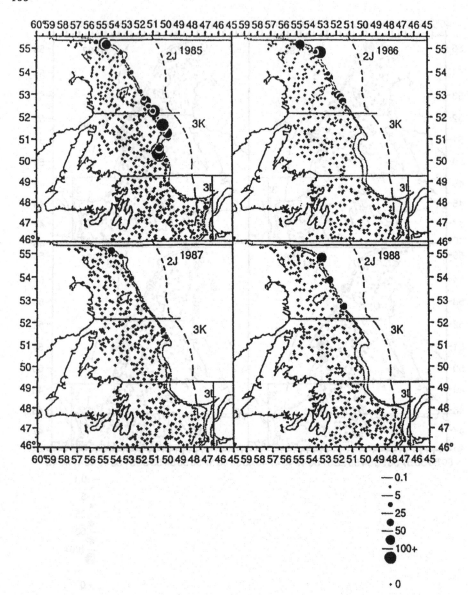

Figure 15: Distribution and trends in catches (kg/30 min. tow) of roundnose grenadier in divisions 2J and 3K as determined from annual Canadian fall stratified random surveys to 1000 m. 1985-88

101

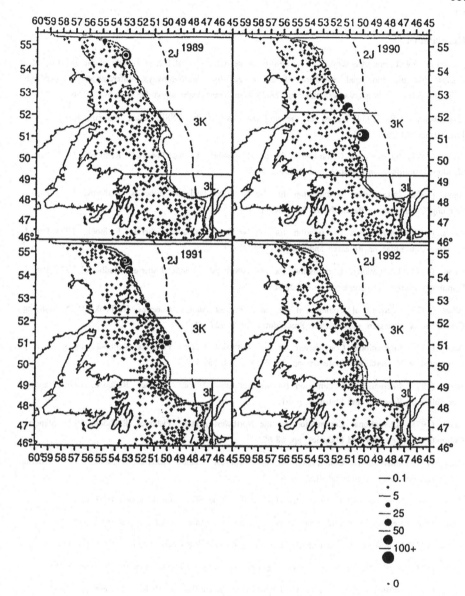

Figure 15: Distribution and trends in catches (kg/30 min. tow) of roundnose grenadier in divisions 2J and 3K as determined from annual Canadian fall stratified random surveys to 1000 m. 1989-92

102

References

Alekseeva, Ye. I., and Alekseev, F. Ye. (1984) The sexual cycles of fish in a study of the structure of a species and the functional structure of the range. In: Vnutrividovaya differentsiatsiya morskikh promyslovykh ryb i bespozvonochynkh. pp. 28-38. Can. Trans. Fish. Aquat. Sci. No. 5402, 15p.

Andriyashev, A.P. (1954) Fishes of the northern seas of the USSR. Israel Program for Scientific Translations. 617 p.

Anon. (1971) International Commission for the Northwest Atlantic Fisheries. Redbook, 1971 Part I, Standing Committee on Research and Statistics. p. 9.

Anon. (1972) International Commission for the Northwest Atlantic Fisheries. Redbook, 1972 Part I, Standing Committee on Research and Statistics. p. 7.

Anon. (1973) International Commission for the Northwest Atlantic Fisheries. Redbook, 1973 Part I, Standing Committee on Research and Statistics. pp. 27-29.

Anon. (1974) International Commission for the Northwest Atlantic Fisheries. Redbook, 1974 Part I, Standing Committee on Research and Statistics. p. 28.

Anon. (1975) International Commission for the Northwest Atlantic Fisheries. Redbook, 1975, Standing Committee on Research and Statistics Proceedings. pp. 25-27, 33.

Anon. (1976) International Commission for the Northwest Atlantic Fisheries. Redbook, 1976, Standing Committee on Research and Statistics. pp. 33, 71, 81, 134, 169-171.

Anon. (1977) International Commission for the Northwest Atlantic Fisheries. Redbook, 1977, Standing Committee on Research and Statistics. p. 61.

Anon. (1978) International Commission for the Northwest Atlantic Fisheries. Redbook, 1978, Standing Committee on Research and Statistics. pp. 68-69.

Anon. (1979) International Commission for the Northwest Atlantic Fisheries. Redbook, 1979, Standing Committee on Research and Statistics. pp. 82-83.

Anon. (1981) Northwest Atlantic Fisheries Organization. Scientific Council Reports 1981. p. 45.

Anon. (1982) Northwest Atlantic Fisheries Organization. Scientific Council Reports 1982. pp.29-30.

Anon. (1983) Northwest Atlantic Fisheries Organization. Scientific Council Reports 1983. pp. 48-49.

Anon. (1984) Northwest Atlantic Fisheries Organization. Scientific Council Reports 1984. pp. 56-57.

Anon. (1985) Northwest Atlantic Fisheries Organization. Scientific Council Reports 1985. pp. 72-73.

Anon. (1986) Northwest Atlantic Fisheries Organization. Scientific Council Reports 1986. pp. 81-82.

Anon. (1988) Northwest Atlantic Fisheries Organization. Scientific Council Reports 1988. pp. 69-70.

Anon. (1990) Northwest Atlantic Fisheries Organization. Scientific Council Reports 1990. pp. 92-94.

Anon. (1993) Northwest Atlantic Fisheries Organization. Scientific Council Reports 1993. pp. 107-109.

Atkinson, D.B. (1979) Roundnose grenadier in ICNAF Subareas 0+1 and 2+3. ICNAF Res. Doc. 79/VI/57. Serial No. 5397. 16 p.

Atkinson, D.B. (1980a) Assessment of roundnose grenadier stocks in NAFO Subareas 0+1 and 2+3. NAFO SCR. Doc. 81/VI/100. Serial No. N155. 13 p.

Atkinson, D.B. (1980b) Weight-length relationships of roundnose grenadier in Subareas 0+1 and 2+3. NAFO SCR Doc. 80/83. Serila No. N137. 5 p.

Atkinson, D.B. (1981a) Assessment of roundnose grenadier in NAFO Subareas 2+3. NAFO SCR. Doc. 81/VI/60. Serial No. N344. 12 p

Atkinson, D.B. (1981b) Partial length as a replacement for total length in measuring grenadiers. J. Northw. Atl. Fish. Sci. 2: 53-56.

Atkinson, D.B. (1982) Status of roundnose grenadier stocks in NAFO Subareas 0+1 and 2+3. NAFO SCR. Doc. 82/VI/55. Serial No. N548. 9 p.

Atkinson, D.B. (1983) Roundnose grenadier stocks in NAFO Subareas 0+1 and 2+3. NAFO SCR. Doc. 83/VI/37. Serial No. N690. 6 p.

Atkinson, D.B. (1984) An update on the status of roundnose grenadier in Subareas 0+1 and 2+3. NAFO SCR. Doc. 84/VI/20. Serial No. N793. 6 p.

Atkinson, D.B. (1985) The roundnose grenadier of Subareas 0+1 and 2+3. NAFO SCR. Doc. 85/46. Serial No. N995. 10 p.

Atkinson, D.B. (1989) Weight-Length relationships of roundnose grenadier (*Coryphaenoides rupestris* Gunn.) in different areas of the North Atlantic. Fisheries Research. 7: 65-72.

Atkinson, D.B. (1993) Some observations on the biomass ans abundance of fish captured during stratified random bottom trawl surveys in NAFO divisions 2J3KL, fall 1981-1991. NAFO SCR Doc. 93/29. Serial No. N2209. 18 p.

Atkinson, D.B. and Power, D. (1986) An update of the status of roundnose grenadier in Subareas 0+1 and 2+3. NAFO SCR. Doc. 86/29. Serial No. N1143. 10 p.

Atkinson, D.B. and Power, D. (1987a) An evaluation of the status of roundnose grenadier in Subareas 0+1 and 2+3. NAFO SCR. Doc. 87/39. Serial No. N1324. 11 p.

Atkinson, D.B. and Power, D. (1987b) The distribution of roughhead and roundnose grenadiers in the Northwest Atlantic. NAFO SCR Doc. 87/73. Serial No. N1398. 28 p.

Atkinson, D.B. and Power, D. (1988) Roundnose grenadier in NAFO SA 0+1 and 2+3. NAFO SCR. Doc. 88/26. Serial No. N1462. 11 p.

Atkinson, D.B. and Power, D. (1989) An update on the status of roundnose grenadier in NAFO subareas 0+1 and 2+3. NAFO SCR. Doc. 89/55. Serial No. N1635. 26 p.

Atkinson, D.B. and Bowering, W.R.. (1987) The distribution and abundance of Greenland halibut, deepwater redfish, golden redfish, roundnose grenadier and roughhead grenadier in Davis Strait. Can. Tech. Rept. Fish. Aquat. Sci. No. 1578. v+29 p.

104

Atkinson, D.B., Power, D., and Kulka, D.W. (1993) The roundnose grenadier (*Coryphaenoides rupestris*) fisheries in NAFO subareas 2+3. NAFO SCR Doc. 93/74. Serial No. N2259. 20 p.

Atkinson, D.B., Bowering, W.R., Parsons, D.G., Horsted, Sv. Aa., and Minet, J.P. (1982) A review of the biology and fisheries for roundnose grenadier, Greenland halibut and northern shrimp in Davis Strait. NAFO Sci. Coun. Studies. 3: 7-27.

Baird, J. and Bowering, W.R. (1986) Biomass estimates for cod and Greenland halibut beyond the Canadian 200-mile Economic Zone in NAFO Division 2J+3KL. NAFO SCR Doc. 86/51. Serial No. N1168. 6 p.

Bajdalinov, A.P., Dorovskikh, R.S., and Stulova, L.I. (1986) On the transition to a new system of measurement of the North Atlantic roundnose grenadier (*Coryphaenoides rupestris*) length. NAFO SCR Doc. 86/61. Serial No. N1178. 8 p.

Bergstad, O.A. and Isaksen, B. (1987) Deepwater resources of the Northeast Atlantic: distribution, abundance and exploitation. NAFO SCR Doc. 87/85, Ser. No. N1388, 40 p.

Berth, U., and Vaske, B. (1980) Report on groundfish survey of Walter Barth in Subarea 2 during autumn 1979. NAFO SCR Doc. 80/VI/102. Serial No. N157, 15 p.

Berth, U., Schultz, N., and Vaske, B. (1979) Report on groundfish survey carried out by the RV Ernst Haeckel in Statistical Area 0, Subarea 2 and Div. 3K during autumn 1978. ICNAF Res. Doc. 79/VI/127. Serial No. 5523, 27 p.

Bigelow, H.B. and Schroeder, W.C. (1953) Fishes of the Gulf of Maine. Fish. Bull., 74. Fish. Bull. U.S. Fish and Wildlife Service. 53: 1-577.

Borovkov, V., Gorchinsky, K., Kovalev, S., Savvatimsky, P., Rikhter. V.A., and Sigaev, I.K. (1993) Russian Research Report for 1992. NAFO SCS Doc. 93/10. Serial No. N2183. 14 p.

Borovkov, V., Kovalev, S., Savvatimsky, P., Rikhter, V.A., and Sigaev, I.K. (1992) Russian Research Report for 1991. NAFO SCS Doc. 92/12. Serial No. N2066. 21 p.

Borovkov, V.A., Bulatova, A. Yu., Chumakov, A.K., Savvatimsky, P.I., and Tevs, I.I. (1989) Bottom water effects on the distribution and density of bottom fish in NAFO Subarea 3. NAFO SCR Doc. 89/87. Serial No. N1671. 16 p.

Borrmann, H. (1976) Preliminary stock assessments of roundnose grenadier in ICNAF Subareas 0+1 and 2+3. ICNAF Res. Doc. 76/VI/27. Serial No. 3807. 12 p.

Borrmann, H. (1978) Stock Assessment of roundnose grenadier in the Northwest Atlantic. ICNAF Res. Doc. 78/VI/54. Serial No. 5220. 8 p.

Bowering, W.R. (1983) By-catch levels of Greenland halibut in the roundnose grenadier directed fishery in NAFO subareas 2+3. NAFO SCR Doc. 83/VI/28. Serial No. N680. 4 p.

Bowering, W.R. and Brodie, W.B. (1994) Greenland halibut (*Reinhardtius hippoglossoides*): A review of the dynamics of its distribution and fisheries off Eastern Canada and Greenland in A.G.Hopper (ed), Deep Water Fisheries of the North Atlantic Oceanic Slope. This volume.

Bridger, J.P. (1978) New deep-water trawling grounds to the west of Britain. Ministry of Agriculture Fisheries and Food, Directorate of Fisheries Research, Laboratory Leaflet No. 41, Lowestoft, 40 p.

Brodie, W.B., Baird, J.W., and Power, D. (1992) Analysis of data from deepwater surveys in Div. 0B, 2GHJ, and 3KLM in 1991. NAFO SCR Doc. 92/82. Serial No. N2137. 8 p.

Burmakin, V.V. (1978) The effect of water temperature on grenadier and halibut catches off Baffin Island. Tr. Polyarn. Nauchno-Issled. Proekt. Morsk. Rybn. Khoz. Okeanogr. XL: 148-152. Can. Trans. Fish. Aquat. Sci. No. 4915, 7p.

Chumakov, A.K. and Postolaky, A.I. (1979) On the USSR fisheries of Greenland halibut and roundnose grenadier in the Davis Strait area. ICNAF Res. Doc. 79/VI/126. Serial No. 5522. 6 p.

Chumakov, A.K. and Savvatimsky, P.I. (1983) On the Greenland halibut by-catch in the directed fishery for roundnose grenadier on the Labrador continental slope and in Davis Strait (NAFO Subareas 0, 1, 2 and 3K). NAFO SCR Doc. 83/IX/91. Serial No. N757, 12 p.

Chumakov, A.K. and Savvatimsky, P.I. (1984) Roundnose grenadier-Greenland halibut ratio in bottom trawl catches taken in NAFO area in 1970-1983. NAFO SCR Doc. 84/VI/37. Serial No. N822. 15 p.

Chumakov, A.K. and Savvatimsky, P.I. (1987) Distribution of Greenland halibut and roundnose grenadier in the Northwest Atlantic in relation to hydrographic conditions in 1968-1986. NAFO SCR Doc. 87/93. Serial No. N1397, 38 p.

Chumakov, A.K. and Savvatimsky, P.I. (1990) Distribution of Greenland halibut (*Reinhardtius hippoglossoides*) and roundnose grenadier (*Coryphaenoides rupestris*) in the Northwest Atlantic in relation to hydrographic conditions in 1968-86. NAFO Sci. Coun. Studies 14: 51-65.

Chumakov, A.K. and Podrazhanskaya, S.G. (1983) Feeding of Greenland halibut in the Northwest Atlantic. NAFO SCR Doc. 83/79. Serial No. N745. 22 p.

Danke, L. (1987) Some Particularities of roundnose grenadier (*Coryphaenoides rupestris* Gunn.) in the north Mid-Atlantic Ridge region. NAFO SCR Doc. 87/78. Serial No. N1378, 10 p.

Dushchenko, V.V. (1980) Preliminary results of investigations on intraspecific groupings of roundnose grenadier in the North Atlantic. ICES C.M. (1980/G:22 Demersal Fish Committee, 12 p.

Dushchenko, V.V. (1983) The relationship between genetic stability and variability in populations of rock grenadier in the North Atlantic. In: Genetics of commercial and cultivated species: Proceedings of the Second All-Union Conference on Genetics, Selective Breeding, and Hybridization of Fish. pp. 34-37. Can. Trans. Fish. Aquat. Sci. No. 5407, 7 p.

Dushchenko, V.V. (1985) The formation of the commercial stock of the north Atlantic rock grenadier. In: Behavior of commercial fishes, P.A. Moiseev, ed., USSR. Collection of scientific works of VNIRO, pp 141-154. Can. Trans. Fish. Aquat. Sci. No. 5340, 17 p.

Dushchenko, V.V. (1988) Intraspecific structure of the rock grenadier, *Coryphaenoides rupestris*, of the North Atlantic in relation to population interchange of juveniles: genetic processes. Voprosy ikhtiologii. 28: 10-21.

Dushchenko, V.V. and Savvatimsky, P.I. (1987) Intraspecific structure of rock grenadier, *Coryphaenoides rupestris* of the North Atlantic: variability of local groups and reasons for their formation. Voprosy ikhtiologii. 27: 784-793.

Ehrich, S. (1983) On the occurrence of some fish species at the slope of Rockall Trough. Arch. Fishwiss. 33: 105-150.

Eliassen, J-E. (1983) Distribution and abundance of roundnose grenadier (*Coryphaenoides rupestris* Gunnerus) (*Gadiformes, Macrouridae*) in northern and mid-Norway. ICES C.M. (1983/G:43 Demersal Fish Committee, 24 p.

Ernst, P. (1984) A contribution to by-catch of Greenland halibut (*Reinhardtius hippoglossoides* Walb.) in the roundnose grenadier (*Coryphaenoides rupestris* Gunn.) directed fishery in NAFO Subarea 2. NAFO SCR Doc. 84/IX/96. Serial No. N891. 8 p.

Forest, A. Minet, J.P., and Perdou, J.B. (1978) Results of groundfish survey on Baffin Island Shelf (ICNAF Statistical Division 0B). ICNAF Res. Doc. 78/VI/45. Serial No. 5207, 9 p.

Gavaris, S. (1980) Use of a multiplicative model to estimate catch rate and effort from commercial data. Can. J. Fish. Aquat. Sci. 37: 2272-2275.

Geistdoerfer, P. (1975) Ecological types of food of *Macrouridae* (Teleosteens Gadiformes) feeding - morphology and histology of the digestive system. The place of *Macrouridae* in the deep-sea food chain. These PhD. Universit, de Paris, Paris, France. 278 p. Can. Trans. Fish. Aquat. Sci. No. 4570. 278 p.

Geistdoerfer, P. (1979a) New data on the reproduction of Macrourids (Teleostei, Gadiformes). Sarsia 64: 109-112.

Geistdoerfer, P. (1979b) Alimentation du grenadier, *Coryphaenoides rupestris*, dans l'Atlantique nord-est. ICES C.M. (1979/G:31 Demersal Fish Committee, 9 p.

Gordon, J.D.M. (1978) The distribution of roundnose grenadier (*Coryphaenoides rupestris* Gunnerus) on the west of Scotland slope. Annls. Biol., Copenh. 34: 225-226.

Grigorev, G.V. (1972) Reproduction of *Macrourus rupestris* Gunner of the northern Atlantic. Trudy PINRO 28: 107-115. Fish. Res. Bd. Can. Trans. Serial No. 2529, 15 p.

Grigorev, G.V. and Serebryakov, V.P. (1981) Eggs of the rock grenadier, *Coryphaenoides rupestris* Gunnerus 1765. J. Northw. Atl. Fish. Sci. 2: 73-74.

Gulland, J.A. (1968) Area reviews on living resources of the world's oceans: Northwest Atlantic. ICNAF Res. Doc. 68/75. Serial No. 2064. 14 p.

Gushchin, A.V. and Podrazhanskaya, S.G. (1984) Feeding of roundnose grenadier, *Coryphaenoides rupestris* Gunn., and its trophic relationships in the North Atlantic. NAFO Sci. Coun. Studies. 7:53-59.

Haedrich, R.L. (1974) Pelagic capture of the epibenthic rattail *Coryphaenoides rupestris*. Deep Sea Research, 21: 977-979.

Hilborn, R. and C.J. Walters. 1992. Quantitative Fisheries Stock Assessment: choice, dynamics and uncertainty. Chapman and Hall, Inc. New York. 570 p.

Houston, K.A. (1983) Food sources for deep-sea fishes of the Newfoundland continental slope. NAFO SCR Doc. 83/IX/89. Serial No. N755. 19 p.

Houston, K.A. and Haedrich, R.L. (1986) Food habits and intestinal parasites of deep demersal fishes from the upper continental slope east of Newfoundland, Northwest Atlantic Ocean. Marine Biology. 92:563-574.

Jensen, J.M. (1976) Length measurement of roundnose grenadier *(Macrourus rupestris).* ICNAF Res. Doc. 76/VI/93. Serial No. 3913. 2 p.

Jørgensen, O. and Akimoto, K. (1990) Results of a stratified random trawl survey in NAFO Subarea 1 in 1989. NAFO SCR Doc. 90/39. Serial No. N1756. 14 p.

Jørgensen, O. and Akimoto, K. (1991) Results of two trawl surveys in NAFO Subarea 1 in 1990. NAFO SCR Doc. 91/50. Serial No. N1933. 14 p.

Konstantinov, K.G. (1980) Note on deep-sea trawling beyond the limits of the Canadian 200-mile zone. NAFO SCR Doc. 80/VI/52. Serial No. N089, 2 p.

Konstantinov, K.G. and Noskov, A.S. (1968) USSR Research Report, 1967. International Commission for the Northwest Atlantic Fisheries. Redbook, 1968 Part II, Reports on Researches in the ICNAF Area in 1967. pp.102-124.

Korotkov, V.K. and Yakimovets, A.V. (1985) Characteristics of rock grenadier accumulations and its behavior response to trawls. In: Behavior of commercial fishes, P.A. Moiseev, ed., USSR. Collection of scientific works of VNIRO, pp 122-130. Can. Trans. Fish. Aquat. Sci. No. 5339, 11 p.

Kosswig, K. (1980) A contribution on the age and growth of roundnose grenadier (*Coryphaenoides rupestris* Gunn.) at West Greenland (NAFO Subarea 1). NAFO SCR Doc. 80/92. Serial No. N147. 5 p.

Kosswig, K. (1986) Investigations on grenadier fish (*Macrourus berglax* and *Coryphaenoides rupestris*) by the Federal Republic of Germany in 1983. Annls. Biol., Copenh. 40: 176-177.

Kotthaus, A. and Krefft, G. (1967) Observations on the distribution of demersal fish on the Iceland-F aroe Ridge in relation to bottom temperatures and depth. ICES Rapp. Proc. Verb. 157: 238-267.

Kulka, D.W. (1985) The effect of changing effort patterns of catch composition in the roundnose grenadier fishery, 1978-1983. NAFO SCR Doc. 85/16. Serial No. N956. (19 p.

Liem, A.H., and Scott, W.B. (1966) Fishes of the Atlantic Coast of Canada. Fish. Res. Bd. Can. Bulletin No. 155, 485 p.

Logvinenko, B.M., Nefedov, G.N., Massal'skaya, L.M., and Polyanskaya, I.B. (1983) A population analysis of rock grenadier based on the genetic polymorphism of nonspecific esterases and myogenes. In: Genetics of commercial and cultivated species: Proceedings of the Second All-Union Conference on Genetics, Selective Breeding, and Hybridization of Fish. pp 29-34. Can. Trans. Fish. Aquat. Sci. No. 5406, 14 p.

Magnússon, J.V. (1977) Some notes on the spawning habits of *Macrouridae* at Iceland. ICES. C.M. (1977/F:49. Demersal Fish Committee, 9 p.

Magnússon, J.V. (1978) Icelandic investigations on grenadier fish (*Coryphaenoides rupestris* Gunnerus and *Macrourus berglax* Lacépède) in 1976. Annls. Biol., Copenh. 33: 174-176.

Magnússon, J.V. (1979) Icelandic investigations on grenadier fish (*Coryphaenoides rupestris* Gunnerus and *Macrourus berglax* Lacépède) in 1977. Annls. Biol., Copenh. 34: 223-225.

Magnússon, J.V. (1980) Icelandic investigations on grenadier fish (*Coryphaenoides rupestris* Gunnerus and *Macrourus berglax* Lacépède) in 1978. Annls. Biol., Copenh. 35: 243-244.

Magnússon, J.V. (1986) Icelandic investigations on grenadier fish (*Macrourus berglax* and *Coryphae-noides rupestris* in 1983. Annls. Biol., Copenh. 40: 176.

Magnússon, J.V. (1987) Grenadier fish in Icelandic waters. NAFO SCR Doc. 87/87. Serial No. N1341,19p.

Markle, D.F. and Musick, J.A. (1974) Benthic-slope fishes found at 900m depth along a transect in the Western N. Atlantic Ocean. Marine Biology 26: 225-233.

Markle, D.F., Dadswell, M.J., and Halliday, R.G. (1988) Demersal fish and decapod crustacean fauna of the upper continental slope off Nova Scotia from LaHave to St. Pierre Banks. Can. J. Zool. 66: 1952-1960.

Marshall, N.B. (1965) Systematic and biological studies of the Macrourid fishes (*Anacanthini-Teleostii*). Deep-Sea research, 12: 299-322.

Marshall, N.B., and Iwamoto, T. (1973) *Genus Coelorhynchus*. In. N.B. Marshall, Family *Macrouridae*. In: Fishes of the western North Atlantic, D.M. Cohen, editor-in-chief. Memoir, Sears Foundation for Marine Research (1) part 6, pp 538-563.

McLellan, T. (1977) Feeding strategies of the macrourids. Deep Sea Research. 24. pp. 1019-1036.

Merrett, N.R. (1978) On the identity and pelagic occurrence of larval and juvenile stages of rattail fishes (Family *Macrouridae*) from 60ø N, 20ø W and 53ø N, 20ø W. Deep-Sea Research 25: 147-160.

Minet, J.P., Forest, A., and Perodou, J.B. (1978) Stratification scheme for ICNAF Statistical Division 0B. ICNAF Res. Doc. 78/64. Serial No. 5232. 2 p.

Nodzynski, J. and Zukowski, C. (1971) Biological and technological characteristics of grenadier family fishes (*Macrouridae*) of the Northwestern Atlantic. Studia I Materialy. Series D. No. 5. 45 p.

Parsons, D.G., Veitch, P.J., and Legge, W.E. (1978) Some characteristics of the roundnose grenadier fisheries in ICNAF Subareas 0+1 and 2+3. ICNAF Res. Doc. 78/VI/47. Serial No. 5209. 9 p.

Parsons, L.S. (1976) Distribution and relative abundance of roundnose, roughhead and common grenadiers in the Northwest Atlantic. ICNAF Sel. Papers No 1, pp 73-88.

Pechenik, L.N. and Troyanovsky, F.M. (1969) Rock Grenadier. Rybnoe Khoziaistvo. 12: 7-9. Can. Trans. Fish. Aquat. Sci. No. 5387, 8p.

Pechenik, L.N. and Troyanovsky, F.M. (1970) Trawling resources on the North-Atlantic continental slope. Murmanskoe Knihnoe Izdatel'stvo, Murmansk. Israel Program for Scientific Translations. (1971. 66 p.

Phleger, C.F. (1971) Biology of Macrourid fishes. Am. Zool. 11: 419-423.

Pinhorn, A.T. (1974) Preliminary estimates of sustainable yield for roundnose grenadier (*Macrourus rupestris*) in ICNAF Subareas 2 and 3. ICNAF Res. Doc. 74/6. Serial No. 3149. 5 p.

Podrazhanskaya, S.G. (1971) Feeding and migrations of the roundnose grenadier, *Macrourus rupestris*, in the Northwest Atlantic and Icelandic waters. ICNAF Redbook 1971(III): 211-220.

Sahrhage, D. (1986) Commercially important grenadiers of the North Atlantic. Reports from the Institute of Deepsea Fisheries, Hamburg, Federal Republic of Germany, No. 37, 81 p. Can. Trans. Fish. Aquat. Sci. No. 5376, 89 p.

Satani, M., Kawahara, S., and Jørgensen, O. (1993) Results of two stratified random bottom trawl surveys off West Greenland in 1992. NAFO SCR Doc. 93/58. Serial No. N2241. 13 p.

Savvatimsky, P.I. (1969) The grenadier of the North Atlantic. Trudy PINRO, 72 p. Fish. Res. Bd. Can. Trans. Ser. No. 2879, 86 p.

Savvatimsky, P.I. (1971a) Studies of the age and growth of roundnose grenadier (Macrourus rupestris Gunn.) in the North Atlantic, 1967-1990. ICNAF Redbook, Part III. pp. 125-138.

Savvatimsky, P.I. (1971b) Determination of the age of grenadiers (Order Macruriformes). Voprosy ikhtiologii. 11: 397-403.

Savvatimsky, P.I. (1972) The age of the rock grenadier in the north-west Atlantic and a possible influence of fisheries on its population numbers. Trudy PINRO 28: 116-127. Fish. Res. Bd. Can. Trans. Ser. No. 2879, 26 p.

Savvatimsky, P.I. (1975) The age and growth rate of grenadiers in the North Atlantic. Trudy PINRO. Fis. and Marine Serv. Trans. Ser. No. 3924. 9 p.

Savvatimsky, P.I. (1977) Prospects of roundnose grenadier fishery in the Northwest Atlantic. ICNAF Res. Doc. 77/VI/30. Serial No. 5055. 17 p.

Savvatimsky, P.I. (1981) On length measurements of roundnose grenadier (Coryphaenoides rupestris) in the Northwest Atlantic. NAFO SCR Doc. 81/VI/20. Serial No. N296. 11 p.

Savvatimsky, P.I. (1982) Reproduction and sex composition of the North Atlantic roundnose grenadier. In: Abundance and mode of life of the Northwest Atlantic commercial fishes, Murmansk, pp 32-53. Can. Trans. Fish. Aquat. Sci. No. 5389, 29 p.

Savvatimsky, P.I. (1984) On correlation between total length and pre-anal length of roundnose grenadier (Coryphaenoides rupestris) in the North Atlantic. NAFO SCR Doc. 84/VI/44. Serial No. N829. 14 p.

Savvatimsky, P.I. (1985) Diurnal vertical migrations of cod and rock grenadier in the Northwest Atlantic. In: Behavior of commercial fishes, P.A. Moiseev, ed., USSR. Collection of scientific works of VNIRO, pp 67-77. Can. Trans. Fish. Aquat. Sci. No. 5338, 13 p.

Savvatimsky, P.I. (1986) Changes in the composition of the bottom fish catches at different depths along the Continental Slope in NAFO Subareas 0, 2 and 3 in 1970-1985. NAFO SCR Doc. 86/67. Serial No. N1184, 26 p.

Savvatimsky, P.I. (1987a) Changes in species composition of trawl catches by depth on the continental slope from Baffin Island to Northeastern Newfoundland, 1970-85. NAFO Sci. Coun. Studies. 11: 43-52.

Savvatimsky, P.I. (1987b) Methods of conversion of roundnose grenadier (Coryphaenoides rupestris Gunnerus, Macrouridae) ante-anal distance into zoological length. NAFO SCR Doc. 87/91. Serial No. N1395. 13 p.

Savvatimsky, P.I. (1988) Possible reasons for variations in roundnose grenadier catches composition in the Northwest Atlantic in 1871-87. NAFO SCR Doc. 88/100. Serial No. N1557. 21 p.

Savvatimsky, P.I. (1989) Dynamics of roundnose grenadier catch in the Northwest Atlantic. NAFO SCR Doc. 89/8. Serial No. N1572. 11 p.

Savvatimsky, P.I. (1990) Variations in catch composition of roundnose grenadier from the Northwest Atlantic during 1971-1989. NAFO SCR Doc. 90/6. Serial No. N1717. 16 p.

110

Savvatimsky, P.I. (1991) Causes of decrease in total catch of roundnose grenadier (*Coryphaenoides rupestris* Gun.) in the Northwest Atlantic in 1979-1990. NAFO SCR Doc. 91/8. Serial No. N1180. 22 p.

Savvatimsky, P.I. (1992) Distribution and biological characteristics of rock grenadier (*Coryphaenoides rupestris*) as shown by trawl surveys in the Northwest Atlantic in 1989-1991. NAFO SCR Doc. 92/2. Serial No. N2051. 12 p.

Savvatimsky, P.I. (1993) Results of investigations of roundnose grenadier in NAFO Subarea 0, 2 and Div. 3K in 1971-1992. NAFO SCR Doc. 93/12. Serial No. N2189. 8 p.

Savvatimsky, P.I. and \Shafran, I.S. (1981) Status of roundnose grenadier stocks and possibilities for their commercial removal in the Northwest Atlantic. NAFO SCR Doc. 81/IX/106. Serial No. N410. 29 p.

Savvatimsky, P.I., and Atkinson, D.B. (1993) Length-weight relationships of roundnose grenadier (*Coryphaenoides rupestris* Gunn.) in different areas of the Northwest Atlantic. NAFO Sci. Coun. Studies. 19: 71-78.

Savvatimsky, P.I., and Shibanov, V.N. (1987) On population structure of roundnose grenadier (*Coryphaenoides rupestris* Gunnerus, *Macrouridae*) in the North Atlantic. NAFO SCR Doc. 87/92. Serial No. N1396, 17 p.

Savvatimsky, P.I., Rudneva, G.B., Danke, L., Muller, H., Atkinson, D.B., and Power, D. (1990) Roundnose grenadier (*Coryphaenoides rupestris*) in NAFO subareas 0+1 and 2+3. NAFO SCR Doc. 90/75. Serial No. N1797. 46 p.

Savvatimsky, P.I., \Kokh, Kh., and \Yernst, P. (1977) Comparison of methods for determining the age of grenadiers (*Macruriformes*, Pisces) from the North Atlantic. Voprosy ikhtiologii. 17:324-327.

Schroeder, W.C. (1955) Report on the results of exploratory otter-trawling along the continental shelf and slope between Nova Scotia and Virginia during the summers of 1952 and 1953. Papers in Marine Biology and Oceanography, Suppl. to vol. 3 of Deep-Sea Research. pp. 358-372.

Travin, V.I. and Pechenik, L.N. (1962) Soviet Fisheries Investigations in the Northwest Atlantic. Yu. Yu. Marti, editor. Israel Program for Scientific Translations, 1963. p 4-54.

Yamada, H., Okada, K., and Jørgensen, O. (1988) West Greenland biomasses estimated from a stratified-random trawl survey in 1987. NAFO SCR Doc. 88/31. Serial No. N1469. 6 p.

Yano, K. and Jørgensen, O. (1992) Results of two stratified random trawl surveys at West Greenland in 1991. NAFO SCR Doc. 92/48. Serial No. 2100. 14 p.

Yatso, A. and O. Jørgensen. (1989a) West Greenland biomasses estimated from a stratified-random trawl survey in 1988. NAFO SCR Doc. 89/30. Serial No. N1607. 7 p.

Yatso, A. and Jrgensen, O. (1989b) Distribution, abundance, size, age, gonad index and stomach contents of Greenland halibut (*Reinhardtius hippoglossoides*) off West Greenland in September/October 1988. NAFO SCR Doc. 89/31. Serial No. N1606. 12 p.

Zaferman, M.L. (1991) On the behavior of the blunt-snouted grenadier, *Coryphaenoides rupestris*, based on underwater observations. Voprosy ikhtiologii. 31: 1028-1033.

Zakharov, G.P., and Mokanu, I.D. (1970) Distribution and biological characteristics of *Macrourus rupestris* of the Davis Strait in August-September, 1969. Reports of PINRO Marine Expeditionary Investigations, 2nd Cruise of R/V Perseus III. (in Russian) cited in Podrazhanskaya (1971).

Zilanov, V.K. (1976) The problem of species composition and value of by-catch obtained in the course of a specialized grenadier (*Macrourus rupestris*) fishery in the Northwest Atlantic. ICNAF Res. Doc. 76/VI/112. Serial No. 3935. 11 p.

Zubchenko, A.V. (1981) Parasitic fauna of some *Macrouridae* in the Northwest Atlantic. J. Northw. Atl. Fish. Sci. 2:67-72.

Zubchenko, A.V. (1984) Parasites of roundnose grenadier (*Macrourus rupestris*) in two areas of the North Atlantic in 1981. Annls. Biol., Copenh. 38: 200-201.

GREENLAND HALIBUT (*Reinhardtius hippoglossoides*) . A REVIEW OF THE DYNAMICS OF ITS DISTRIBUTION AND FISHERIES OFF EASTERN CANADA AND GREENLAND

W.R. BOWERING and W.B. BRODIE

Northwest Atlantic Fisheries Center
P.O. Box 5667, St. John's, Newfoundland.
Canada A1C 5X1

ABSTRACT

Greenland halibut is distributed throughout the deep waters of the northwest Atlantic from as far north as 78° N to as far south as the Grand Bank and Flemish Cap with no apparent break in the continuity of its distribution along the continental slope. It is also found in the fjords of West Greenland and the deepwater bays of eastern Newfoundland and the Gulf of St. Lawrence. The biomass of the stock has been declining since the early 1980's although catches during recent years have reached unprecedented levels. This is mainly due to increased fishing effort primarily in Flemish Pass, Davis Strait and at West Greenland.

1. General Distribution

Throughout the Northwest Atlantic, small concentrations of Greenland halibut (*Reinhardtius hippoglossoides*) have been found as far north as Smith Sound (78° latitude) based upon catches during the Godthab expedition of 1928 as reported by Smidt (1969a), to as far south as the southern tip of the Scotian Shelf (Templeman 1973) and in the Bay of Fundy, Nova Scotia (Barrett 1968). More recently, individual specimens have been reported also on Georges Bank (Scott and Scott 1988) and even off the coast of New Jersey (Robbins and Ray 1986). In the northern Canadian waters they are found in the deep waters (to 1500 m) of the Davis Strait area off Baffin Island (Templeman 1973; Bowering and Chumakov 1989) and extend into Cumberland Sound where they are reported to depths of 1125 m (Crawford 1992). Recent investigations using deepsea longlines off West Greenland by Boje and Hareide (1993) found that Greenland halibut could be caught as deep as 2200 m. Greenland halibut have also been reported well into Hudson Strait and on the east side of Ungava Bay (Dunbar and Hildebrand 1952; Hudon 1990). Off the Canadian coast

A. G. Hopper (ed.), Deep-Water Fisheries of the North Atlantic Oceanic Slope, 113–160.
© 1995 Kluwer Academic Publishers.

they are located throughout the entire range including the Gulf of St. Lawrence (Bowering 1982); Fortune Bay and St. Pierre Bank (Templeman 1973) as well as Flemish Cap (Bowering and Baird 1980; Vazquez 1993). In more recent years, as commercial exploitation and scientific investigation cover progressively deeper water it is likely that the known distribution of this species as well as other deepwater species will expand.

Greenland halibut are also distributed along the deep slopes of West Greenland between Canada and Greenland where that part of the continental slope runs east to west rather than north to south. They are prevalent in the deep fjords all along the coast of West Greenland to the far north and are especially abundant in the Disko Bay area, Umanak district and around Jakobshavn (Smidt 1969a; Riget and Boje 1988).

Although Greenland halibut is a Pleuronectiforme it is not typical of a flatfish that is almost completely associated with the sea bottom and leads a rather passive and sedentary life. It is actually a very active swimmer and has been caught occasionally at the surface over oceanic depths in salmon drift nets off southern Greenland (Christensen and Lear 1977). It has a voracious appetite and can feed on animals on the bottom and in the water column which can be large relative to its own size (Bowering and Lilly 1992). In addition, it exhibits strong diel behaviour patterns from the bottom to the water column as demonstrated by Chumakov (1970) for the Icelandic waters and Bowering and Parsons (1986) for Labrador.

2. Spawning and Migrations

It has been the prevailing theory that most spawning of Greenland halibut in the northwest Atlantic takes place in Davis Strait during winter-spring at depths from 650-1000 m (Templeman 1973) to depths well beyond 1000 m (Bowering and Chumakov 1989). The actual location is believed to be south of the Greenland-Canadian cascade between Greenland and Baffin Island at approximately 67° N latitude where bottom temperatures are usually warmer than at shallower depths (Atkinson and Bowering 1987). The eggs and small larvae produced here are generally found in depths of 600-1000 m. The larvae later rise near the ocean surface and are trapped by currents where they are transported to and along the west coast of Greenland to northern Davis Strait. Where the current turns in northern Davis Strait many larvae are transported southward to the banks off Baffin Island (Bowering and Chumakov 1989). Many larvae caught in these currents continue to drift further southward where they colonize the deep channels around the banks of Labrador and eastern Newfoundland as well as the deepwater bays on the Newfoundland east coast. While a separate spawning stock of Greenland halibut is located in the Gulf of St. Lawrence it is also believed that there is mixing from the north by eggs or larval drift through the Strait of Belle Isle at the northern tip of Newfoundland separating the island from Labrador (Bowering 1982).

As Greenland halibut get older in the Labrador-eastern Newfoundland area they move progressively offshore to the deep edges of the continental slope and as they approach maturity they generally migrate northward presumably to Davis Strait where they spawn. This hypothesis was first proposed by Templeman (1973) and again by Chumakov (1975). There have been a variety of studies since then that support it, such as the tagging results reported in Bowering (1984) which showed several tag returns from northern Labrador as well as the middle of Davis Strait from fish that were tagged off Newfoundland. These returns were considered significant given that the numbers of fish tagged were low and the amount of commercial fishing effort in the respective areas was also very low. In addition, Bowering (1983) in studying age, growth and sexual maturity observed that Greenland halibut mature at a faster rate in the more northerly NAFO (Northwest Atlantic Fisheries Organisation) divisions. It was concluded, however, that this observation was not a result of actual real differences in maturity rates but more a result of migration patterns of maturing fish i.e. "if a spawning migration northward occurs, then it would be expected that the proportion of maturing individuals in catches at any particular size (age) up to 100% would be greater moving progressively northward. This would result in a shifting of the maturity curve to the left going north and subsequently producing lower values of M_{50} (age or size at which 50% of the population matures)" (from Bowering and Brodie 1991). It should be pointed out, however, that when most of the above studies were conducted the amount of commercial effort as well as research effort in depths greater than 1000 m was very limited. In recent years there has been more intensive research work as well as commercial fishing effort (by both trawlers and gillnetters) in depths beyond 1000 m (in some cases as deep as 1700 m). These operations are beginning to provide additional information on the subject of spawning locations and intensity especially along the continental slope of Labrador and eastern Newfoundland including the area of the Flemish Pass east of the Grand Bank. Data gathered from deepwater gillnet fisheries operating in depths beyond 1100 m in NAFO Divisions 3K (Bowering and Brodie 1993) and 2GH (unpublished data of the Northwest Atlantic Fisheries Center, St. John's, Newfoundland, Canada) showed the presence of mature fish at various stages including pre-spawning, spawning and post-spawning. The proportions of mature fish in the catches were, nevertheless, higher in the north than in the south. Junquera and Zamarro (1992) also found spawning Greenland halibut in the Spanish commercial catches during 1991 in the area of the Flemish Pass particularly at depths beyond 900 m. They observed peak spawning in August although only 20% of mature females had any hydrated (ready to spawn) eggs. Throughout the rest of the year the percentage of mature females with hydrated eggs averaged about 3% per month. Overall, these observations are not inconsistent with the hypothesis that most mature Greenland halibut migrate to the far north for spawning. It does suggest, however, that there is also localized spawning (but to a much lesser degree) all along the deep slopes of the continental shelf throughout the range of distribution becoming more diminished towards the southern extremes.

Within the Gulf of St. Lawrence, the resident part of the stock of Greenland halibut spawns in the deepest part of the Laurentian Channel off southwestern Newfoundland. After spawning they disperse throughout the channel areas of the Gulf, however, most appear to concentrate well inside the mouth of the St. Lawrence River. Here they become the object of a rather intensive gillnet fishery during summer by fishermen mainly from Quebec (Bowering 1982).

In the West Greenland area young Greenland halibut are particularly abundant in the area of Disko Bay and in the offshore area to the north of Store Hellefiske Bank (Riget and Boje 1988). Smidt (1969a) reported high densities of age 1 Greenland halibut to the west of Disko Island and this area has been regarded as a very important nursery area. Results from many research surveys (for example, Atkinson and Bowering (1987)) and studies on by-catch and discards in the northern shrimp fishery (NAFO 1992) have corroborated this observation. Farther north in the offshore area west of Umanak (71° N) Riget and Boje (1988) reported that the Danish research vessel *Dana* found large numbers of young Greenland halibut in shrimp trawl catches during July of 1971. As well, considerable numbers of juvenile Greenland halibut are being caught in the newly-developed shrimp fishery between 71° and 73° N (NAFO 1992). Considering the results of Smidt (1969) together with those of Riget and Boje (1988), the main areas of distribution of young stages of Greenland halibut at West Greenland appear to be from 68° N to at least 73° N although the full extent of northward distribution remains unknown. This agrees with Jensen's (1935) conclusions of spawning of Greenland halibut in Davis Strait south of 67° and Smidt's (1969a) results concerning the distribution of pelagic larvae with densest occurrence between 62° 32' N and 66° 15' N.

It has been hypothesized by Riget and Boje (1988) that the high abundance of juvenile Greenland halibut in certain coastal areas at southwest Greenland, on the other hand, cannot be explained by the passive transport of larvae by currents from spawning in Davis Strait since water masses from that area are unlikely to be transported to the southernmost areas of the West Greenland coast. Consequently, it is their view that Greenland halibut in these areas do not originate from Davis Strait. Because pelagic 0-group Greenland halibut are found in abundance from spawning at southeast Greenland and are caught in the Irminger current which flows into southwest Greenland it is considered possible that Greenland halibut in this area may, in fact, originate from the East Greenland-Icelandic stock.

3. Stock Structure

In the Northwest Atlantic Greenland halibut has been managed as three separate units: (1) Baffin Island-West Greenland (NAFO Subareas 0 and 1); (2) Labrador-eastern Newfoundland (NAFO Subarea 2 and Divisions 3KL); (3) Gulf of St. Lawrence stock (NAFO Divisions 4RST) (Bowering and Chumakov 1989)(Fig. 1). Because of the commercial importance of these resources, interest by both the coastal states (Canada and Greenland) and the variety of nations fishing these Greenland halibut populations, knowledge of the actual biological stock structure has become an important issue for fisheries management, especially since 1990.

Templeman (1970) conducted the first investigation into stock identification of Greenland halibut in the northwest Atlantic by performing univariate analyses of certain meristic characters especially vertebral averages on samples collected from

West Greenland to the southern Grand Bank and including the northern Gulf of St. Lawrence. He found no significant differences throughout the area investigated with the exception of the Gulf of St. Lawrence which was different from all other areas. It was the conclusion of Templeman (1970) that vertebral averages alone were not particularly useful in the separation of Greenland halibut stocks in the northwest Atlantic with the possible exception of the Gulf of St. Lawrence population. There have been many laboratory experiments conducted to examine the formation of meristic characters especially when interest in stock identification first began. Gabriel (1944), Heuts (1949) and Taning (1944,1952) indicated from their respective experiments that although such characters are genotypically selected, the individual phenotype itself can be significantly influenced by a number of environmental parameters such as ocean temperature, salinity and even light condition. As a consequence, the phenocritical period during which meristic characters are determined in teleostean fishes has been a topic of considerable debate and scientific pursuit. A variety of results of investigations along these lines for a number of species is summarized in Templeman and Pitt (1961).

A study of stock delineation of Greenland halibut in the northwest Atlantic by analysing the prevalence of piroplasm and trypanosome (blood protozoan parasites) infections as biological tags was conducted by Khan et al. (1982) from samples collected throughout the range similar to that described above from Templeman (1970). Results of the analyses indicated that Greenland halibut from Davis Strait, northern Labrador, northeast Newfoundland shelf and the northern Grand Bank were all similar in prevalence of the respective parasites, however, samples from southern Labrador were different from all other areas and might represent an isolated population. Considering its location, however, it was believed to be more representative of a cline in the prevalence data. Data from the Gulf of St. Lawrence were quite distinct from all areas north and east of Newfoundland and the authors considered it strong evidence of a separate stock in that area. Similar conclusions were drawn more recently by Arthur and Albert (1993) based upon examination of infestation rates for a variety of completely different parasites. Misra and Bowering (1984) conducted a multivariate analysis on nine meristic characters of the same samples collected for the parasite study and reached the same conclusions as Khan et al. (1982) including the observation that samples from southern Labrador were different than those from other areas. It was considered that in light of the similarities between the two independent methods it would appear that Greenland halibut in the area of Hamilton Bank may be more isolated that previously thought. A third study using the same samples was carried out by Bowering (1988a) who used multivariate statistical techniques to examine morphometric characteristics. The results of this study were vastly different from the previous two described above in that every sample was significantly different from every other sample which could be construed that there are a large number of Greenland halibut stocks throughout the area. Considering the overwhelming evidence against such a conclusion it is believed that there is a north-south cline in the morphometric characteristics and that the samples were taken far enough apart to give statistically significant differences.

Stock delineation studies on Greenland halibut were performed by Fairbairn (1981) by analysing allele and genotype frequencies from 16 electrophoretically detectable protein loci, from tissue samples collected throughout the range from Davis Strait to the northern Grand Bank. Based on her results, she concluded that Greenland halibut in the Gulf of St. Lawrence supported a separate stock from eastern Newfoundland although the stock was not completely isolated since it showed similarities to fish from the Labrador area. She also indicated from her results that Greenland halibut from Davis Strait to the northern Grand Bank form a single genetically homogenous stock.

A variety of biological studies examining characteristics such as age, growth and sexual maturity (Bowering 1983) and distribution patterns (Bowering and Chumakov 1989) also reached conclusions similar to many of the investigations discussed above. In particular, that there is a single stock from Davis Strait to the Grand Bank as well as a self sustaining stock in the Gulf of St. Lawrence with some mixture from the Labrador region. The strength of these observations is enhanced by the diversity of methods used to study the issue of stock identification. This is consistent with the summary of a special session of the NAFO Scientific Council, held in Canada in 1982 and entitled "Stock Discrimination in Marine Fishes and Invertebrates of the Northwest Atlantic". It was concluded that *"The contributions to the meeting illustrated many different areas of expertise, and it was clear, that in some instances, that conclusions derived from one methodology needed support from other independent sources to carry conviction. This emphasized the need for a multidiciplinary approach toward stock discrimination studies, particularly the involvement of population geneticists."* (NAIF 1982).

The only area off eastern Canada not considered with respect to stock structure of Greenland halibut is in the area of the Flemish Cap in general, and the Flemish Pass in particular, where new fisheries have been developed in recent years. This will be discussed in detail later in this paper. It would appear then that the stock structure of Greenland halibut from the offshore area of Davis Strait southward has been reasonably well defined.

Recent literature on stock delineation investigations from the inshore areas of the West Greenland region leaves some doubts as to the genetic relationship of Greenland halibut in these areas with those in Davis Strait. Riget and Boje (1989) examined Greenland halibut length frequency data from Japanese-Greenland research vessel surveys collected off West Greenland and found that average size increased with depth. They observed that this increase occurred from the banks in an offshore southward direction which they attributed to a migration towards the spawning area in Davis Strait south of 67° N. However, they also observed a similar migration towards the deep West Greenland fjords where the fish grow to a considerable size and constitute the main resource of the commercial fishery in West Greenland. Although intensive spawning has not been seen in the fjords, some large females with ripe and running gonads have been observed which indicate that there might be at least some localized spawning in the fjords. Riget and Boje(1989) also reported that results of tagging experiments do not support the hypothesis of a spawning migration out of the fjords to the Davis Strait and, in fact, the only long distance migrations recorded were six specimens tagged in southern West Greenland and recaptured in the Greenland-Iceland area (Boje 1993). On the other hand, the lack of returns from offshore areas may in part be explained by very low levels of commercial fishing effort. The authors also suggested that there was very little convincing evidence to indicate migration back into the fjords after spawning and that fish captured several years later, which were likely mature at time of tagging, were probably fish that remained in the fjords without migrating at all.

A stock identification study of Greenland halibut was carried out by Riget et al. (1992) in the northwest Atlantic based on meristics and frequencies of electrophoreti-cally detectable protein loci. Samples for this study were collected from off eastern Newfoundland, Davis Strait, three West Greenland fjords and Denmark Strait off East Greenland. According to this study, analyses of genetic variation indicated a low level of divergence among samples in the offshore areas including Denmark Strait. However, a lack of homogeneity in allelic frequencies suggested that there was more than one spawning stock. An analysis of meristic characters also showed minor differences among samples from offshore areas, while significant differences occurred among West Greenland fjords. The authors viewed this as supporting evidence that Greenland halibut in the fjords of West Greenland may be somewhat isolated from the Davis Strait stock. The results of this study, however, could neither support nor reject the hypothesis that Greenland halibut in the southernmost fjords of West Greenland are to some extent recruited from the East Greenland-Iceland area.

4. History of the Fishery

4.1 HISTORICAL PERSPECTIVE (EASTERN CANADA)

The first record of Greenland halibut being sold off Canada appeared in the Newfoundland customs returns for 1857 in which year about 900 kgs was exported (Lear 1970). The quantity of Greenland halibut exploited remained very low until 1916 when nearly 600 tonne was exported from Newfoundland. The landings from 1916 until the early 1960's fluctuated between 250-1000 tonne annually. The traditional fishery used longlines with capelin, squid and herring for bait and the main areas of operation were Trinity, Notre Dame and White Bays off the northeast coast of Newfoundland as well as Fortune Bay on the south coast of Newfoundland (Lear 1970). The Greenland halibut caught in these early fisheries were generally salted in barrels and exported mainly to logging camps in eastern Canada with lesser quantities going to the United States, British West Indies and the French island of St. Pierre (Lear 1970). The fishery was mostly limited to the cooler seasons since the method of curing did not allow it to be maintained in good condition for a long period of time largely because of the high fat content in Greenland halibut. According to Lear (1970) the first shipment of fresh Greenland halibut was exported to the United States in 1901 after being packed in barrels in chopped ice. At this time small quantities of Greenland halibut were being smoked and exported to the United States and Canada. Catches in the early 1960's began to increase substantially as a result of Greenland halibut being utilized for fresh frozen fillets and fish blocks most of which were exported to the United States with smaller quantities being shipped to Europe. At the same time larger fish were still being frozen after gutting and trimming and exported to Germany where they were smoked. The demand for salted Greenland halibut from this period onward remained very low.

As an experimental venture in 1960 synthetic gillnets were introduced to the groundfish fishery (Fleming 1964) including Greenland halibut. The use of this highly efficient gear in the Greenland halibut fishery increased so rapidly in the deepwater bays of the Newfoundland east coast that by 1967 the use of the traditional longlines was all but phased out completely. The first synthetic nets were made of ulstron twine but by late 1967 it was replaced by monofilament nylon which remains in use to this day. As a result of the introduction of these nets fishing effort increased dramatically in the east coast bays of Newfoundland starting with Trinity Bay where catches peaked at 11,000 tonne in 1967 but by 1971 had declined to just over 2,000 tonne. Between 1965 and 1967 catch rates in Trinity Bay declined by a factor of seven and much of the effort moved to Bonavista Bay after 1967 and intensified in Notre Dame Bay and White Bay in 1968 although catches in these bays were far below those of Trinity Bay. Catch rates in these bays, nevertheless, suffered somewhat similar fates as did those in Trinity Bay in very few years (Lear and Pitt 1971). As catches in the bays declined substantially gillnetters moved further offshore and fished the deep channels between the banks where localized concentrations of Greenland halibut could be found.

Landings increased substantially in the offshore areas as well after 1966 by large trawlers primarily from Poland, the former Soviet Union (USSR) and German Democratic Republic (GDR). The USSR and Polish fisheries for Greenland halibut began as incidental fisheries in conjunction with the deepwater fishery for redfish as it progressed to depths of 600-1000 m (Templeman 1973). In addition, the catches of Greenland halibut increased even more by the USSR and the GDR as these countries pursued the fishery for roundnose grenadier. According to Pechenik and Troyanovskii (1970), sometimes Greenland halibut in these earlier years were also the subject of directed fisheries. It was noted by Templeman (1973) that the greatest proportion of the catches by these European nations was taken from NAFO Division 3K. Catches by country from 1963 to 1992 for the Labrador and eastern Newfoundland area are shown in Table 1.

In NAFO Subarea 0, catches were first recorded in 1968 by the USSR largely as a by-catch in the developing roundnose grenadier fishery (ICNAF 1975). By 1972, the GDR also began to fish in Subarea 0 for both Greenland halibut and roundnose grenadier. For many years since that time, catches have fluctuated in this area for a variety of reasons such as adverse ice conditions, variability in catch rates, redeployment of effort to other fisheries and changes in management regulations as a result of extension of coastal jurisdiction. However, in more recent years interest in this area by Canadian fishermen has expanded substantially. This will be explored in further detail later in this paper. Catches by country from 1968 in this area are presented in Table 2.

In the Gulf of St. Lawrence there has been a small fishery for Greenland halibut for many years largely as a by-catch of other groundfish fisheries. In the late 1970's, however, catches increased considerably as a result of several successive good year-classes which brought about a substantial increase in the stock biomass in the Gulf area. As these year-classes passed through the fishery and the population, catches declined again. Catch levels within the Gulf of St. Lawrence since then tend to fluctuate around the presence of good and poor year-classes with little change in effort. When good year-classes occur in the Gulf area they tend to be coincidental with good year-classes outside the Gulf. Catches from the Gulf of St. Lawrence since 1965 are presented in Table 3.

4.2 HISTORICAL PERSPECTIVE (WEST GREENLAND)

According to Smidt (1969b), the Greenland halibut fishery at West Greenland is of *"ancient vintage"* and until the 1920's when milder ocean climate brought cod in abundance to West Greenland waters it was the most important fishery. The following excerpt from Rink (1852) as reported in Smidt (1969b) best describes the earliest description of the development of the Greenland halibut fishery at West Greenland:

" it is characterized by occurring at great depths and is caught almost exclusively in the ice fjords, or where large icebergs pass by and ground themselves, and finally only in the most cold months of the year. These fish have been known to occur in various areas, partly by being caught, partly by seeing them being chased near the surface of the water by 'blacksides' (seals) and finding them in the stomachs of seals, and in white whales. But they are being caught only in two locations, namely in Jacobshavn's Isefjord and in Omenak Fjord and in each of these waters on certain banks. In Jacobshavn's Isefjord these banks are inside the mouth of the fjord and inside the shoal which is always full of the very biggest icebergs; just outside the mouth there is a fishing bank which is used less, however. In the month of January, holes are made in the ice in certain locations and the fishermen jig at the extraordinary depth of 350, yes even 380 fathoms, it is claimed. Ordinary thin line is used, but preferably whalebone (the fibre from baleen which in earlier days was used to weave fishing lines and nets) since in this case it would be much easier to notice the jerk when the fish bites. Thus much time is saved in hauling in when one is not sure if it is a bite or not."

Later Rink writes *" In the Omenak Fjord there are many Greenland halibut banks in various locations; each of the settlements have one or more banks less than 1/2 mile away with the exception of the outer part of the fjord near Niakornak where they occur very sparingly. The fishery in Omenak Fjord is carried out at a depth of only a little over 200 fathoms and here the chances are also much less for having the banks packed with icebergs; but the fish are generally somewhat smaller and not as plentiful as they are in Jacobshavn's Isefjord."*

In the beginning, Greenland halibut was fished only for local consumption by the Greenlanders themselves and their sled dogs. However, during the 1890's Greenland halibut was bought from Greenlanders in Jacobshavn by the Danish colonial administration to be salted for export (Smidt 1969b) which was the beginning of the

122

Greenland halibut commercial fishery. During the early 1900's there were a number of exploratory fishing expeditions for Greenland halibut experimenting with longlines. For example, in Jacobshavn by the vessels *Karen* and *Havgasin* in 1906; in Lichtenau Fjord by the Danish expeditionary vessel *Tjalfe* in 1909. During the ensuing few years the Greenlandic fishermen were taught to use longlines (Smidt 1969b). With the success of these exploratory fishing trials Jacobshavn district and Lichtenau Fjord developed into the major centers for the Greenland halibut fishery where most of the fish bought was salted in barrels and exported to Denmark.

Catches in traditional areas fluctuated up to the late 1950's mainly due to a diversion of effort to the cod fishing which was much easier and gave a better financial return (Smidt 1969b). The fishery for Greenland halibut expanded during the late 1950's and early 1960's to other West Greenland districts with the establishment of freezing plants and heated salting houses. The fishery remained very much in its existing form for the next 20 years and catches fluctuated over this period as well because of decreasing interests in certain districts, increasingly poor ice conditions in northern regions and the influence of the shrimp fishery. By the 1990's, however, the cod stock at West Greenland had collapsed (Hovgard 1993) and once again interest in the fishery for Greenland halibut began to increase considerably and has again become the most important groundfish fishery at West Greenland. Catches at West Greenland (NAFO Subarea 1) since 1963 can be seen in Table 4.

5. Recent Catches and Quota Management

5.1 NAFO SUBAREA 2 AND DIVISIONS 3KL (MN)

Since the directed fishery for Greenland halibut began in the early-1960's in the deepwater bays of eastern Newfoundland, catches increased from fairly low levels in the early-1960s to over 36,000 tonne by 1969 and ranged from 24,000 tonne to 39,000 tonne over the next 15 years (Table 1). From 1985 to 1989, catches exceeded 20,000 tonne only in 1987. In 1990, an intensive fishery for Greenland halibut developed in the deepwater area of the NAFO Regulatory area near the boundary of Divisions 3L and 3M in the areas known as the Sackville Spur and Flemish Pass. The development of this fishery quickly resulted in increased catches to about 47,000 tonne by 1990. It was estimated that the catch in 1991 was at least as high as 65,000 tonne although some estimates put the catch at nearer 75,000 tonne. Catches during 1992 remained high and are believed to have been in the order of about 63,000 tonne of which more than 10,000 tonne were estimated as non-reported. The major participants in this fishery have been EU/Spain and EU/Portugal, who are members of NAFO, as well as some non-member countries such as Panama. Catches listed as "Subarea 3 Outside" in Table 1 include all non-Canadian catches during recent years and are illustrated in Fig. 2 and 3 for comparison with traditional fishing areas.

Up until 1990, Canada, USSR, GDR, and Poland were usually the main participants in the fishery, although Portugal and Japan have become increasingly involved in the fishery since 1984. USSR/Russia catches were about 1,100 tonne in 1988-90 but increased to 8,200 tonne in 1991, the highest level since 1975. Most of this catch in

TABLE 1. Greenland halibut landings (metric tons) by year and country for Subarea 2 and Division 3K and Lfrom 1963 to 1992. Catches from Div 3M included for 1992.

Country	63	64	65	66	67	68	69	70	71
Canada	776	1757	8082	16209	16604	13322	11553	10706	9408
Fed. Rep. Germany	10	35	·	355	42	4	202	13	·
Poland	691	1834	939	1114	3296	5806	5406	8266	5234
Iceland	·	·	·	·	·	·	·	·	2
Norway	·	·	·	·	·	·	4	·	·
USSR	125	302	479	242	4287	8732	9268	7384	9094
Romania	·	·	·	·	·	·	40	225	7
German Dem. Republic[b]	·	·	·	1324	1415	4122	10014	·	647
Denmark (Far)	·	·	·	·	·	·	·	·	·
Spain	·	·	·	·	·	·	·	·	·
UK	·	·	·	·	·	·	·	·	·
Denmark (Grlnd)	·	·	·	·	·	·	·	·	·
Portugal	·	·	·	·	·	·	·	·	·
France (M)	·	·	·	·	·	·	·	·	·
France (SP)	·	·	·	·	·	·	·	·	·
Japan	·	·	·	·	·	·	·	·	·
Other	·	·	·	·	·	·	·	·	·
Total	1602	3928	9500	19244	25644	31986	36488	26594	24392

Table 1. (Cont'd.)

Country	72	73	74	75	76	77	78	79	80
Canada	8952	6840	5745	7807	9306	17967	24692	29940	31774
Fed. Rep. Germany	86	707	515	622	927	755	1022	15	55
Poland	6986	9060	7105	8447	5942	5998	5215	1813	203
Iceland	·	·	·	·	·	·	·	·	·
Norway	1389	501	117	-	6	15	3	8	1
USSR	10183	8652	9650	9439	6799	4308	5632	1961	238
Romania	120	80	·	·	·	·	·	·	·
German Dem. Republic[b]	402	1681	2701	2025	1512	1953	1636	178	316
Denmark (Far)	970	950	4	·	·	350	268	·	·
Spain	3	·	·	·	6	·	·	4	·
UK	731	201	1112	62	1	476	53	110	22
Denmark (Grlnd)	·	65	2	·	·	·	·	·	·
Portugal	·	207	161	231	73	119	·	38	21
France (M)	·	·	5	·	·	·	·	·	·
France (SP)	·	·	6	48	32	·	5	1	·
Japan	·	·	·	·	·	·	3	·	12
Other	·	·	·	·	·	·	·	·	·
Total	29822	28944	27123	28681	24598	31941	38532	34068	32642

124

Table 1. (Cont'd.)

Country	Year								
	81	82	83	84	85	86	87	88	89
Canada	24125	19248	19031	17283	12277	8213	13450	8451	11919
Fed. Rep. Germany	-	57	2	9	482	15	1	-	5
Poland	1806	1111	5258	943	460	177	1001	904	360
Iceland	-	-	-	-	-	-	-	-	-
Norway	-	-	15	18	1	-	-	-	-
USSR	3325	1471	937	440	149	770	6716	1063	1058
Romania	-	-	-	-	-	-	-	-	-
German Dem. Republic[b]	1350	2487	2587	2498	1850	1868	3268	2246	1726
Denmark (Far)	-	-	-	-	193	451	2877	740	730
Spain	-	-	-	-	-	-	107	15	22
UK	-	1	-	3	-	-	-	-	-
Denmark (Grlnd)	-	-	-	-	-	-	-	-	-
Portugal	16	1818	-	2612	2940	3107	1390	4118	3168
France (M)	-	-	-	-	-	-	-	-	-
France (SP)	-	7	-	-	-	-	-	-	-
Japan	60	14	-	1003	258	1277	2128	1506	477
Other	-	-	9	-	-	-	-	-	-
Total	30682	26206	27839	24809	18610	15878	30938	19043	19465

Table 1. (Cont'd.)

Country	Year		
	90[a]	91[a]	92[a]
EEC[d]	-	13388	-
Canada	9863	10942	6935
Fed. Rep. Germany	-	-	-
Poland	360	-	-
Iceland	-	-	-
Norway	-	-	-
USSR	1161	8190	-
Romania	-	-	-
German Dem. Republic[b]	12	10	42
Denmark (Fra)	571	753	759
Spain	4685	-	34520
UK	-	-	9
Denmark (Grlnd)	-	-	10
Portugal	10637	-	10539
France (M)	-	-	-
France (SP)	-	-	-
Japan	1662	1990	1882
Other	18115[c]	29715[c]	49
Total	46796	64989	54745

[a]Provisional.
[b]Includes catches recorded as Germany from 1991 and 1992.
[c]Includes catches by EEC and some non-members of NAFO.
[d]No breakdown available for EEC (Sp) and EEC (Port).

1991 was taken in Division 2H. Increase in Russian catches are largely a result of Russian vessels fishing under charter by Canadian companies fishing Canadian allocations. Catches by Canadian vessels peaked in 1980 at just over 31,000 tonne while the largest non-Canadian catches before 1990 occurred in 1969-70. In most years, most of the catch has come from Div. 3K and 3L, with catches from Div. 2G and 2H usually being relatively low (detailed breakdown in Brodie and Baird (1992)).

Canadian catches are taken mainly by gillnet and have been around 7,000-10,000 tonne in recent years, down from a peak of about 28,000 tonne in 1980. The 1991 gillnet catch of 3,500 tonne was the lowest in the time series. As previously reported, the traditional gillnet fishery has been conducted by relatively small vessels fishing in the deepwater channels near the Newfoundland and Labrador coast as well as the Newfoundland east coast deepwater bays. However, this component of the fishery has been declining rapidly in recent years due to the lack of raw material in the area (see Bowering and Power 1993). The Canadian gillnet catches in the last couple of years represent mainly those of a newly developed fishery along the deep edge of the continental slope especially in Div. 3K. Canadian otter trawl catches peaked at about 8,000 tonne in 1982, declined to less than 1,000 tonne in 1988, then increased to about 7,400 tonne in 1991 which is the highest level since 1982. In 1992 , otter trawl catches were less than half that of 1991 due to low catch rates.

The first total allowable catch (TAC) or quota for this resource was introduced in 1974 by ICNAF (International Commission for the Northwest Atlantic Fisheries) at a level of 40,000 tonne (ICNAF 1974). It was reduced in 1976 to 30,000 tonne and remained there until it was increased slightly to 35,000 tonne in 1980 (ICNAF 1975; NAFO 1979). It was further increased from 35,000 tonne in 1980 to 55,000 tonne in 1981-84, 75,000 tonne in 1985, and 100,000 tonne in 1986 (ICNAF 1979; NAFO 1980, 1984, 1985). These increases in TAC's were the result of research vessel information on estimates of stock biomass which indicated both high levels of fishable biomass as well as prospects of several better than average recruiting year-classes. After observing a major reduction in stock biomass from the late 1970's to the late 1980's of about 50% the TAC was reduced to 50,000 tonne in 1990 (NAFO 1989) and this level has been maintained since that time despite the substantive declines in stock size throughout the normal range of observed historical stock distribution. Although a precise catch level could not be advised by the NAFO Scientific Council for this resource for 1994 (NAFO 1993), the Canadian government set the TAC at 25,000 tonne for 1994 within its jurisdictional limits based on a recommendation of the Fisheries Resource Conservation Council, an advisory body established in 1993. Nevertheless, it is intended that this precautionary level of catch should include catches from this stock outside the Canadian zone as well.

5.2 NAFO SUBAREAS 0 AND 1

During the period 1981-89 Greenland halibut catches have remained relatively stable in NAFO Subareas 0 and 1 with an average annual catch of about 9,000 tonne (Tables 2 and 4). From 1989, catches increased substantially to 20,000 tonne in 1990, 22,000 tonne in 1991 and nearly 29,000 tonne in 1992. From 1989 to 1990 the increase in

TABLE 2.. Nominal catches (metric tons) of Greenland halibut from NAFO Subarea 0 from 1968 to 1992

Country	1968	69	70	71	72	73	74	75	76	77	78	79	80
Canada	-	-	-	-	-	-	-	-	-	-	-	-	136
Fed. Rep. Germany	-	-	-	-	-	-	-	-	-	-	-	94	-
Denmark (Far)	-	-	-	-	-	-	-	825	916	930	550	50	60
Japan	-	-	-	-	-	-	-	-	-	-	-	-	-
Norway	-	-	-	-	-	-	24	-	-	-	146	-	-
USSR	1443	813	215	1298	9397	1218	861	455	3990	2967	-	987	1546
German Dem. Republic	-	-	-	-	563	-	-	-	-	240	-	-	-
Denmark (Grlnd)	-	-	-	-	-	912	4	288	-	-	-	-	1
Total	1443	813	215	1298	9960	2130	889	1568	4906	4137	696	1131	1743

Country	1981	82	83	84	85	86	87	88	89	90*	91*	92*
Canada	-	-	-	-	-	-	-	-	-	6194	256	3826
Fed. Rep. Germany	-	-	-	-	335	-	-	-	-	-	-	-
Denmark (Far)	170	337	765	370	525	240	388	963	698	2540	2350	476
Japan	-	-	-	-	-	-	-	-	-	-	1016	235
Norway	-	-	-	-	-	-	-	-	-	-	3959	-
USSR	3626	3468	3772	109	179	32	-	59	29	-	3203	7169
German Dem. Republic	-	-	-	-	-	-	-	-	-	-	-	-
Denmark (Grlnd)	-	-	-	-	-	-	-	-	-	-	-	-
Total	3796	3805	4537	479	1039	272	388	1024	727	10242	10784	11706

'Provisional

TABLE 3. Nominal catches (metric tons) of Greenland halibut from NAFO Divisions 4R, S andT from 1965 to 1992

Country	1965	66	67	68	69	70	71	72	73	74	75
Canada	24	365	365	689	802	1112	954	670	763	1011	1534
Others	-	-	-	-	-	-	-	13	-	-	10
Total	24	365	365	689	802	1112	954	683	763	1011	1544

Country	1976	77	78	79	80	81	82	83	84	85	86
Canada	1994	3962	6247	8779	7006	3176	2269	1104	2126	2332	6469
Others	25	-	-	12	-	-	-	-	-	-	-
Total	2019	3962	6247	8791	7006	3176	2269	1104	2126	2332	6469

Country	1987	88	89	90*	91*	92*
Canada	11069	8027	4958	2441	2293	3423
Others	-	-	1	4	-	-
Total	11069	8027	4959	2445	2293	3423

'Provisional

TABLE 4. Nominal catches (metric tons) of Greenland halibut from NAFO Subarea 1 from 1963 to 1992

Country	1963	64	65	66	67	68	69	70	71	72
Fed. Rep.	252	167	-	88	258	133	68	13	16	136
Germany	-	-	18	2	-	-	1	-	38	442
Denmark (Far)	-	-	-	-	-	-	-	-	-	-
Japan	-	13	1	2	-	-	19	-	1168	12
Norway	-	-	2	-	-	42	123	444	545	245
USSR	-	-	-	31	235	137	8	-	112	-
German Dem.	2671	-	3045	2573	1834	1568	1477	1212	1159	2950
Republic	-	-	-	-	-	-	-	-	-	20
Denmark (Grlnd)	-	-	3	-	-	-	1	-	-	135
Denmark (Mnlnd)	-	-	-	-	-	-	-	-	-	-
Poland	-	-	-	-	-	-	-	-	-	-
Portugal	-	-	-	-	-	-	-	-	-	5
UK										
Total	2923	180	3069	2696	2327	1880	1697	1669	3038	3945

Country	1973	74	75	76	77	78	79	80	81	82
Fed. Rep.	65	2	24	93	590	4965	12784	1174	10	9
Germany	-	-	-	34	77	2	-	-	-	-
Denmark (Far)	-	-	-	-	-	-	-	-	-	-
Japan	35	12	37	7	11	5	-	-	-	-
Norway	2913	8650	19775	6944	1389	-	-	-	-	-
USSR	754	601	56	160	335	-	-	-	-	-
German Dem.	3655	4054	3436	3546	6110	5985	5273	5355	5755	5397
Republic	5	-	2	3	-	-	-	-	-	-
Denmark (Grlnd)	-	-	-	-	-	-	-	-	-	-
Denmark (Mnlnd)	-	32	41	95	-	-	-	-	-	-
Poland	1	18	9	-	-	-	-	-	-	-
Portugal										
UK										
Total	7428	13369	23380	10882	8512	10957	18057	6529	5765	5406

Country	1983	84	85	86	87	88	89	90*	91*	92*
Fed. Rep.	14	15	-	-	-	-	-	-	-	-
Germany	-	-	-	-	-	-	-	131	73	213
Denmark (Far)	-	26	5	-	906	1581	1300	861	571	1629
Japan	-	2	-	-	-	-	-	-	-	1309
Norway	-	-	-	-	-	-	-	-	-	6
USSR	-	-	-	-	-	-	-	-	-	-
German Dem.	4136	6509	9127	8705	8668	7003	7492	8352	10209	1218
Republic	-	-	-	-	-	-	-	-	-	1
Denmark (Grlnd)	-	-	-	-	-	-	-	-	-	-
Denmark (Mnlnd)	-	-	-	-	-	-	-	-	-	-
Poland	-	-	-	-	-	-	-	-	-	-
Portugal										-
UK										
Total	4150	6552	9132	8705	9574	8584	8792	9344	10853	16795

*Provisional

catch was mainly due to a new developing offshore trawler fishery in Division 0B by Canada while the increase in catch from 1991 to 1992 was a result of a general increase in the offshore catch in Subarea 0 and both the inshore and offshore catch in Subarea 1 (Jorgensen and Boje 1993).

The annual catches in Division 0B in the early 1980's declined from an average of 4,000 tonne to less than 1,000 tonne by 1989. Catches increased rapidly to 10,500 tonne in 1990 and 1991 and further to 11,700 in 1992. The fishery generally takes place only in the second half of the year as it is dependent on suitable ice conditions. Nearly all the catch has recently been taken by Canadian, Russian and Japanese trawlers, although a lesser proportion of the catch was taken by Canadian and Faroese longliners using gillnets and longlines, respectively. Many juvenile Greenland halibut are also captured and discarded as a by-catch of the shrimp fishery.

Commercial catches of Greenland halibut in Subarea 1 averaged about 5,500 tonne during the early 1980's, increased to 9,000 tonne in 1985 and remained at that level until 1990 (Jorgensen and Boje 1993). In 1991, the catch increased to near 11,000 tonne and by 1992 reached a level of almost 17,000 tonne. Although both inshore and offshore catches increased, more than 70% was taken in the fjords, nearly all of which was reported from Division 1A. According to Jorgensen and Boje (1993) the use of gillnets in the inshore fishery became common practice in the mid 1980's and during 1986-89 gillnets and longlines accounted equally for the effort applied to the fishery. During 1990 and 1991, longlines comprised more than 80% of the inshore catch while in 1992 gillnets accounted for about 60% of the inshore catch. It was also observed that in recent years the inshore fishery for Greenland halibut had become less of a seasonal nature with catch and effort spread more evenly throughout the year (Jorgensen and Boje 1993). Catches in the offshore area of Subarea 1 are usually taken in the southern Divisions of 1C and 1D mostly by trawlers from Japan, Norway and Russia. Some catch is also taken by longliners from both the Faroe Islands and Norway. The increase in the offshore fisheries are mainly attributable to the introduction of Norwegian trawl fisheries and increasing allocations to Japan.

The first total allowable catch imposed on Greenland halibut in Subareas 0 and 1 was introduced under the auspices of ICNAF in 1976 at a precautionary level of 20,000 tonne which remained in effect through 1977. For 1978, ICNAF increased the TAC to 25,000 tonne which corresponded to the highest annual catch recorded to that time (ICNAF, 1975). Although the status of this resource has been reviewed annually since that time through ICNAF and subsequently NAFO, the TAC has remained at an annual level of 25,000 tonne up to and including 1994. The main reason for this has been the lack of adequate scientific information in order to advise a more precise catch level represented by the reference fishing mortality desired.

6. Geographic Distribution From Research Vessel Surveys

6.1 DAVIS STRAIT AND LABRADOR

Comprehensive research vessel surveys directed primarily towards Greenland halibut in the continental slope area of Davis Strait and Labrador by the former Soviet Union and more recently the Russian Federation have been conducted since the late 1970's. In addition to these, the former German Democratic Republic (GDR) also conducted surveys towards Greenland halibut but covered some additional area. Since the geographic distribution has been relatively stable from 1979-86 only data from 1979 is presented for illustrative purposes (Fig. 4) by combining the results of both Russian and German data based on those published in Bowering and Chumakov (1989). It should be pointed out at this time that the following observations are intended to describe geographic distribution only and are not intended to reflect numerical changes in stock size from year to year. This will be dealt with in the following section on trends in biomass.

According to results from these surveys, Greenland halibut are concentrated all along the slope of the continental shelf from the northern part of Division 0B southward to the northern extent of Division 2J (Fig. 4). Where fishing sets occurred towards Hudson Strait at the northern tip of Labrador, Greenland halibut were also caught in similar quantities as along the shelf edge. In 1978, a Greenland halibut survey was also conducted by the GDR only, from Davis Strait to eastern Newfoundland (Fig. 5). Observations from this survey also indicate consistent distribution along the slope of the continental shelf (Berth et al. 1979). As well, larger catches were experienced in all the deepwater channels from Hudson Strait to eastern Newfoundland. In the summer of 1978, a Canadian survey was conducted in Davis Strait directed towards pink shrimp *(Pandalus sp.)* where Greenland halibut comprises a significant by-catch. This survey extended farther north than the Greenland halibut surveys of Russia and the GDR and also included sets in Cumberland Sound, inner Hudson Strait and Ungava Bay but did not cover the deep continental slope area in a significant way (Fig. 6). This survey also indicated some catch of Greenland halibut in nearly all sets. Interestingly, some of the larger catches occurred in Cumberland Sound, east Ungava Bay and more particularly at the border of Divisions 0A and 0B near where most spawning is believed to take place (Templeman 1973; Chumakov 1975; Bowering and Chumakov 1989). A dedicated deepwater survey directed for Greenland halibut and roundnose grenadier was conducted in Davis Strait during the summer of 1986 by Canada. Consistently large catches were experienced along the deep slope of Divisions 0B and 1C (Fig. 7). Large catches were also observed along the Divisions 0A and 1B border in the similar locations as shown by the results of the shrimp survey in 1978. Clearly the catch levels in the 1986 survey on the tops of the shallower banks are proportionately much lower than those shown in the shrimp survey. The probable reason is that the shrimp survey was conducted with much smaller mesh gear and would be more efficient at capturing smaller fish which are more abundant in shallow water.

By 1987 and including 1988, the Russian surveys for Greenland halibut in the north had expanded to include shallower depth zones as well the continental slope area in Subarea 1 to 67° North latitude in Division 1B, which is a continuation of the east-west slope in Subarea 0. The results of the 1988 survey are illustrated in Fig. 8 which are representative of both 1987 and 1988 (Chumakov et al. 1988). The distribution of Greenland halibut along the continental slope during 1987 and 1988 is

130

similar to that described in the above surveys. Additionally, Greenland halibut were caught in shallower waters although in lesser quantities than in deeper water. Larger catches were taken consistently across the slope area of both Divisions 0B and 1C and 1D. There were very few sets in which there was no catch of Greenland halibut.

In general terms, it would appear that Greenland halibut are widely distributed throughout the entire range of these surveys with the larger catches occurring in the deeper areas (greater than 750 m) as indicated by Bowering and Chumakov (1989). More significantly, there would seem to be no break in the continuity of the distribution over this range which is consistent with studies on stock structure that suggest that all Greenland halibut in these areas, at least in the offshore, comprise a single population .

6.2 SOUTH LABRADOR, NE. NEWFOUNDLAND,GRAND BANK AND FLEMISH CAP

USSR/Russian groundfish surveys have been conducted off northeast Newfoundland, the Grand Bank and Flemish Cap to depths of about 750 m for many years primarily directed towards cod, American plaice, yellowtail flounder and redfish. From 1987-90, data on Greenland halibut catches during these surveys were provided in Brodie et al. (1991). Survey coverage during each of these surveys was generally comprehensive and Greenland halibut distribution consistent among years. Therefore, the results of the 1987 survey are presented in (Fig. 9) as generally representative of distribution. Greenland halibut were caught in almost all sets in Division 3K, along the edge of the Grand Bank and around the deeper part of Flemish Cap covered by the survey. Unfortunately, there was no coverage in the area of the Flemish Pass where the largest fishery for Greenland halibut worldwide is currently occurring. The largest catches clearly occurred in Division 3K especially along the continental slope in the southern part of the division and the deep channel off the Newfoundland northeast coast similar to results of the GDR survey in 1978 discussed previously. Unlike the shelf off Labrador, there were virtually no Greenland halibut caught in any of the sets over the shallower areas of both the Grand Bank and Flemish Cap. During 1978-85, surveys were conducted on Flemish Cap by Canada in winter. These surveys also caught Greenland halibut in the deeper areas surveyed, however, catches were not as abundant or as frequent as indicated in the Russian surveys (unpublished data of the Northwest Atlantic Fisheries centre, St. John's, Newfoundland, Canada). In 1991, as a result of the large fishery being conducted in the area of the Flemish Pass, Canada conducted a survey in deep water (750-1500 m) from Division 3K south into 3L and east into 3M in order to gather information on distribution and population structure in this area. Results are shown in Fig. 10 as published in Bowering and Power (1993). The gap in survey coverage along the 1000 m contour in Division 3L was due to poor trawlable bottom. However, catches of the sets that were possible in this area suggest that there is little reason to expect a significant break in the continuity of the distribution along the slope from Division 3K into 3L, 3M and possibly south into 3N as indicated by the Russian surveys (Fig. 9). The highest catches experienced in the survey occurred in the northern part of Division 3K. Catch levels were lower in southern 3K and northern Division 3L but were consistently much higher on average in the proximity of the Division 3L and 3M boundary. It is notable that every set made had some catch of Greenland halibut.

During the period 1978-93, Canada has been conducting autumn groundfish surveys in NAFO Divisions 2J and 3K and in Division 3L since 1981. Up until the late 1980's, the geographic distribution remained relatively stable as illustrated in Fig. 11 for the years 1978-81. Profound changes in distribution took place during the 1990's and this is demonstrated in Fig. 12 for 1990-93. Both figures have been adopted from Bowering and Power (1993). For the earlier period Greenland halibut were relatively abundant in the deep channels running between the fishing banks, particularly in Divisions 2J and 3K (Fig. 11). Similar observations were made in the GDR survey of 1978. Catches were also high along the slope of the continental shelf throughout the extent of Division 2J and including the northern portion of Division 3K. Bowering and Chumakov (1989) observed that the largest catches in fall-winter surveys in these divisions from 1977-86 were taken in depths of 401 to 500 m and 751 to 1000 m. By the late 1980's Greenland halibut began to disappear rapidly from Division 2J (Bowering and Power 1993) followed by Division 3K in 1990 (Fig. 12). By 1992, catches in Divisions 2J and 3K were extremely low with little or no area of concentration. In 1993, however, there were some increased catches again in the area of the deep channels off northeastern Newfoundland and distribution was more similar to that seen in 1991.

Throughout the period of the surveys, no large catches were observed in Division 3L, nevertheless, most Greenland halibut were caught near the "nose" of the Grand Bank in the area known as the Sackville Spur. Highest catches in Division 3L came during the earlier period when surveys only covered depths up to 366 m whereas during the last three years when coverage was complete to over 700 m catches were negligible.

7. Biomass Trends

Research vessel surveys are the primary sources of information on trends in the abundance and biomass of Greenland halibut in the Northwest Atlantic. Attempts to use methods such as sequential population analysis to estimate population size have not been particularly successful (Bowering et al. 1990; Chumakov et al. 1990), due mainly to incomplete input data and an inability to quantify migration rates (NAFO 1990). In addition, catch-per-unit-effort data are either lacking from many fisheries, or too sporadic to provide a reasonable index of total stock abundance (NAFO 1992). Thus for many areas, research vessel surveys provide the only data on Greenland halibut distribution and abundance.

Given the vast area inhabited by Greenland halibut, it is not surprising that there is no single index of abundance covering the entire stock range. What does exist is a series of separate surveys, from a number of countries, covering various parts of the distribution of Greenland halibut as noted in the previous section. Most of the surveys have been done annually, although in some areas they were less frequent, and in a few cases were one-time only. The majority of the surveys used a stratified random design,

132

with stratification schemes as outlined by Doubleday (1981) and Bowering (1987a,b). As was noted in the discussion of distribution, many of these surveys do not cover the entire depth range inhabited by Greenland halibut, resulting in a lack of information from areas deeper than 1000 m in most NAFO Divisions.

For the surveys discussed below, the index of stock size calculated was the biomass, often referred to as the minimum trawlable biomass or swept area biomass. This is calculated from the stratified mean catch weight per tow, the area swept by the trawl during each tow, and the total stratified area covered by the survey (for details on the calculations, see Smith and Somerton, 1981). For presentation, the available survey data have been grouped by NAFO Division as follows: Divisions 0A, 0B and Subarea 1; Divisions 2G and 2H; Divisions 2J, 3K and 3L; and Division 3M.

7.1 DIVISIONS OA AND OB, SUBAREA 1

The former USSR has conducted surveys in Division 0B since 1979 usually to a depth of 1500 m, with some of the recent surveys being done as joint USSR/German efforts. A plot of the biomass estimates from these surveys is shown in Fig. 13 (from Fig.4 of Gorchinsky, 1993). These data show a sharp drop in biomass from 1986 to 1987, from a relatively stable level around 120,000 tonne to about 37,000 tonne. Since 1987, the biomass has ranged between 37,000 and 79,000 tonne. A survey by France in 1977 gave a biomass estimate of 78,000 tonne for most of the area in Division 0B down to 700 m (Atkinson et al. 1982), a survey by the former GDR in 1978, to a maximum depth of 850 m, produced a biomass estimate of 90,000 tonne (Berth et al. 1979), and a Canadian survey in 1986, to depths of 1250 m (Bowering 1987c) gave a biomass estimate of approximately 125,000 tonne for Division 0B, which is very close to the USSR estimate in the same year. Survey data from Division 0A are sparse and there are no time series which can be used to determine trends in abundance in this area.

In Divisions 1BCD, the USSR/German surveys produced biomass estimates of 56,000, 47,000, and 88,000 tonne in 1987, 1988, and 1990 respectively, and there were no surveys in 1989, 1991, and 1992 (Jorgensen and Boje 1993). Surveys done jointly by Japan and Greenland, to depths of 1500 m in some areas, produced relatively stable biomass estimates in Divisions 1BCD from 1987 to 1992 in the range of 53,000 to 77,000 tonne. Division 1A was not completely covered by these surveys, and no data from Divisions 1EF were shown by Jorgensen and Boje (1993).

7.2 DIVISIONS 2GH

Survey data from this area are relatively scarce; Canada has surveyed the area sporadically since 1977, the former German Democratic Republic conducted surveys in 1978 and 1979, and the former USSR has attempted surveys of the area in the fall-winter period of each year since 1978. However, in many of the surveys, there are substantial gaps in coverage and/or few sets in all depth ranges.

Post-stratified estimates of biomass from Canadian summer-fall surveys in Division 2GH in 1978, 1979, and 1981 ranged from 71,000 to 122,000 tonne (Bowering 1988b), despite the fact that there were no sets deeper than 750 m in any of the three surveys. The values obtained in the GDR surveys of 1978 and 1979 (a maximum depth of 850 m) were also within this range (Berth et al. 1979; Berth and Vaske 1980). Biomass estimates from Canadian surveys in 1987 and 1988 were between 37,000 and 39,000 tonne, even though these surveys were expanded to cover depths down to 1250 m. A survey in 1991 in Division 2H to 500 m suggests that a further decline in biomass has occurred, at least in the shallower depths (Brodie and Baird 1992).

USSR surveys (maximum depth of 1500 m) also show a substantial decrease in the biomass of Greenland halibut from the early to late 1980's. Comparing years where the survey coverage was relatively complete in both divisions, eg. 1982 and 1987, these surveys suggest a decline in biomass in Divisions 2GH from over 200,000 tonne in the earlier period to about 30,000 tonne in the latter period (Gorchinsky 1993). The biomass estimates in the late 1980's are consistent with those from the Canadian surveys, although the USSR surveys suggest an even steeper decline in biomass than is evident from the Canadian data. Surveys in the early 1990's suggest the biomass has remained relatively stable at the lower level (Gorchinsky 1993).

Canadian shrimp surveys in Hopedale Channel in Division 2H from 1984-88 and 1990 provided information on the strength of recruiting year-classes (Brodie 1991). A Canadian line transect survey in 1991 gave post-stratified biomass estimates of Greenland halibut of 13,500 tonne in Division 2G and 16,000 tonne in Division 2H, for the areas between 1000 and 1500 m (Bowering et al. 1993). These values also suggest a further decline in biomass from the levels observed in 1987-88.

7.3 DIVISIONS 2J, 3K, AND 3L

There are several series of stratified-random survey data from these areas: Canada from 1978-93 in Divisions 2J3K, and 1981-93 in Division 3L; and the former USSR from 1983-84 and 1986-90 in Division 3K, and 1987-91 in Division 3L.

The Canadian surveys in Division 2J and 3K were done between October and December each year and generally covered all depths to 1000 m. Peak biomass values of over 100,000 tonne in each of these divisions occurred in the early to mid-1980's, after which time the biomass declined very sharply in both divisions (Fig. 14). Recent estimates of biomass in both areas are about 10-20 % of the earlier values. A line transect survey in 1991, which was post-stratified, gave biomass estimates in Divisions 2J and 3K respectively of 13,000 and 22,000 tonne for the area from 1000 to 1500 m. This compares with the estimates from 100 to 1000 m of 11,000 and 36,000 tonne for Division 2J and 3K respectively in 1991 (Bowering et al. 1993).

In Division 3L, the surveys were also conducted late in each year, although the survey coverage only extended to 732 m since 1990 on a regular basis. Prior to 1990, regular coverage only extended to a depth of 366 m. In this division, the biomass also peaked in the mid-1980's (around 20,000 tonne), but has declined to less than a quarter of that value in 1993 (Fig. 14). As a further indication of the importance of the deepwater areas to the abundance of Greenland halibut, a post-stratified line transect survey done by Canada in 1991 produced a biomass estimate of 18,000 tonne for the area in Division 3L between 750 and 1500 m (approx 4,400 square nautical miles) (Bowering et al. 1993). This compares with the value of the stratified random survey in Division 3L in 1991 of 7,300 tonne, which covered the area out to 732 m (approx. 39,000 square nautical miles).

Russian surveys in Division 3K, carried out in the spring-summer period down to a depth of 1000 m (Rikhter et al. 1991), show a decline in biomass from the mid 1980's to 1990 (Fig. 14), although the results were somewhat more variable than those from the Canadian surveys (Brodie et al. 1991). Nonetheless, in 3 of the last 4 years in this series, the biomass estimate from the Russian spring-summer survey was within fifteen percent of the estimate from the Canadian fall survey. In Division 3L, biomass values ranged from 1,800 tonne in 1990 to 7,700 tonne in 1991, and there was no trend in the data from 1987-91. With the exception of the 1991 point, these biomass estimates from spring-summer are somewhat lower than those from the Canadian fall surveys in the same years (Fig. 14).

There are some data from other surveys in the Divisions 2J3KL area, such as Canadian spring surveys in Division 3L from 1971-93 (Bowering and Brodie 1988), surveys by the former German Democratic Republic in Division 2J in 1978-79 and Division 3K in 1978 (Berth et al. 1979, Berth and Vaske 1980), fall-winter surveys in Division 2J and 3K by the former USSR in 1980 and 1982, and surveys by the former Federal Republic of Germany in Division 2J from 1972-83 (Messtorff 1984). However, the Canadian and FRG surveys did not cover many of the deeper areas, and the coverage in the GDR surveys was relatively sparse. Therefore their usefulness as indices of abundance for Greenland halibut is severely limited. In addition to these, there were Canadian shrimp surveys in Division 2J in Hawke and Cartwright Channels during the periods 1984-88 and 1988-90 respectively. Data on by-catches of Greenland halibut have been used to compare relative strengths of year-classes at young ages in these areas (Bowering and Brodie 1988; Brodie 1991).

7.4 DIVISION 3M

There are three series of survey data available from the Flemish Cap in Division 3M, all of which covered depths to 732 m: Canadian from 1978-85, USSR/Russia from 1987-91, and EU from 1988-92. The Canadian surveys, which were done in January-February of each year, indicated a relatively stable biomass of Greenland halibut around 2,000 tonne from 1978-85 (Fig. 15). The USSR/Russian surveys showed an increase from about that level in 1987-88 to over 4,000 tonne in 1989, followed by a sharp decline to very low levels in 1990 and 1991 (Fig. 15). A possible explanation for the 1991 value could be a change in vessel in that year as well as a change in timing, from June-July in 1987-90 to April-May in 1991. This, however, can not account for the decline in 1990. The EU surveys, which used different vessels in 1989 and 1990 from the one used in 1988, 1991, and 1992, were done during the period of late June to early August of each year. These surveys gave biomass estimates of between 4,300 and 8,600 tonne, and indicated an increase from 1989 to 1992 (Fig. 15). A 1991 Canadian survey of the western and northern areas of the Flemish Cap, including the Flemish Pass, covering mainly depths from 750 to 1500 m (Fig. 10), produced a biomass estimate of about 23,000 tonne for the surveyed area in Division 3M (Bowering et al. 1993).

Given the development of large scale fisheries in the Flemish Pass and Flemish Cap areas in the early 1990's (Bowering et al. 1993), it is generally believed that the biomass of Greenland halibut in these areas was higher at this time than in the early to mid 1980's. The survey results from Canada and EU in Fig. 15 support this hypothesis. However, the data from the Russian surveys in 1990 and 1991 are difficult to explain in this context, particularly since the EU and Russian surveys in 1990 occurred at almost exactly the same time.

7.5 OVERALL TREND IN BIOMASS

It is clear from the survey data examined that there was a substantial decline in the estimate of trawlable biomass of Greenland halibut during the 1980's. Russian surveys suggest that this decline occurred in Divisions 0B and 2GH in the early to middle part of the decade, which agrees with the results from Canadian surveys in Divisions 2GH. In Divisions 2J3K, Canadian surveys indicate that the biomass decline was most severe from the mid 1980's through the early 1990's. From these areas alone (Divisions 0B, 2GHJ, 3K) the combined biomass estimates from the surveys in 1982 and 1983 were around 600,000 tonne. Even if the estimates from the Canadian surveys in Divisions 2GH from 1978-81 are used instead of the higher Russian estimates, and the 0B estimates from 1984-85 are used instead of the peak 1982-83 values, a conservative estimate of the cumulative biomass from Divisions 0B, 2GHJ, 3K was still in excess of 370,000 tonne. It should also be noted that the surveys in Divisions 2J3K did not cover depths beyond 1000 m. In 1991 and 1992, the combined estimate from the same surveys, which covered the same areas, had fallen to about 80,000-100,000 tonne. The total biomass in 1991, from Division 0B south to Division 3M, covering virtually all areas out to 1500 m, was estimated from a number of surveys to be just over 200,000 tonne (Bowering et al. 1993).

These declines in biomass from the surveys have been substantiated by available data from fishery-based indices. For example, there have been major reductions in gillnet catches in virtually all 'traditional' fishery areas off the northeast coast of Newfoundland. In 1993, the Scientific Council of NAFO stated that "there has been probably a redistribution of the Greenland halibut resource and that a substantial part of the stock component being exploited in the regulatory Area of Division 3L, 3M and 3N is likely to have originated in divisions to the north, at least from Division 2J and 3KL" (NAFO 1993). In any case, it is clear that the total estimated biomass of the Greenland halibut stock complex in the early 1990's was substantially lower than it had been about ten years earlier.

8. Possible Factors Affecting Change in Distribution and Biomass

Based on the information presented and the literature reviewed, this resource has evidently exhibited major changes in geographic distribution and substantial, yet systematic, reductions in stock size. It would appear that the changes in geographic distribution are clearly apparent in the southern area of Division 2J and Subarea 3 whereas in the more northerly region, any major shifts in distribution are not so clear. On the other hand, it is quite clear that there were extremely sharp reductions in biomass throughout all of Subareas 0,2 and 3 and these seemed to occur around the early 1980's in the north and about the mid to late 1980's in the south. Given the level of decline in biomass and the abruptness at which it apparently occurred, it is difficult to attribute it, however, to commercial fishing activity alone. The cumulative levels of commercial catch (Tables 1,2 and 4) during the particular periods in question are far below the magnitudes of the declines even if under the unlikely assumptions of zero growth and recruitment. Observations of this nature are very typical of many other groundfish species in this area especially in Divisions 2J , 3K and 3L. Atkinson (1992) examined for trends in biomass and abundance of a wide variety of groundfish species caught during the Canadian fall surveys. He found that there were declining trends in biomass for all species or species groups investigated. Certain species were commercially fished , some heavily and some lightly, whereas many species examined were of no commercial importance. Nevertheless, all species declined systematically during the recent time period. Observations on the pelagic species, capelin (Mallotus villosus) also exhibited massive declines in abundance since about 1989, based upon the results of offshore acoustic surveys, especially in Divisions 2J and 3K although fishing mortality on this species was very low (Miller and Lilly 1991; Miller 1992, 1993). It seems reasonable to conclude, therefore, that while commercial fishing may have exacerbated recent declines in the stock size for some species, large declines would nevertheless have occurred. It would also seem from the apparent timing of the declines that they probably began in the north and moved progressively southward, at least for Greenland halibut. This observation is also rather consistent with certain other species that are under investigation, for example, cod. Cod stocks at West Greenland started to collapse in the mid to late 1980's especially when anticipated good recruitment began to disappear (Hovgaard 1993). The causes for the disappearance of cod at West Greenland are still not fully explained, but to some degree the 1984 and 1985 year-classes may have moved back to the Iceland area where they were spawned. Similar events have occurred with a number of cod stocks in the southern part of the northwest Atlantic, however, they appear to have occurred somewhat later. While there is much investigation into the causes of the declines, no firm conclusions have yet been reached.

In recent years, as a result of collapsing groundfish stocks in the northwest Atlantic many studies have been undertaken to evaluate changes in ocean climate and possible associated effects. It is most notable that the early 1990's exhibited some of the most severe climatic conditions experienced in the area particularly as indicated by low sea bottom temperatures on the continental shelf, extreme surface ice conditions and an increasing volume of the cold intermediate layer (Colbourne et. al 1993). Many are convinced that the severity of ocean climate played a significant role in declining stocks possibly through increased natural mortality, but it has been difficult to identify conclusively the underlying mechanisms which ultimately caused the declines. Lilly (1994), in trying to relate the collapse of the cod stock in Divisions 2J, 3K and 3L in the 1990's with water temperature, concluded that his examination of the data could not support the hypothesis that the disappearance of cod from these areas was directly related to low bottom temperature. He suggested that although many data indices reveal that temperatures were far below normal in 1990-92, it was unclear that conditions were any more severe than in 1972-75 and 1983-85, two previous cold periods that did not seem to affect cod distribution and abundance in these same areas. Given that Greenland halibut occupy much deeper water (where temperature is more stable)(Bowering and Chumakov 1989) than shelf species such as cod, it is unlikely that the change in water temperature alone directly would have contributed to significant changes in its distribution and abundance. Nevertheless, the change in temperature may have acted as a catalyst to some other unknown parameter which negatively affects Greenland halibut and other species.

It has been shown by Bowering and Lilly (1992) that Greenland halibut feed heavily on capelin especially in Divisions 2J and 3K to the point that it is the main prey item. The coincidental disappearance of capelin in the offshore areas of these divisions along with substantive declines in Greenland halibut in the same areas could suggest that there may be some relationship between the two events. Detailed information on feeding of Greenland halibut in this area during the 1990's, however, is not available at this time to allow for a more complete evaluation of any relationship.

As stated above, it was concluded by the NAFO Scientific Council in its 1993 report (NAFO 1993) that much of the Greenland halibut located and fished heavily during the 1990's in the area of the Flemish Pass probably originated in Divisions 2J, 3K and 3L inside the Canadian 200 mile fishery zone. Additionally, there was at least one Greenland halibut tagged in Labrador in 1981 that was recaptured in the Flemish Pass in 1990 as reported by Portugal. It is uncertain as to what the biomass of Greenland halibut was in the area either when the fishery began or at present, however, considering cumulative catches it would suggest that biomass may have been fairly high especially at depths greater than 900 m. What is more uncertain is why these fish would have moved to this particular area and occupy such great depths. Examination of stomach contents from the commercial catches of Spain in 1992 indicated that the percentage of empty stomachs ranged from 62% in August to 84% in December (Rodriquez-Marin et. al 1993). These data were collected from fishing sets at depths ranging from 720-1533 m at an average depth of 979 m. The main prey group was fish, dominated by blue hake *(Antimora rostrata)*, Macrouridae, Gadidae, and Pleuronectidae, comprising 39% of the diet. In addition, decapod crustaceans (mainly Pandalus) accounted for 22% and squid *(Illex illecebrosus)* 32%. About 10% of the diet was made up of waste products, especially Greenland halibut heads. It is unlikely that Greenland halibut would have followed any of this prey to the area of the Flemish Pass. Clearly capelin was not an important food item here, therefore, it would not appear that Greenland halibut were chasing them, although capelin were caught on

Flemish Cap during 1993 (J. Carscadden, Northwest Atlantic Fisheries centre, St. John's, Newfoundland, Canada, pers. comm.). Prior to then there were no records of capelin ever caught here. It is of interest to observe that given the significance of food items such as shrimp, Greenland halibut is likely to make distant vertical migrations in search of food. The significance of shrimp is not surprising considering the extremely rapid increase in its biomass recently, in Division 3M (NAFO 1993). Despite all the observations and hypotheses, however, it remains difficult to explain precisely why Greenland halibut may have moved here. Nevertheless, Greenland halibut appear to have remained in this general area since a very high level of effort continues to be exerted here. According to recent catch reports there may even be further migration southward as catches are becoming proportionately higher along the continental slope of Division 3N. Preliminary reports indicate that for 1993, nearly 58,000 tonne of Greenland halibut were caught outside the Canadian fishery zone in Subarea 3 of which about 21,000 tonne was taken in Division 3N.

9. New Fishery Developments

In recent years, there have been a variety of fishery developments and expansions for Greenland halibut in the Northwest Atlantic as a result of a number of events as follows: a) reduced availability of other groundfish such as cod in both Canadian and Greenland waters, b) increased fishing effort by foreign nations outside Canada's 200 mile fishery zone in the deep waters of the Flemish Pass, c) perceptions of underutilized resources in the Canadian offshore Arctic and d) increased effort from local communities in inshore areas of Baffin Island. The major developments that are of significant concern for the well-being of the Greenland halibut resource overall are discussed in further detail.

9.1 THE INTERNATIONAL FISHERY IN THE FLEMISH PASS

Throughout the 1970's and 1980's, there was little or no directed fishing effort for Greenland halibut in the deepwater areas of Divsions 3LM by any nations, as indicated by the catch levels (Fig. 2 and 3). However, in 1990, catches by Spain, Portugal and some non-NAFO members such as Panama rose sharply in the NAFO Regulatory Area in Divisions 3LM. Some of the catch was taken by large freezer trawlers new to the area, which had the capability of fishing in the depths required to catch Greenland halibut, ie. 1000 to 1500 m. Most of the catch was from the area around the Flemish Pass north to the Sackville Spur, and NAFO considered that catches from Division 3M were from the Subarea 2 and Division 3KL stock and should therefore be included in the assessment of that resource (NAFO 1991). It was estimated that approximately 35,000 tonne of Greenland halibut was caught in the fishery in this area in 1990 (Table 1).

The fishery continued to escalate in 1991, resulting in an increase in catch from the Division 3LM area to a value in the range of 45,000 to 65,000 tonne, there being some doubt as to the actual nominal catch (NAFO 1992). Much of the increase came from an influx of Spanish vessels, with some ships of < 600 gross registered tonnage (GRT) and some of 600 - 1000 GRT joining a fleet of larger vessels over 1000 GRT.

The fishery occurred at depths from 800 to 1500 m (Junquera et al. 1992). In 1992, the catch from this area was again around 45,000 tonne, with about three-quarters of this being caught by Spain (Table 1). From 1991 to 1993, there has been some expansion of the fishery into the adjacent deepwater slopes of Division 3N (NAFO 1993).

The Scientific Council of NAFO expressed concern about the rapid development of this fishery on a number of occasions (NAFO 1992; 1993). The fishery was outside Canada's 200 mile fishery zone, there was no management of the fishery by NAFO, and questions were raised on the sustainability of catches in this relatively small area of 50,000 tonne annually. There was little in the way of biological information on Greenland halibut in this area, and the declines in biomass in other areas were noted with concern. Data from Portuguese trawlers involved in this fishery "showed a gradual decline in catch rate from 1989 to 1991" (NAFO 1992). An analysis of Spanish catch rate data in 1993 was generally inconclusive in terms of a trend in stock size (de Cardenas et al. 1993). A Canadian survey covered much of the fishery area in 1991, and was discussed in detail above. Data from a similar survey conducted in February and March 1994 were not available for analysis at this time. Given the conclusion of the NAFO Scientific Council that there probably had been some redistribution of the resource from the areas north of Division 3L to the area of the fishery in Divisions 3LMN (NAFO 1993), the impact of this large totally unregulated fishery on the well-being of the Greenland halibut resource in the northwest Atlantic could be disastrous. Despite the serious concerns, NAFO Scientific Council in 1993 was unable to advise on an appropriate catch level for this fishery or for the stock as a whole due to an inadequate database.

9.2 THE CANADIAN OFFSHORE FISHERY IN THE DAVIS STRAIT

As major groundfish resources dwindled during the late 1980's in the northwest Atlantic, the Canadian government began to look seriously at other groundfish stocks that were not fully subscribed in relation to their respective catch quotas. These stocks were considered to be "underutilized" and thus a program was put in place to encourage the exploitation of such resources. Greenland halibut in Davis Strait and to a lesser extent northern Labrador were included in this development program. Quotas were established for the program by taking back enterprise allocations from members of the fishing industry who were not utilizing them and putting them into a pool of surplus quota. Proponents were able to apply for a share of the surplus quotas according to a set of rules that generally intended to maximize Canadian employment opportunities associated with harvesting and processing the catch. As a result of this program, introduced in 1990, catches of Greenland halibut in these areas began to increase substantially during the 1990's, as detailed earlier (see Table 2), and now comprise one of the most significant groundfish fisheries in Atlantic Canada. Although this fishery is well regulated with a much lower exploitation rate than the fishery in Flemish Pass, it is, nevertheless, substantial and highly concentrated in a rather localized area. It is, therefore, subject to similar concerns as those expressed for the Flemish Pass fishery (NAFO 1993).

It is sometimes argued that because catches in this area do not exceed the existing quotas there should be no cause for alarm. However, what is rarely considered in making this argument is the biological validity of the existing quota. In the case of the total allowable catch in place for the Greenland halibut fishery in Davis Strait, the quota of 25,000 tonne has existed for the past 16 years. This should not be misconstrued as representing stability in the stock status, but rather a result of insufficient data to advise a more appropriate catch level. Additionally, with very low catch levels until recent years there was indeed little cause for concern. Based on survey biomass indices (Fig. 13) and declining catch rates (Unpublished data of the Northwest Atlantic Fisheries centre, St. John's, Newfoundland, Canada), at least on the Canadian side of Davis Strait, precautionary considerations should be seriously taken into account in any further development of this fishery.

9.3 THE CANADIAN DEEP-WATER GILLNET FISHERY

With the virtual disappearance of Greenland halibut from many of the traditional fishery areas by the late 1980's, some gillnet fishermen in Newfoundland and Labrador began to fish for this species in the deepwater areas of the continental slope. Exploratory work in the summers of 1989-91 off the northeast coast of Newfoundland concluded that large areas in Divisions 2J, 3K and 3L had commercial potential for vessels in the 16 to 20 metre length range and that these vessels were able to fish for Greenland halibut with gillnets in depths as great as 1300 metres (Yetman and Staubitzer 1990). The distance from land to these areas ranged from 140-180 nautical miles compared to a maximum of about 50-60 nautical miles for the traditional fishing areas.

Technical and financial assistance was provided to some fishermen by the Department of Fisheries and Oceans in Canada, with the result that catches in the deepwater gillnet fishery increased from negligible amounts in 1990 to about 900 tonne in 1991. An influx of effort in 1992 caused the catch in this fishery to increase further to about 3,000 tonne, 70 % of which was caught in Division 3K. A study of this fishery (Melindy and Flight 1992) showed that some vessels carried as many as 800 gillnets, most of which had a mesh size between 190 and 203 mm, and were typically fished for periods of 7 to 13 days between hauls. This compares to about 80-150 nets per vessel fished for 1-4 days during the earlier years of the traditional fishery. Twenty-two vessels participated in this study and according to the fishing logs, these vessels made 160 trips and hauled over 59,000 nets during nearly 900 days. Results indicated a steady decline in the mean monthly catch rate over the fishing season (May to October) in 1992 and a corresponding increase in the mean fishing time per net. Data collected from this fishery in 1993 generally indicate a decline in catch rate in the southern areas (Divisions 3KL). Many of the vessels that were involved in the 1992 fishery did not participate during 1993 because catches were too low for economic viability. Vessels that did fish, did so with much reduced catch rates and a shorter season. However, there was some expansion of the fishery into the deepwater areas of the continental slope in Divisions 2GH, where catch rates were often 3-4 times higher per net than in the south (Canadian Department of Fisheries and Oceans, unpublished data).

Biological sampling of the catches in 1992 and 1993 showed that females comprised 60 to 98% of the catches by number, generally increasing with depth fished. In the northern areas in 1993, 97% of the fish examined were female, compared to 89% in the south. In the north, 60% of the females examined were determined to be sexually immature, compared to 75% in Divisions 3KL. These percentages varied on a sample by sample basis, depending on season and depth.

The effects on the resource of this fishery are largely unknown, given its short time span, the paucity of fishery-independent data from the area, and the suspected southward migration of Greenland halibut in recent years. Despite the relatively low catches from this fishery, concerns have been expressed about the rapid escalation of gillnet effort, noting the effects on stock abundance of the increased gillnet catches in the deepwater bays in the 1960's which were discussed in an earlier section of this paper. The concentration of this fishery on females is also of concern from a biological viewpoint, although little is known of stock size versus recruitment relationships in this stock.

10. Management Concerns

It is somewhat ironic that at the time when every available index of stock size for Greenland halibut in the entire Northwest Atlantic is declining, catches have reached unprecedented levels largely under the guise of "developmental fisheries". In fact, the term represents little more than a displacement of fishing effort from other groundfish fisheries that are either closed to fishing or no longer economically viable, from West Greenland to the Flemish Cap. With the recognition that Greenland halibut from Davis Strait to the Newfoundland Grand Bank and Flemish Cap consists of a single interbreeding stock it is clear that excessive fishing on any component of the resource will have a detrimental effect on the resource as a whole. This concern has been raised by the NAFO Scientific Council on many occasions and it is more critical now than at any time previous.

With most groundfish stocks in the northwest Atlantic in various states of depletion there has been and will likely continue to be increasing pressure to expand fishing activity towards these so called "underutilized" stocks and species especially those that occupy the deeper slope of the continental shelf such as Greenland halibut. It is most disconcerting that the resources in these areas are the ones for which we have the least knowledge. Nevertheless, new fisheries tend to develop and escalate before sufficient information is obtained to offer any meaningful advice on the type and level of exploitation that is both biologically appropriate and economically sustainable. With a fishing industry that is hurting badly due to lack of raw material, there is also the perception by many that there is a panacea of deepsea resources that is going to make up for the devastating shortfall of resource in traditional groundfish fisheries. While there may be opportunities to develop some deepsea fisheries it is painfully obvious that the amount of fish inhabiting a very narrow edge on the continental slope could never be expected to compete in quantity with the normal level of fishery resources occupying the many hundreds of thousands of square nautical miles of continental shelf adjacent to it. It would seem that the short life of "newly developed fisheries" described above would critically attest to that. Unless serious consideration is given to these observations and the necessary action taken from both a management and scientific point of view, the Greenland halibut resource in the northwest Atlantic will continue to decline; the consequences of which will be as undesirable as they have been for the fisheries they are attempting to replace.

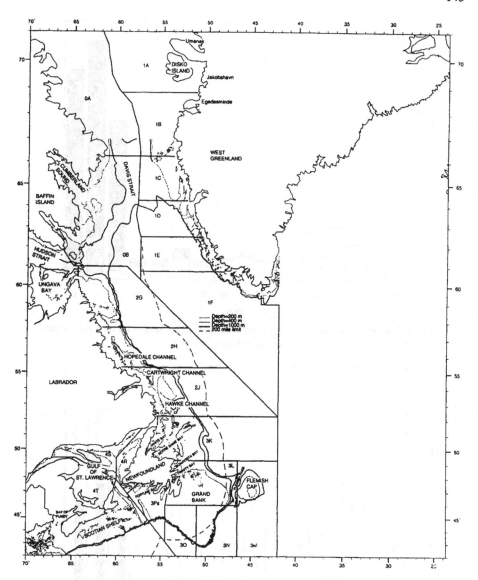

Figure 1 Map showing major areas and place names mentioned in the text

144

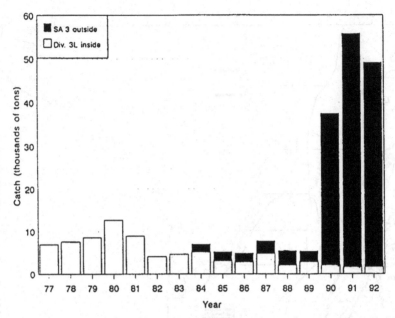

Figure 2 . Catch of Greenland halibut inside 200 miles (Div 3L) compared to the catch outside 200 miles (Subarea 3) from 1977 to1992

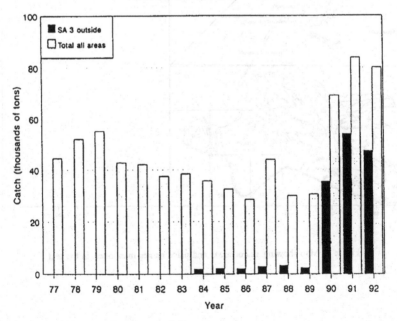

Figure 3. Catch of Greenland halibut from all areas compared to the portion of the catch outside 200 miles (Subarea 3) from 1977 to 1992

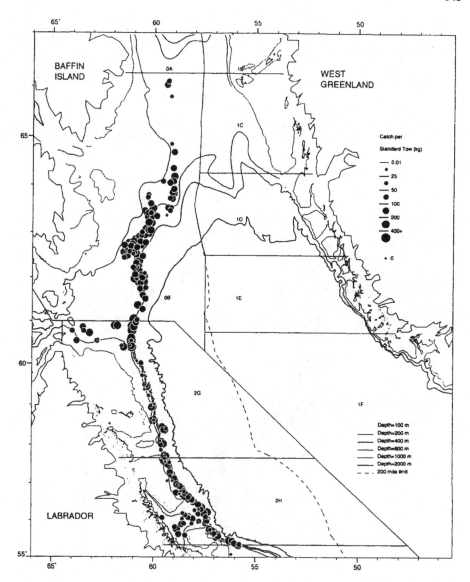

Figure 4. Catch per standard tow (kg) of Greenland halibut during the autumn of 1979 in the Davis Strait and Northern Labrador by the Russian research vessel *Suloy* and the GDR vessel *Walther Barth* combined. All survey tows standardised to 1.75 nautical miles.

146

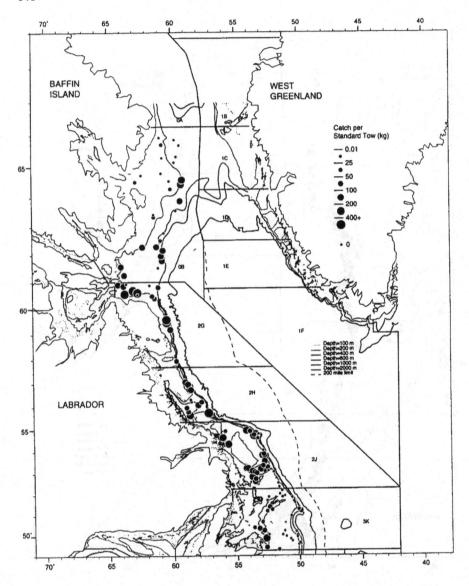

Figure 5. Catch per standard tow (kg) of Greenland halibut during the summer of 1978 in the Davis Strait, Labrador and North Eastern Newfoundland by the German research vessel *Ernst Haeckel*. All survey tows standardised to 1.75 nautical miles.

147

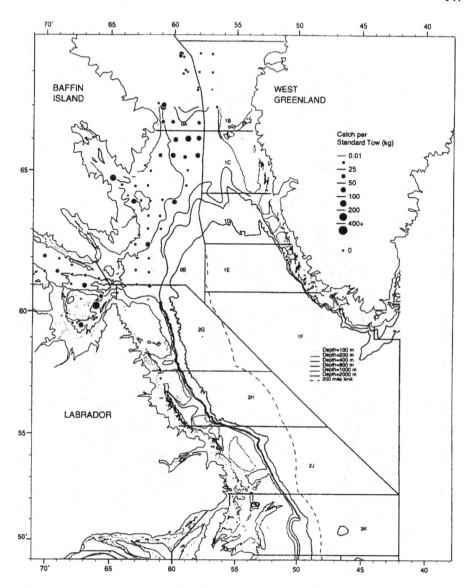

Figure 6. Catch per standard tow (kg) of Greenland halibut during the summer of 1978 in the Davis Strait and the Hudson Strait by the Canadian research vessel *Canso Condor*. All survey tows standardised to 1.75 nautical miles.

148

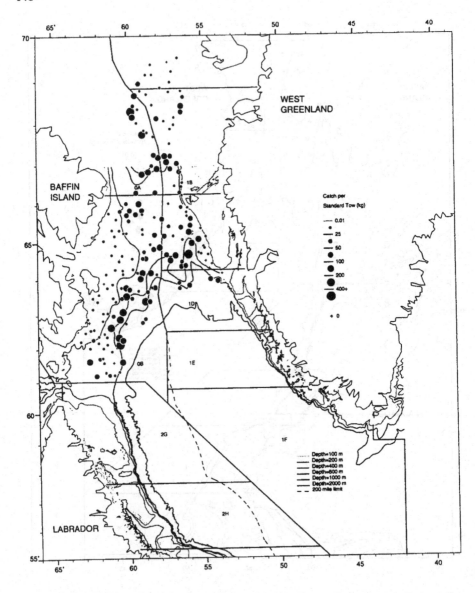

Figure 7. Catch per standard tow (kg) of Greenland halibut during the summer of 1986 in the Davis Strait by the Canadian research vessel *Gadus Atlantica*. All survey tows standardised to 1.75 nautical miles.

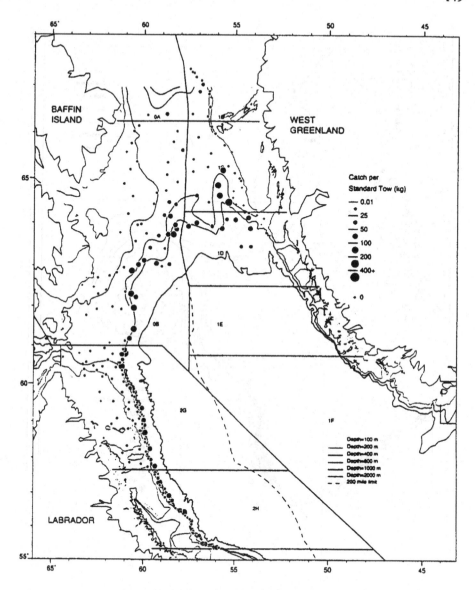

Figure 8. Catch per standard tow (kg) of Greenland halibut during the autmn of 1988 in the Davis Strait and Northern Labrador by the Russian research vessel *K. Shaitanov*. All survey tows standardised to 1.75 nautical miles.

150

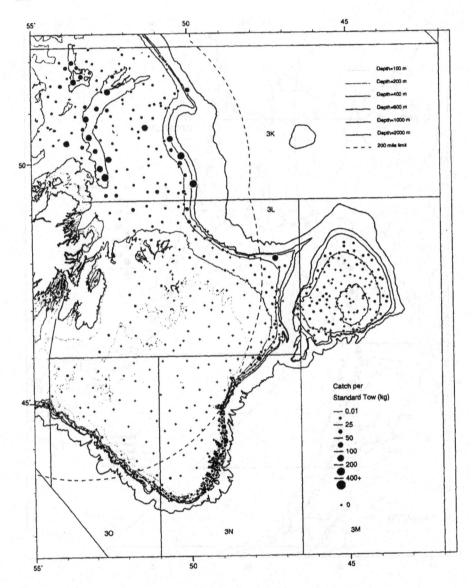

Figure 9. Catch per standard tow (kg) of Greenland halibut during the spring of 1987 at Eastern Newfoundland, Grand Bank and Flemish Cap by the Russian research vessel *Persey*. All survey tows standardised to 1.75 nautical miles.

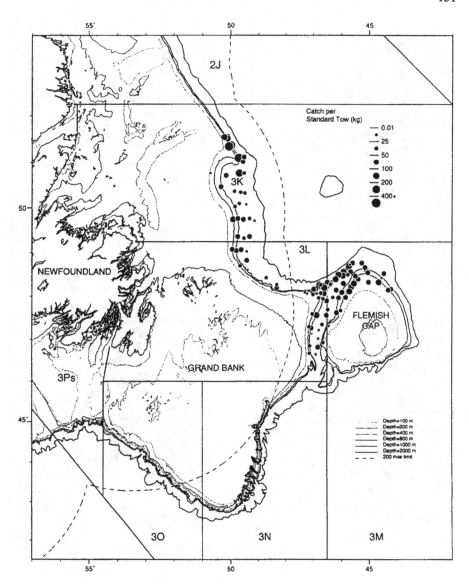

Figure 10. Distribution of Greenland halibut catches from a Greenland halibut directed deep-water Canadian survey by the *Cape Adair*, Summer 1991, to NAFO Divisions 3K,L and M

152

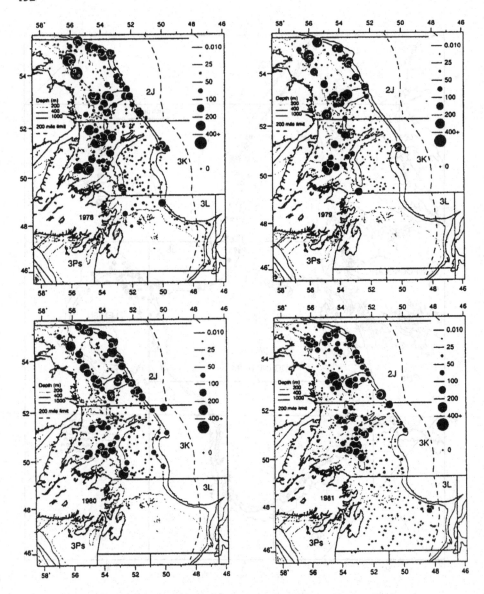

Figure. 11. Distribution of Greenland halibut catches from 1978 to 1981 Canadian autumn
surveys to NAFO Divisions 2J, 3K and 3L by the Canadian research vessel *Gadus
Atlantica*. All survey tows standardised to 1.75 nautical miles.

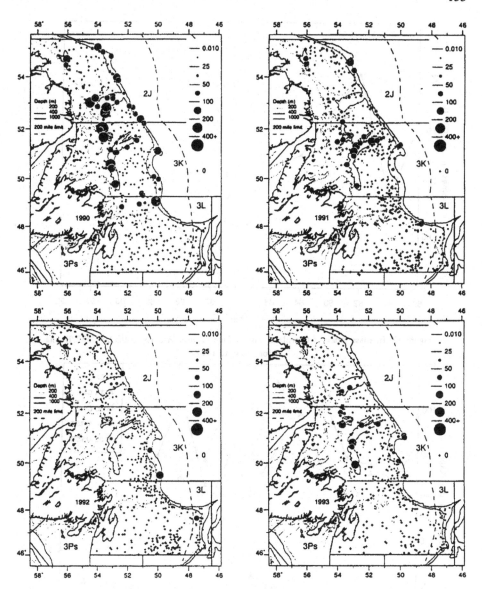

Figure. 12. Distribution of Greenland halibut catches from1990 to 1993 Canadian autumn surveys to NAFO Divisions 2J, 3K and 3L by the Canadian research vessel *Gadus Atlantica*. All survey tows standardised to 1.75 nautical miles.

154

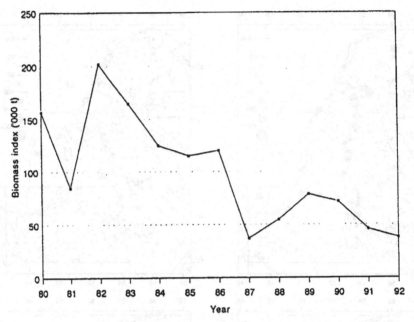

Figure 13. Biomass index of Greenland halibut from surveys in Division 0B by USSR/Russia
from 1980 to 1992

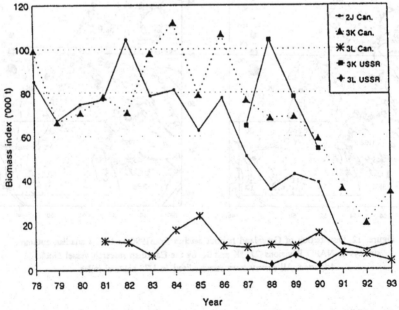

Figure 14. Biomass indices of Greenland halibut from stratified random surveys in Divisions 2J, 3K and
3L by Canada, and Divisions 3K and 3 L by USSR/Russia

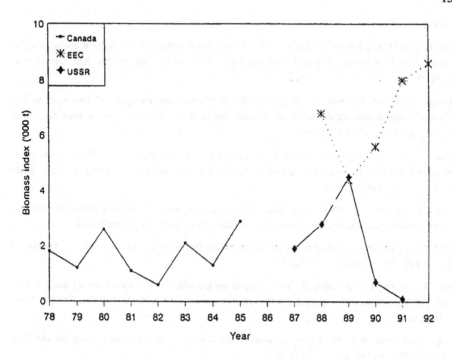

Figure 15. Biomass indices of Greenland halibut from stratified random surveys at Flemish Cap by Canada (1975 to 1985) and the EU(EEC) (1988 to 1992)

156

References

Atkinson, D.B. (1992). Some observations on the biomass and abundance of fish captured during stratified random bottom trawl surveys in NAFO Divisions 2J3KL, fall 1981-91. Can. Atl. Sci. Adv. Comm., Res. Doc. 92/72, 42p.

Atkinson, D. B. and Bowering, W. R. (1987). The distribution and abundance of Greenland halibut, deepwater redfish, golden redfish, roundnose grenadier and roughhead grenadier in Davis Strait. Can. Tech. Rep. Fish. Aquat. Sci. 1578: v + 29 p.

Atkinson, D. B., Bowering, W. R., Parsons D. G., Horsted Sv. Aa., and Minet J. P. (1982). A review of the biology and Fisheries for roundnose grenadier, Greenland halibut and northern shrimp in Davis Strait. NAFO Sci. Coun. Studies, 3: 7-27.

Arthur, J.R. and Albert, E. (1993). Use of parasites for separating stocks of Greenland halibut (Reinhardtius hippoglossoides) in the Canadian Northwest Atlantic. Can. J. Fish. Aquat. Sci. 50: 2175-2181.

Barrett, B. E. (1968). First occurence of Greenland halibut (Reinhardtius hippoglossoides) in the Bay of Fundy. J. Fish. Res. Board Can. 25: 2721-2722.

Berth, U., Schultz N., and Vaske, B. (1979). Report on groundfish survey carried out by the RV Ernst Haeckel in Statistical Area 0, Subarea 2 and Div. 3K during autumn 1978. ICNAF Res. Doc. 79/VI/127, Ser. No. 5523, 27 p.

Berth, U. and Vaske, B. (1980). Report on survey of the Walther Barth in Subarea 2 during autumn 1979. NAFO SCR Doc. 80/102, Ser. No. N157, 15 p.

Boje, J. (1993). Migrations of Greenland halibut in the Northwest Atlantic from tagging experiments in West Greenland 1986-1989. ICES C.M.1993/G:65.

Boje, J. and Hareide, N.R. (1993). Trial deepwater longline fishery in the Davis Strait, May-June, 1992. NAFO SCR Doc. 93/53, Serial No. N2236, 6p.

Bowering, W. R. (1982). Population dynamics of Greenland halibut in the Gulf of St. Lawrence. J. Northw. Atl. Fish. Sci. 3: 141-147.

Bowering, W. R. (1983). Age, growth, and sexual maturity of Greenland halibut, Reinhardtius hippoglossoides (Walbaum), in the Canadian Northwest Atlantic. Fish. Bull. 81: 599-611.

Bowering, W. R. (1984). Migrations of Greenland halibut, Reinhardtius hippoglossoides, in the Northwest Atlantic from tagging in the Labrador-Newfoundland region. J. Northw. Atl. Fish. Sci. 5: 85-91.

Bowering, W. R. (1987a). A newly developed stratification scheme for NAFO Divisions 2G and 2H. NAFO SCR Doc. 87/23, Ser. No. N1306, 6 p.

Bowering, W. R. (1987b). A newly developed stratification scheme for selected areas in NAFO Subareas 0 and 1. NAFO SCR Doc. 87/25, Ser. No. N1308, 5 p.

Bowering, W. R. (1987c). Distribution and abundance of Greenland halibut in Davis Strait (NAFO Subareas 0 and 1) from a Canadian research vessel survey in 1986. NAFO SCR Doc. 87/22, Ser. No. N1305, 10 p.

Bowering, W. R. (1988a). An analysis of morphometric characters of Greenland halibut (Reinhardtius hippoglossoides) in the northwest Atlantic using a multivariate analysis of covariance. Can. J. Fish. Aquat. Sci. 45: 580-585.

Bowering, W. R. (1988b). Biomass estimates of Greenland halibut in NAFO Div. 2GH from post-stratified and stratified Canadian groundfish surveys. NAFO SCR Doc. 88/40, Ser. No. N1480, 12 p.

Bowering, W. R. and Baird, J. (1980). Greenland halibut and witch flounder on the Flemish Cap (Division 3M). NAFO SCR Doc. 80/85, Ser. No. N139. 11 p.

Bowering, W. R. and Brodie, W. B. (1988). A review of the status of the Greenland halibut resources in NAFO Subarea 2 and Divisions 3K and 3L. NAFO SCR Doc. 88/69, Ser. No. N1512, 23 p.

Bowering, W. R. and Brodie, W. B. (1991). Distribution of commercial flatfishes in the Newfoundland-Labrador region of the Canadian northwest Atlantic and changes in certain biological parameters since exploitation. Neth. J. Sea Research 27 (3/4): 407-422 (1991).

Bowering, W. R., Brodie, W. B., and Baird, J. W. (1990). An assessment of the Greenland halibut stock component in NAFO Subarea 2 and Divisions 3K and 3L. NAFO SCR Doc. 90/51 (revised), Ser. No. N1772, 21 p.

Bowering, W. R., Brodie, W. B., and D. Power. (1993). An evaluation of the status of the Greenland halibut resource in NAFO Subarea 2 and Divisions 3KLMN. NAFO SCR Doc. 93/75, Ser. No. N2260, 29 p.

Bowering, W. R. and Chumakov, A. K. (1989). Distribution and relative abundance of Greenland halibut (Reinhardtius hippoglossoides (Walbaum)) in the Canadian Northwest Atlantic from Davis Strait to the northern Grand Bank. Fish. Res. 7: 301-328.

Bowering, W. R. and G. R. Lilly. (1992). Greenland halibut (Reinhardtius hippoglossoides) off southern Labrador and northeastern Newfoundland (Northwest Atlantic) feed primarily on capelin (Mallotus villosus). Neth. J. Sea Research 29 (1-3): 211-222 (1992).

Bowering, W. R. and Parsons, D. G. (1986). Diel variability in trawl catches of Greenland halibut (Reinhardtius hippoglossoides) from the channels off coastal Labrador and implications for resource management. N. Am. J. Fish. Manage. Vol.6, No. 2: p149-155.

Brodie, W. B. (1991). An assessment of Greenland halibut in SA 2 + Divisions 3KL. NAFO SCR Doc. 91/88, Ser. No. N1972, 29 p.

Brodie, W. B. and Baird, J. W. (1992). Data for the assessment of Greenland halibut in SA2 + Divisions 3KLM. NAFO SCR Doc. 92/81, Ser. No. N2136, 16 p.

Brodie, W. B. , Chumakov, A. K., and Bowering W. R. (1991). Update of abundance and biomass estimates of witch flounder in Divisions 3NO and Greenland halibut in Divisions 3KL from USSR surveys in 1987-90. NAFO SCR Doc. 91/56, Serial No. N1940, 7p.

de Cardenas, E., Junquera S., and Vazquez, A. (1993). Abundance indices of Greenland halibut in deepwater fishing zones on NAFO Divisions 3LMN. NAFO SCR Doc. 93/61, Serial No. N2244, 8p.

Christensen, O. and Lear, W. H.. (1977). Bycatches in salmon driftnets at West Greenland in 1972. Medd. Gronl. 205(5): 38 p.

Chumakov, A. K. (1970). The Greenland halibut (Reinhardtius hippoglossoides) in the Iceland area - The halibut fisheries and tagging. Tr. Polyarn. Naucho-Issled. Proetkn. Inst. Morsk. Rybn. Khoz. (1970.

Chumakov, A. K. (1975). Localities of Greenland halibut stocks in the Northwest Atlantic. Tr. Polyarn. Naucho-Issled. Proektn. Inst. Morsk. Rybn. Khoz. Okeanogr. 35: 203-209, 1975.

Chumakov, A. K., Rikhter, V. A., and Sigaev, I. K. (1989). USSR research report for 1988. NAFO SCS Doc. 89/8, Serial No. N1585, 23p.

Chumakov, A. K., Rudneva, G. B., Ernst, P., and Muller, H. (1990). Status of Greenland halibut (*Reinhardtius hippoglossoides*) stocks and feasible yield in NAFO Subareas 0, 1, and Div. 2GH. NAFO SCR Doc. 90/52, Ser. No. N1773, 18 p.

Colbourne, E., Narayanan S., and Prinsenberg, S. (1993). Climate change and environmental conditions in NAFO Divisions 2J3KL. Int. Coun. Explor. Sea. Cod and Climate Change Symposium Poster No. 7.

Crawford, R. E. (1992). Life history of the Davis Strait Greenland halibut, with reference to the Cumberland Sound fishery. Can. Man. Rep. Fish. Aquat. Sci. No. 2130, 19 p.

Doubleday, W. G. (1981). Manual on groundfish surveys in the Northwest Atlantic. NAFO Sci. Coun. Studies, No. 2, 55 p.

Dunbar, M. J., and Hildebrand, H. H. (1952). Contribution to the study of fishes of Ungava Bay. J. Fish. Res. Board Can. 9: 83-128.

Fairbairn, D. J. (1981). Biochemical genetic analysis of population differentiation in Greenland halibut (*Reinhardtius hippoglossoides*) from the Northwest Atlantic, Gulf of St. Lawrence, and Bering Sea. Can. J. Fish. Aquat. Sci. 38: 669-677.

Fleming, A. M. (1964). The inshore cod fishery with nylon gillnets. Fish. Res. Bd. Canada, Biol. Stn. Circ., St. John's, Nfld. No. 11, p 2-3.

Gabriel, M. L. (1944). Factors affecting the number and form of vertebrae in *(Fundulus heteroclitus)*. J. Exp. Zool. 95: 105-147.

Gorchinsky, K. V. (1993). Results from Greenland halibut assessment in Divisions 0B, 2GH by the data from 1992 trawl survey. NAFO SCR Doc. 93/15, Ser. No. N2192, 7 p.

Heuts, M. J. (1949). Racial divergence in finray variation patterns in *(Gasterosteus aculeatus)*. J. Genet. 49: 183-191.

Hovgard, H. (1993). The fluctuations in cod *(Gadus morhua)* fisheries off West Greenland in the twentieth century. NAFO Sci. Coun. Studies 18: 43-45.

Hudon, C. (1990). Distribution of shrimp and fish by-catch assemblages in the Canadian eastern Arctic in relation to water circulation. Can. J. Fish. Aquat. Sci. 47: 1710-1723.

ICNAF. (1974). Redbook, pp. 82.

ICNAF. (1975). Stat. Bull. Int. Comm. Northw. Atl. Fish., Vol. 23.

ICNAF. (1979). Redbook, pp. 81-82.

Jensen, A. S. (1935). The Greenland halibut *(Reinhardtius hippoglossoides)* its development and migrations. K. Dan. Vidensk. Selsk. Skr. 9Rk. 6: 1-32.

Jorgensen, O. and Boje, J. (1993). An assessment of the Greenland halibut stock component in NAFO Subareas 0+1. NAFO SCR. Doc. 93/80, Ser. No. N2265, 6 p.

Junquera, S., Iglesias, S. and Cardenas, E. de (1992). Spanish fishery of Greenland halibut *(Reinhardtius hippoglossoides)* in 1990-91. NAFO SCR Doc. 92/28, N2075, 14 p.

Junquera, S. and Zamarro, J. (1992). Sexual maturity and spawning of the Greenland halibut *(Reinhardtius hippoglossoides)* from Flemish Pass area. NAFO SCR Doc. 92/41, Ser. No. N2092, 10 p.

Khan, R. A., Dawe, M., Bowering, R., and Misra, R. K. (1982). Blood protozoa as an aid for separating stocks of Greenland halibut, *Reinhardtius hippoglossoides*, in the northwestern Atlantic. Can. J. Fish. Aquat. Sci. 39: 1317-1322.

Lear, W. H. (1970). The biology and fishery of the Greenland halibut *(Reinhardtius hippoglossoides* Walbaum) in the Newfoundland area. M.Sc. thesis, Memorial Univ. Newfoundland, St. John's, Nfld. 132 p.

Lear, W. H. and Pitt, T. K. (1971). Distribution of Greenland halibut in the Newfoundland-Labrador area in relation to depth and temperature. Fish. Res. Board Can. Tech. Rep. 274: 34 p.

Melindy, S. and Flight, J. (1992). Development of a deepwater turbot fishery by inshore gillnetters, 1992. Canada/Newfoundland Inshore Fisheries Development Agreement. iv + 30 p.

Messtorff, J. (1984). Research vessel survey results for cod in Division 2J as obtained by R/V *Anton Dohrn* and R/V *Walther Herwig* in autumn, 1972-83. NAFO SCR Doc. 84/91, Ser. No. N882, 7 p.

Miller, D. S. (1992). Observations and studies on SA2 + Div. 3K capelin in 1991. Can. Atl. Sci. Adv. Comm., Res. Doc. 92/15, 18p.

Miller, D. S. (1993). Observations and studies on SA2 + Div. 3K capelin in 1992. Canada, Dept. Fish. and Oceans Atlantic Fisheries Res. Doc. 93/10, 10p.

Miller, D. S. and Lilly, G. R. (1991). Observations and studies on SA2 + 3K capelin in 1990. Can. Atl. Sci. Adv. Comm., Res. Doc. 91/11, 30p.

Misra, R. K. and Bowering, W. R. (1984). Stock delineation of Greenland halibut in the Northwest Atlantic using a recently developed, multivariate statistical analysis based on meristic characters. N. Am. J. Fish. Manage. 4A: 390-398.

NAFO. (1980). Scientific Council Reports, pp. 84.

NAFO. (1982). Scientific Council Reports, pp. 77.

NAFO. (1984). Scientific Council Reports, pp. 55-56.

NAFO. (1985). Scientific Council Reports, pp. 71-72.

NAFO. (1989). Scientific Council Reports, pp. 81-82.

NAFO. (1990). Scientific Council Reports, pp. 91.

NAFO. (1992). Scientific Council Reports, pp. 135.

NAFO. (1993). Scientific Council Reports, pp. 99-103, 154.

Pechenik, L. N. and Troyanovskii, F. M. (1970). Trawling Resources of the North-Atlantic continental slope. Murmanskoe Knizhnoe Izdatel'stvo, Murmansk 1970. Translated from Russian by Israel Program for Scientific Translations. Jerusalem 1971: 66 p.

Riget, F. and Boje, J. (1988). Distribution and abundance of young Greenland halibut *(Reinhardtius hippoglossoides)* in West Greenland waters. NAFO Sci. Coun. Studies 12: 7-12.

160

Riget, F. and Boje, J. (1989). Fishery and some biological aspects of Greenland halibut *(Reinhardtius hippoglossoides)* in West Greenland waters. NAFO Sci. Coun. Studies 13: 41-52.

Riget, F., Boje, J., and Simonsen, V. (1992). Analysis of meristic characters and genetic differentiation in Greenland halibut *(Reinhardtius hippoglossoides)* in the Northwest Atlantic. J. Northw. Atl. Fish. Sci., Vol. 12: 7-14.

Rikhter, V. A., Sigaev, I. K., Borovkov, V., Kovalev, S., and Savvatimsky, P. (1991). USSR research report for 1990. NAFO SCS DOC. 91/5, Ser. No. N1889, 25 p.

Rink, H. (1852). De danske Handelsdistrikter i Nordgronland, deres geografiske Beskaffenhed og produktive Erhvervskilder. Forste Deel. - In: Gronland geografisk og statistisk beskrevet, 1, (1857).

Robbins, C. R. and Ray, G. C. (1986). A field guide to Atlantic coast fishes of North America. Houghton Mifflin Co., Boston. 354 p.

Rodriquez-Marin, E., Puzon, A., and Paz, J. (1993). Greenland halibut *(Reinhardtius hippoglossoides)* Feeding in Flemish Pass - NAFO Divisions 3LM. NAFO SCR Doc. 93/18, Serial No. N2195, 9p.

Scott, W. B. and Scott, M. G. (1988). Atlantic fishes of Canada. Can. Bull. Fish. Aquat. Sci. 219: 731 p.

Smidt, E. (1969a). The Greenland halibut *(Reinhardtius hippoglossoides)*, biology and exploitation in Greenland waters. Medd. Dan. Fisk. Havunders. 6: 79-148.

Smidt, E. (1969b). The Greenland halibut. Translated from Tidsskrifted Gronland, p259-366, by the Translation Bureau, Multilingual Serv. Div., Dept. of Sec. of State of Canada, Trans. Ser. No. 3094.

Smith, S. J. and Somerton, G. D. (1981). STRAP: A user-oriented computer analysis system for groundfish research trawl survey data. Can. Tech. Rep. Fish. Aquat. Sci. 1030: iv + 66 p.

Taning, A. V. (1944). Experiments on meristic and other characters in fishes. Medd. Komm. Dan. Fisk. Havunders. Bd. XI, Fiskeri, p. 1-66.

Taning, A. V. (1952). Experimental study on meristic characters in fishes. Biol. Rev. 27: 169-193.

Templeman, W. (1970). Vertebral and other meristic characteristics of Greenland halibut, *Reinhardtius hippoglossoides*, from the Northwest Atlantic. J. Fish. Res. Board Can. 27: 1549-1562.

Templeman, W. (1973). Distribution and abundance of the Greenland halibut, *Reinhardtius hippoglossoides* (Walbaum), in the Northwest Atlantic. ICNAF Res. Bull. 10: 83-98.

Templeman, W. and Pitt, T. K. (1961). Vertebral numbers of redfish, *Sebastes marinus* (L.), in the Northwest Atlantic, 1947-1954. ICNAF Spec. Publ. 3: 56-89 [also Rapp. P.-V. Reun. Cons. Int. Explor. Mer 150: 56-89].

Vazquez, A. (1993). Results of bottom trawl survey of Flemish Cap in July 1992. NAFO SCR Doc. 93/19, Ser. No. N2196, 22 p.

Yetman, L. and Staubitzer, D. (1990). Deepwater turbot survey, Northeast coast of Newfoundland, 1990. Project Report FDD 1990-148, v + 25 p.

THE DISTRIBUTION, RELATIVE ABUNDANCE, AND BIOLOGY OF THE DEEP- SEA FISHES OF THE ICELANDIC SLOPE AND REYKJANES RIDGE

JUTTA V. MAGNÚSSON & JAKOB MAGNÚSSON
Marine Research Institute,
PO. Box 1390
1210 Reykjavik
Iceland.

ABSTRACT

This working paper describes the distribution and in some cases the relative abundance of several deep-sea fish species at Iceland. The information given is mainly based on deep sea hauls carried out during numerous research cruises since 1976.

The main species dealt with are: roundnose and roughhead grenadiers, blue ling, orange roughy, black scabbard fish, smooth-head, Rhinochimæridae, several dogfishes and sharks, e.g. black dogfish and Portuguese shark.

Biological information is given for some of the species. The roundnose grenadier is considered to be a self-sustained stock at Iceland. The spawning pattern of the blue ling is discussed and a possible connection between blue ling stocks at Iceland and the Faroes is mentioned. It is most likely that the orange roughy spawns at Iceland.

An extract from a recent deep-sea cruise to the Reykjanes Ridge, in March 1993 is included.

A. G. Hopper (ed.), Deep-Water Fisheries of the North Atlantic Oceanic Slope, 161–199.
© *1995 Kluwer Academic Publishers.*

1. Introduction

The deep waters surrounding Iceland are inhabited by a great variety and number of
fish species but only few of them are subject to commercial fishing. However, there
is potential in some of the other species.

In this document, the definition "deep sea fish" refers to those species which have
their main abundance in waters 500 m and deeper. It should be mentioned that deep
sea fish like Greenland halibut (*Reinhardtius hippoglossoides*) and deep sea redfish
(*Sebastes mentella*) are already subjected to heavy exploitation and are thoroughly
dealt with elsewhere, and are therefore not discussed here. Neither is the greater
silver smelt (*Argentina silus*) for which the life-cycle in Icelandic waters is relatively
well known, but which is also abundant in waters shallower than 500 m depth.

In this paper the main emphasis is on those deep sea fish believed to have some
potential in future fisheries. A few species are already exploited to some extent but
most of them not at all. These species are: roundnose and roughhead grenadiers, blue
ling, orange roughy, North Atlantic codling, smooth-head, black scabbard fish, some
Rhinochimaeridae, several dogfishes and sharks.

The material used partly dates back to 1975 and 1976 and covers the period up to
and including 1993. In the beginning of this time period little attention was paid to
the sampling of biological information on those deep sea species which were not of
commercial interest at that time. Records, however, were usually kept on the
occurrence but a great part of measurements and other biological observations
presented in the paper derive mainly from the last decade. Some of the latest
information was obtained from a deep sea survey to the Reykjanes Ridge, in March
1993, and this is presented in a separate section. A brief note on preliminary results
of a new study on the scattering layers in the Irminger Sea is also given.

It should be pointed out that this paper has been prepared as a contribution for the
Workshop and thus, gives an overview of observations almost entirely based on
investigations carried out by Iceland, with restricted references to the background
literature.

The distribution of tows from which the samples derive is shown in Fig. 1. All
species mentioned in the paper are listed in Table 1.

2. Physical factors of the Icelandic slope

2.1 NOTES ON THE WATER TEMPERATURE

Off the south coast of Iceland, the bottom water temperatures range from 7 to 5deg.C
in 600-1000 m depth. The conditions are mostly stable in these depths and do not
show great seasonal fluctuations. Slight differences may occur between years but in
general terms, the bottom temperature in 1000 m depth is about 5deg.C. Likewise,
temperatures in 600 m depth are around 7deg.C. In this depth, the temperature may
fluctuate slightly more than in 1000 m depth i.e. mainly between 6 and 7deg.C. Most
of the species discussed in this paper are recorded with their greatest abundance in

this depth range and range of water temperature.

The temperatures are slightly lower in the western slope area in these depths. In 600m depth the temperature is about 6 deg C but in 1000 m depth it is about 4.5 deg.C.

2.2 BOTTOM CONDITIONS FOR TRAWLING ON THE SLOPE

The bottom conditions in the deep zone of the Icelandic slope are very variable. The western slope is in general quite suitable for bottom trawling. In the northern part of it, however, the rocky bottom may cause difficulties. In the southern part, on the other hand, trawling may be difficult because of too muddy a bottom. The Reykjanes Ridge itself is in general a very difficult area for bottom trawling mainly because of the extremely irregular bottom topography with its countless peaks and rifts. On both sides one may encounter huge quantities of sponges in some places and loose mud in others. It is possible to trawl to some extent on the eastern slope of the Reykjanes Ridge closest to the shelf, but with dangerous obstacles in between tows. The deep water area from the Reykjanes Ridge to east of the Westman Islands (approximately longtitude 20° W) is in general suitable for trawling. In deep hauls (>1000 m) mud may in some places cause difficulties. As reflected in the distribution charts (fig.3 to 16) of the various species, the deep water zone off the south east coast has been considered unsuitable for trawling. The deep zone below approximately 500 m is covered with coral. Recently, well equipped powerful trawlers have started to venture into this coral belt, mainly in search for orange roughy, but with limited results. The western slope of the Iceland-Faroe Ridge is suitable for trawling but in great depths and there are dangerous areas for the trawl gear. Many unknown ship wrecks from World War II are still causing damage to gear.

3. Species

3.1 GRENADIERS (*Macrouridae*)

The information given here on the grenadiers is mainly based on a conference paper (Magnússon 1987). The sea areas mentioned are shown in figure 2. Several species of *Macrouridae* are common in deep waters off Iceland. Only two of them have potential commercial value, i.e. the roundnose grenadier (*Coryphaenoides rupestris* Gunnerus) and the roughhead grenadier (*Macrourus berglax,* Lacépède). Both species have been subject to commercial fishing particularly in the 1960s and early 1970s. The roundnose grenadier was caught by the Soviet fleet mainly in the 1960s (Pechenik & Troyanovskii, 1970) off the south and south east coasts, and the roughhead grenadier by the international fleet in connection with the fishery for Greenland halibut off the west and north west coasts in the 1960s and early 1970s. The magnitude of these fisheries is not known. There has not been any directed grenadier fishery in Iceland, although Icelandic trawlers do, however, occasionally make big catches of roundnose grenadier when engaged in the fishery for other species.

164

TABLE 1. List of species. (English names in alphabetical order)

ENGLISH	LATIN	ICELANDIC
Birdbeak dogfish	Deania calceus	Faltrefur
Black dogfish	Centroscyllium fabricii	Svarthásfur
Black scabbard fish	Aphanopus carbo	Stinglax
Blue ling	Molva dypterygia	Blálanga
Boa dragonfish	Stomias boa ferox	Marsnákur
Capelin	Mallotus villosus	Loðna
Deep sea redfish	Sebastes mentella	Djúpkarfi
Deep sea cat shark	Apristurus laurussonii	Gíslaháfur
Eelpouts	Lycodes spp.	Mjórar
Goitre black smelt	Bathylagus euryops	Skjár/Blálax
Greater lantern shark	Etmopterus princeps	Dökkháfur
Greater silver smelt	Argentina silus	Stóri gulllax
Greenland halibut	Reinhardtius hippoglossoides	Grálúða
Greenland shark	Somniosus microcephalus	Hákarl
Halibut	Hippoglossus hippoglossus	Lúða
Knifenose chimaera	Rhinochimaera atlantica	Trjónufiskur
Lantern shark	Etmopterus spinax	Litli loðháfur
Leafscale gulper shark	Centrophorus squamosus	Rauðháfur
Longnose velvet dogfish	Centroscymnus crepidater	Þorsteinsháfur
Longnosed chimaera	Hariotta raleighana	Langnefur
Loosejaw	Malacosteus niger	Kolbíldur
North Atlantic codling	Lepidion eques	Bláriddari
Orange roughy	Hoplostethus atlanticus	Búrfiskur
Portuguese shark	Centroscymnus coelolepis	Gljáfiskur
Roughhead grenadier	Macrourus berglax	Snarphali
Roundnose grenadier	Coryphaenoides rupestris	Sl. langhali
Sea tadpole	Careproctus reinhardti	Hveljusogfiskur
Shortnosed snipe eel	Serrivomer beani	Trjónuáll
Silver rocking	Onogadus argentatus	Rauða sævesla
Smooth-head	Alepocephalus bairdii	Gjölnir
Viperfish	Chauliodus sloani	Slóans gelgja

3.1.1 Roundnose grenadier *(Coryphaenoides rupestris)*. Figure 3.

The roundnose grenadier is widely distributed and is common off the south and west coasts of Iceland (Fig. 3). It is also common on the western slope of the Iceland-Faroe Ridge. Although there are no records of roundnose grenadier from the slope of the continental shelf off the southeast coast of Iceland in our material, it is assumed that there, the roundnose grenadier is in fact very common. There are even indications that the greatest concentrations might be found in this area. Most of the reports on big catches of up to 30 tons in one haul(personal communication with captains of Icelandic trawlers) in relatively shallow waters are from the southeast corner of Iceland, in winter and early spring. Additionally, the Soviet fleet reportedly obtained good catches in this area in the 1960s. The reason for the lack of observations from the southeast area in our material is that the sea bed in this particular region is unsuitable for trawling at the depths which are considered optimal for the abundance of roundnose grenadier.

From the north and northeast coasts there are no records of roundnose grenadier. Only few specimens have been recorded off East Greenland.

An example of the abundance according to depth is indicated in Table 2 which shows the average catch per hour by area and depth regardless of time. This is based on material from several cruises. In all areas, the average catch is best in depths greater than 800 m. The samples from depths less than 800 m are rather few except for the southwest area. The biggest average catches were obtained in depths greater than 1000 m but it should be noted that in these depths, a great part of the catch consisted of small juvenile fish.

TABLE 2 Roundnose grenadier. Catch per hr (kg) and no of hauls by area and depth(Records of the years 1974-1985 regardless of season)

Area	Northwest		West		Southwest-South	
Depth (m)	No.h.	Catch	No.h.	Catch	No.h	Catch
600	-	-	2	21	26	43
600-800	2	7	3	51	59	136
800-1000	8	224	25	278	38	186
1000	-	-	13	284	21	235
Total hauls	10	-	43	-	144	-
Average catch		180		251		147

Young fish are defined in this paper as fish of 20 cm total length and smaller. Young stages of roundnose grenadier are found spread over a wide area along the continental slope. None, however, have been recorded off East Greenland, and they seem to be rare on the Iceland-Faroe Ridge. Off the northwest and west coasts of Iceland (north of latitude 65° N) they are frequently found, but only in small numbers. They are most abundant off the southwest coast in particular in the area from the Reykjanes Ridge north to 64° N (Fig. 17). Small roundnose grenadier are also abundant at some locations between the Westman Islands and the Reykjanes Ridge. Young fish are seldom found in depths less than 800 m but more often deeper than 900 m on both sides of the Reykjanes Ridge, i.e. within an area showing the greatest depth distribution and highest densities of young stages. The area, can thus be characterized as a nursery ground for the roundnose grenadier. Small roundnose grenadier are particularly frequent in catches taken in depths greater than 1000 m. In these depths, the small roundnose grenadier show the widest area distribution.

In the following text table the frequency of catches in which young roundnose grenadier are present in different depths is shown:

Depth (m)	No of hauls	No of hauls with young	%
<600	29	1	3.5
600-800	66	8	13.3
800-1000	80	37	46.3
>1000	40	28	70.0

166

Some observations on the feeding habits were carried out. *Euphausids* are by far the most common food species mentioned both in frequency and quantity. Other deep sea crustaceans such as unidentified deep sea shrimps were not uncommon. Less common were small squid, *medusae, myctophids spp.* and several other species of the deep sea community. In Table 6, the results are given from a random sample of 132 fish taken in 700 to 730 m depth, with a length range of 38 to 96 cm total length. All fish were at maturity stages I and II. According to this sample, the feeding habits of males and females are very similar. Usually the stomachs of individual fish contained only one food component, and only occasionally the diet was more variable.

TABLE 3. Roundnose. Mean length at age (1984-1985)
for males and females separately (Snout/analfin)

	MALES		FEMALES	
AGE	NO	1/2 CM	NO	1/2 CM
8	-	-	3	19.65
9	1	22.02	2	21.74
10	7	23.68	5	23.80
11	6	25.29	6	25.83
12	13	26.86	24	27.84
13	27	28.39	26	29.82
14	30	29.88	26	31.78
15	14	31.33	15	33.72
16	15	32.76	22	35.65
17	7	34.16	9	37.55
18	3	35.54	6	39.45
n	123		144	

TABLE 4. Age composition of roundnose
grenadier in Icelandic waters 1984-1987

	MALES		FEMALES	
AGE	NO	%	NO	%
8	-	-	105	2.6
9	63	2.6	34	0.8
10	261	10.9	106	2.6
11	80	3.3	109	2.7
12	177	7.4	513	12.7
13	467	19.4	678	16.8
14	595	24.8	611	15.1
15	297	12.4	416	10.3
16	259	10.8	891	22.1
17	157	6.5	285	7.1
18	48	2.0	291	7.2
Total	2404	100.1	4039	100.0

The roundnose grenadier were observed in temperatures ranging from 2.5 deg.C to 6.9 deg.C in Icelandic waters regardless of area and depth. However, in the area of greatest abundance, i.e. in the southwest and west area, they are most common in temperatures between 4deg.C and 6 deg.C in depths greater than 800 m. Young stages were most abundant in temperatures about and below 5 deg.C. On the western slope of the Iceland-Faroe Ridge, temperatures in which the roundnose grenadier were observed varied greatly, i.e. from 2.5 to 6.9 deg.C and there, they were not linked to certain depths as they were in the southwest area.

Spawning roundnose grenadier appear extremely seldom in the bottom trawl catches which leads to the assumption that the roundnose grenadier spawn off the bottom i.e. bathypelagic and are therefore inaccessible to bottom trawls. The few direct observations of spawning specimens are spread over several months. Spawning males were observed during all months except in January, June and August and spawning females were observed in February to July and in September. These observations were made regardless of area, and they indicate that spawning may take place in Icelandic waters to some extent more or less throughout the year.

Since the observations available on spawning fish (stage III)[1] only gave limited information on the spawning habits, the distribution of mature fish of stages II and IV was then studied on a monthly basis. For both males and females, the two stages were observed throughout the year, but the proportion differed considerably according to season. Thus, it is obvious that at least some spawning takes place throughout the year in Icelandic waters. There seems to be, however, a more intensive spawning period during the winter months which might be the main spawning season. There also might be a second period of intensive spawning in the summer and consequently, the possibility of the existence of two spawning stocks, i.e. winter and summer spawners should not be excluded. The main spawning might take place in somewhat different times in the different areas at Iceland.

The proportion of mature fish at length revealed that 50% maturity is reached by males at an approximately anal-fin length of 15.0 cm while the females reach this point at an anal-fin. length of about 17.8 cm.

The roundnose grenadier in Icelandic waters shows a different growth rate for males and females. The growth rate of the females is considerably higher than that of the males (Table 3). With reference to the length at maturity , the males mature (50%) at an age of about 14 years but the females of about 16 years. Ageing of young roundnose grenadier has not been carried out in Iceland, but an age/length key used on samples from 1984-1987 showed that the majority of the bigger fish were between the ages of 12 and 16 years (Table 4). The calculated length/weight relationship for the sexes combined is shown in Fig. 18.

The length/weight relationship is very similar for males and females while the relationship between weight and age, show distinct differences; the female being heavier at the same age (Table 5).

---

[1] Maturity stages: I = juvenile, II = ripening, III = spawning, IV = newly spent

TABLE 5. Roundnose. Weight at age for males and females separately

AGE	MALES (gr)	FEMALES (gr)
8	-	215
9	267	286
10	329	368
11	398	462
12	475	569
13	557	690
14	646	825
15	741	973
16	844	1135
17	953	1313
18	1068	1506

TABLE 6. Roundnose. Information on the stomach content of 132 fishes (45 males and 87 females)

Stomach	MALES no	MALES %	FEMALES no	FEMALES %	TOTAL no	TOTAL %
Empty	3	6.7	7	8.0	10	7.6
Everted	15	33.3	20	23.0	35	26.5
With content	27	60.0	60	69.0	87	65.9
Total	45		87		132	

Stomach content	no	%	no	%	no	%
Euphausids	24	85.7	56	93.3	80	90.9
Shrimps	-	-	1	1.7	1	1.1
Fish	1	3.6	1	1.7	2	2.3
Not recognizable	3	10.7	2	3.3	5	5.7

3.1.2 Roughhead grenadier (Macrourus berglax). Figure 4

In general terms, the roughhead grenadier was found all around Iceland and had thus a much wider distribution in Icelandic waters than the roundnose grenadier (Fig. 4). It was most abundant in the Dohrn Bank-Víkuráll area west of Iceland, but in the other areas, in general it was only recorded in minor quantities usually as single specimens.

In Icelandic waters the roughhead grenadier has been recorded in 295 m to 1260 m depth . The depth range was, however, variable for the different areas but it was most common in depths from 600 m to 900 m, in temperatures from 4 to 5 deg.C. In the Dohrn Bank area where the roughhead grenadier was most abundant, it has been recorded in depths less than 300 m down to 1000 m ,and especially between 700-800 m, in temperatures from 3 to 5 deg.C.

Contrary to the roundnose grenadier, young stages of roughhead grenadier have not been observed in Icelandic waters, and only in very limited numbers off East Greenland.

The main spawning area of roughhead grenadier is the Dohrn Bank-Víkuráll area and was described in 1978 (Magnússon 1978) based on findings of spawning females and of bathypelagic eggs. Additional information collected in later years support these findings.

The spawning takes mainly place in February to May,but a few specimens collected at different localities in the southwest area of Iceland, in September, indicated that some spawning was still going on at that time.

The length/weight relationship for both sexes combined is shown in Fig. 19. There are considerable differences in length/weight for males and females (Fig. 20).

Only few specimens of roughhead grenadier in Icelandic waters have been aged. The range was 9 to 19 years and the corresponding length range 46 to 83 cm total length. This range in age is similar to that observed for the roundnose grenadier.

Observations on the feeding habits of the roughhead grenadier showed that the diet is quite different from that of the roundnose grenadier. Several species of small deep sea fish seem to be preferred by the roughhead grenadier but *Euphausids* and other crustaceans as well as squid and *medusae* were also recorded.

Fish species, such as *Bathylagus euryops, Mallotus villosus, Stomias boa ferox, Malacosteus niger, Myctophidae spp., Onogadus argentatus, Careproctus reinhardti and Lycodes spp.*were identified in the stomach contents.

A notable number of roughhead grenadier were infested by *Sphyrion lumpi*, a parasite not observed on the roundnose grenadier but quite common on redfish in particular from the Irminger Sea oceanic stock.

3.1.3. Conclusions on grenadiers
Young stages of roundnose grenadier are very abundant in Icelandic waters and their densities indicate rather extensive nursery areas. Since the roundnose grenadier spawns in Icelandic waters and all sizes and maturity stages of this species have been observed, it can be assumed that the roundnose grenadier at Iceland should be considered as a separate population or stock.

Accumulations of young stages of roughhead grenadier have not been observed in Icelandic waters so far as it is known. Spawning roughhead grenadier and bathypelagic eggs of the roughhead grenadier were observed in the Dohrn Bank area (Magnússon 1978) and young stages further south on the East Greenland shelf suggesting a possible drift of eggs and fry southwards along the East Greenland coast and even around Cape Farewell.

3.2. BLUE LING *(Molva dypterygia)* Figure 5

In Icelandic waters the blue ling inhabits the relatively warm deep water off the south

170

and west coasts. It is scarce off the north coast where only exclusively small specimens have been recorded, and it is absent off the northeast coast (Fig. 5).

Blue ling is also found along the shelf slope off East Greenland, from Dohrn Bank south to approximately. 63° N but only in the Dohrn Bank area is it common.

Blue ling is reported from various depths. In general, the young and small fish are found in shallower waters (less than 500 m) than the big fish. Generally, blue ling are most abundant in depths from 700 to 900 m, although in certain areas they have been observed at greater depths.

Until 1978, blue ling were only obtained as by-catch mainly in the redfish fishery . In 1979, a directed fishery for blue ling was commenced and continued for some years during a very limited period of time each year mainly in the month of March. The basis for this fishery was the discovery of a spawning area of blue ling at a very restricted locality near the Westman Islands, on and around a small steep hill near the base of the slope mostly in depths of 500 to 800 m, in temperatures of 5-6 deg.C (Fig. 21). The catches of blue ling increased rapidly and reached 8000 tonnes in 1980 which was the highest catch of blue ling ever landed from Icelandic grounds. For the following three years, the catches remained high but decreased rapidly after 1983 and went back to the low by-catch level in 1985 (Fig. 22). It is noteworthy that after 9 years since the decline in the catches no significant accumulation of spawning blue ling has been observed in this area.

TABLE 7. Blue ling. Mean length by age for the years 1978-1982

Age Group	N	Ml (cm)	Increase	N	Ml (cm)	Increase	N	Ml (cm)	Increase
2	1	(33)					1	(33)	
3	2	39.50	6.50	1	(38.)		1	(41)	(8)
4	4	46.50	7.00	1	(44)	(6)	3	47.30	6.30
5	9	55.67	9.17	4	52.50	8.50	5	58.20	10.90
6	17	60.71	5.04	6	59.80	7.30	11	61.20	3.0
7	28	65.72	5.01	12	64.30	4.50	12	67.10	5.9
8	44	70.16	4.44	14	69.40	5.10	20	71.00	3.90
9	96	75.25	5.09	41	74.06	4.66	42	76.87	5.87
10	119	80.82	5.57	44	80.84	6.78	41	82.01	5.14
11	240	85.20	4.38	111	83.83	2.99	55	88.02	6.01
12	253	91.24	6.04	70	88.58	4.75	65	96.24	8.22
13	284	98.83	7.59	70	94.83	6.25	100	102.78	6.54
14	205	104.58	5.75	47	99.80	4.97	83	108.34	5.56
15	85	112.29	7.71	9	105.13	5.33	51	114.36	6.02
16	49	119.35	7.06				30	121.04	6.68
17	32	124.21	4.86	1	(115)		24	125.42	4.38
18	9	129.67	5.46				8	129.63	4.21
19	8	136.13	6.46				5	136.62	6.99
20+	7	137.86	1.73				6	138.50	1.88
	1492	93.31	5.52	431	85.58	5.56	563	98.38	5.86

Until recently, this restricted location was the only known spawning area of blue ling in Icelandic waters. Several indications and circumstances (e.g. the distribution of 0-group blue ling) led to the assumption that there had to be another, or other areas of spawning concentrations of blue ling elsewhere. Some years ago, it was observed that French trawlers were fishing for blue ling in a southerly area of the Reykjanes Ridge and in 1993, the existence of spawning concentrations of blue ling in that area was confirmed. Once again, the location was an underwater hill of very limited size. In this new area, the fishing took mainly place in 800 to 900 m depth.

The size of blue ling in Icelandic waters is very variable. As early as 1982, it was pointed out that there are distinct differences in the length of blue ling in different sub-areas (Magnússon 1982). Thus, the smallest sizes of fish, and also the greatest amounts of small blue ling were observed in the southwest-subarea (apart from the few specimens off the north coast as mentioned previously) indicating a nursery ground in the shallower part of this area. The largest specimens were recorded off the east coast, but there, no small blue ling were observed. A new updated evaluation of the length distribution of blue ling at Iceland confirmed the former findings. Figure 23 shows this data and includes an additional sub-area (south-west) which is the southern Reykjanes Ridge. Small blue ling are also absent here.

The length distribution of blue ling from the recently discovered spawning locality on the Reykjanes Ridge is shown in Fig. 24. It is based on two samples taken early and late March, 1993. The sample taken in early March consisted almost entirely of males (97,7%). The sample taken later in the same month was not sexed but presumably the second peak at the longer lengths represents the females. This might indicate that the males accumulate in the spawning area earlier than the females.

Blue ling belong to the slow-growing species. In Icelandic waters, the average annual growth increase for both sexes is 5,52 cm for 3 years to 20 years or more of age. However, the growth increase for the 2-3 first years of life is presumably much higher judging from the size of 0-group fish (Table 7).

There is a distinct difference in the growth of males and females. Females grow faster and live longer than the males (Table 7). Males are also smaller and younger when reaching maturity (50% retention) at about 75 cm in size, whilst females mature at about 88 cm. Then, according to the given length-age relationship in Table 7, males will be about 9 years and females about 11 years old.

In Fig. 25, the age composition of spawning blue ling off the Westman Islands is shown which demonstrates the age composition of the spawning stock at the beginning of its exploitation.

Blue ling show differences in growth which quite obviously depend on the area of descendance i.e. growth patterns seem to slow down in the northerly direction. Thus, for example, 10 years old females in Icelandic waters are about 82 cm in size, at the Faroes about 93 cm and in the Shetland area about 95 cm (Thomas, 1980). This when taken together with the fact that all stages of the lifecycle of blue ling have been observed in Icelandic waters , including spawning and nursery areas, leads to the conclusion that the blue ling in the Icelandic - East Greenland area is a separate stock although there might exist several populations. As long as there is no proof for spawning areas of blue ling elsewhere in the Icelandic - East Greenland region the blue ling at East Greenland can be regarded as a part of the population which spawns near the Westman Islands and/or on the Reykjanes Ridge.

On the other hand, the blue ling in the area east of Iceland, and on the Iceland-Faroe Ridge are most probably of another origin. As has been said previously small blue ling were never observed in the sub-area E (fig.2). The Iceland - Faroe Ridge seems to be a kind of barrier for the distribution of the Icelandic blue ling stock eastwards. Therefore, it is presumed that the very big blue ling off east Iceland derives from spawning places near the Faroes and the Shetland Islands.

It is not yet known whether there are other areas where there are spawning concentrations of blue ling besides those already discussed, but this possibility should not be excluded. It is, however, an unanswered question why the recruits do not return to the spawning area of their descendance.

3.3. NORTH ATLANTIC CODLING (*Lepidon eques*) Figure 6

The North Atlantic codling is very common in Icelandic waters and quite abundant mainly in the slope area off the southwest coast. Records with 500 to over 1000 specimens per station are not uncommon. Its distribution reaches from the Víkuráll area in the west to the Iceland Faroe Ridge but it is completely absent off the north and east coasts (Fig. 6). It has been observed from depths less than 500 m to depths greater than 1100 m but it is most abundant in 600 to 700 m depth.

An example of the length distribution is given in Fig. 26 for the sexes separately. There are no obvious differences in size between the sexes. The bulk of the North Atlantic codling is of 26 to 36 cm in size. It should be noted that the smallest specimens derive from shallower stations. The most shallow station at which this species was observed was at 330 m depth.

Apparently, the main spawning takes place in February-March and the maturity is reached at 31.14 cm (50% retention).

3.4 ORANGE ROUGHY (*Hoplostethus atlanticus*) Figure 7

The first record of orange roughy in Icelandic waters was off the southeast coast and dates back to 1949. Since then, it was reported from the southwest to southeast coasts from time to time usually as single specimens. But it was not until 1991 that this species was fished commercially, although in minor quantities only. The known distribution is shown in Fig. 7. Orange roughy in the Icelandic area inhabits mainly depths greater than 600 m, and they are most frequent between 800 and 1000 m depth.

Compared with orange roughy of New Zealand and Australia the species seems to be much larger at Iceland. Both total length (T) and standard length (S) were measured, the conversion factor being

$$S = T - (T \times 0.1882).$$

There is a considerable difference in size between sexes (Fig. 27). The average length for males was 57.52 cm (T) and for females it was 60.17 cm (T). There are also differences between the sexes in the weight (Fig. 28). The length-weight relationship was:

$$W = 0.0381 \times L^{2.808} \text{ (males) and}$$

$$W = 0.0598 \times L^{2.678} \text{ (females)}.$$

In common with the New Zealand and Australian orange roughy, the species at Iceland seems to be a winter spawner. Spawning specimens begin to appear in the samples taken in November and they are still observed as late as March, but at that time about one third of the sample had just become spent (stage IV) while the majority were of stage II (recovering). Samples from April to August are not yet available.

Otoliths of orange roughy were examined and ageing was tried on whole otoliths with transparent light. Ageing in this way seems to be impossible on such large specimens. However, the panel of 5 readers agreed that the otoliths were readable up to 16-20 years. At that point there were changes observed which excluded an ongoing reliable reading. One wonders whether this might indicate the onset of the first spawning season.

The stomachs were also examined. The content was mostly unrecognizable. However, the food components which could be analysed were mainly crustaceans and remnants of fish e.g. *Myctophids*.

3.5 BLACK SCABBARD FISH (*Aphanopus carbo*) Figure 8

The distribution and relative abundance of the black scabbard fish is shown in Fig. 8. The species is distributed over the western and southern slope area as well as over the Reykjanes Ridge and the western slope of the Iceland-Faroe Ridge. Black scabbard fish have rarely been observed in less than 500 m depth. They are common in 600 to 1000 m depth, and particularly between 800 to 900 m according to the available material.

The length range of the black scabbard fish in Icelandic waters was rather wide. There are records from 36 cm to 115 cm but the bulk of the fish (75%) was in the range from 90 to 105 cm. Differences in the size of male and female were observed (Fig. 29); the average length for males being 94.30 cm and for females 99.80 cm. On the other hand, no significant differences between the sexes could be observed in the length/weight relationship (Fig. 30). Little is known about the spawning habits in this area. However, the observations of the cruises in March 1993 (Section 4.) showed that the bulk of the mature fish was newly spent (stage IV) although some of them were of stage II (recovering). Since also a spawning specimen was observed it is most likely that this species spawns in the Reykjanes Ridge area. The unity of the maturity stage IV at this time of the year indicates a seasonal spawning time probably during January to March.

Black scabbard fish is believed to have some potential for the commercial fishery since it is already well known in the markets of countries like Portugal and Spain. In the Icelandic area, the bottom trawl might not be the most suitable gear for this fishery because the species is known to live partly off the bottom. Deep sea long-lining might be more suitable but trials have not yet been undertaken.

3.6 SMOOTH-HEAD (*Alepocephalus bairdii*) Figure 9.

The smooth-head is common off the south and west slopes of Iceland and it is quite abundant in some areas (Fig. 9). The depth in which the smooth-head was recorded ranged from less than 500 m down to over 1100m, but mostly it was found in depths over 900 m in particular over 1000 m. The length range of this species was rather wide, i.e. from 8 cm to 68 cm but also specimens up to 90 cm have been measured (Fig. 31). The bulk of the fish was between 35 and 54 cm. No noteworthy differences in the length distribution of males and females were observed. The smallest specimens of 8 to 20 cm were recorded on both sides of the Reykjanes Ridge.

The length-weight relationship was studied on 130 specimens, in the range 24 to 67 cm in size. The average weight ranged from 105 to 2000 g. with the overall average weight being 658 g.

In March, the majority of the mature smooth-head were newly spent (stage IV). Spawning specimens have also been observed and the eggs are rather large, e.g. from a 69 cm female smooth-head, the diameter of the ovarian eggs was 3,8 mm.

Summarizing these observations they indicate the existence of separate stocks in the Icelandic area.

3.7 KNIFENOSE CHIMAERA (*Rhinochimaera atlantica*) Figure 11

This species has a similar distribution to most of the other deep sea fishes but the abundance is more restricted to the slope areas between the Reykjanes Ridge and the Westman Islands (Fig. 11). It is common in depths of 600 to 1000 m. Knifenose chimaera were observed in lengths of 31 to 140 cm but the bulk was between 109 and 115 cm (average length 105.67 cm). Ripe males were observed in March. There were sexual differences in length and weight: the females being larger and heavier than the males. The average length of mature males was 112.44 cm and the average weight 1277 g. while the average length of females was 120.50 cm and the average weight 3866 g.

3.8 LONGNOSED-CHIMAERA (*Hariotta raleighana*) Figure 12.

Longnose chimaera were more restricted to the southwest area and the Reykjanes Ridge than most of the other deep sea species (Fig. 12). It was most abundant in depths greater than 1100 m which might explain the low number of records and the apparently restricted distribution. The length ranged from 19 to 112 cm. The average length of males was 83.58 cm and of females 93.17 cm respectively. The average weight for both sexes was 942 g. but mature females over 100 cm in length weighed between 2340 g. and 2495 g. The largest male of 91 cm length weighed 1160 g. Ripe males were observed in March.

3.9 PORTUGUESE SHARK (*Centroscymnus coelolepsis*) Figure 10

This species is one of the biggest deep sea sharks in Icelandic waters. It is quite common off the west, southwest and south coasts of Iceland (Fig. 10). Although not caught in great quantities, ten specimens and more in one haul are not uncommon. The Portuguese shark has been recorded in depths shallower than 500 m, but generally the Portuguese shark inhabits deeper waters, and was most abundant in 900 to 1000 m depth.

Information on length and weight is given in Table 8 and Fig. 32.

3.10 BLACK DOGFISH (*Centroscyllium fabricii*) Figure 13

The black dogfish is probably the most common deep sea dogfish in Icelandic waters and it is widely distributed in the deep water zone off the west, southwest and south coasts, and at the western slope of the Iceland-Faroe Ridge (Fig. 13). It is very common in great depths and in particular below 600 m and even beyond 1100 m.

There are differences in the length range of males and females. Females were rather evenly distributed from approximately 48 cm to 81 cm, whilst the length distribution of the males showed a distinct peak in the 64 to 66 cm range (Fig. 33). The overall length was from 19 to 91 cm and the average weight from 25 g. (19 cm) to 4265 g. (86 cm) with the overall average weight of 1210 g. At one station at the eastern side of the Reykjanes Ridge at a depth of 925 m spawning females were observed during March. The young of two females (both 78 cm in length, 2160 g. and 2600 g. in weight, respectively) were counted and measured. One of the females released 17 young and the other one 5, all in a length range of 13 to 17 cm.

The black dogfish is considered to have potential for the commercial fishery because of its abundance and distribution, but much will depend on its processing possibilities.

3.11 GREATER LANTERN SHARK (*Etmopterus princeps*) Figure 14

The greater lantern shark was most abundant in the Reykjanes Ridge area (Fig. 14). It was common in depths greater than 700 m and in particular from 900 to 1000 m depth. The length range was from 12 cm to 89 cm with the average length 58.22 cm. See also Table 8. There were sexual differences in the length distribution with two distinct peaks at 61-63 cm for males and 70-72 cm for females (Fig. 34). The average weight ranged from 40 g. (12 cm) to 3360 g. (89 cm) and the overall average weight was 1277 g.

3.12 BIRDBEAK DOGFISH (*Deania calceus*) Figure 15

Although the birdbeak dogfish is found along the slope west off Iceland it was most common off the southwest coast and the Reykjanes Ridge (Fig. 15). It was most frequently caught in depths greater than 700 m. Information on the length and weight ranges as well as on average length and weight is given in Table 8.

3.13 LONGNOSE VELVET DOGFISH (*Centroscymnus crepidater*) Figure 16

The longnose velvet dogfish was distributed off the S coast and on the Reykjanes

Ridge (Fig. 16). It occurred in over a great range of depth but was most common in 700 to 800 m depth. Only a limited number of sexed measurements were available but the length distribution showed a remarkable peculiarity with the males being much smaller than the females with a distinct peak at the 61 to 63 cm group and no overlapping with the length of females which showed a distinct peak at the 76 to 78 cm group (Fig. 35). Information on the weight is given in Table 8.

TABLE 8. Overview over length (cm)/weight(g) of dogfishes and sharks

SPECIES	NO.	LENGTH RANGE	AV. TOTAL	MALE	FEMALE	WEIGHT RANGE	AV. WEIGHT
Deep sea cat shark	160	13-86	47.29			5-2520	529
Leafscale gulper shark	5	110-140	127.40			12200-15500	13336
Black dogfish	1278	16-106	60.39	60.52	60.42	15-2545	1210
Portuguese shark	91	70-120	102.86	90.14	106.56	1180-13090	8336
Longnose velvet dogfish	277	18-90	64.84	62.61	79.32	35-3265	1361
Birdbeak dogfish	192	27-114	88.23	80.13	86.65	1245-6415	3056
Greater lantern shark	443	12-89	58.22	60.34	67.33	40-3360	1277

4. The deep sea survey to the Reykjanes Ridge, March 1993

4.1. THE OBJECTIVES

The task of this survey was to examine the possibilities of deep sea fishing outside the conventional fishing grounds in the Reykjanes Ridge area mainly within the EEZ of Iceland. The survey was carried out by two vessels: the Research Vessel Bjarni Sæmundsson, and the chartered freezer trawler Sjóli, a vessel of 57 m length, and 2991 hp. Conventional bottom trawls were used by both the vessels and the cod-ends were lined with fine-meshed net.The main objectives of the cruise were:-

To examine the possibility of bottom trawling in the area

To obtain information on the availability and catchability of fish species in the area.

Species composition.

To collect environmental and biological information.

The trawling was carried out in depths between 550 m and 1500 m, but mainly in depths of 800 to 1000 m. Because of the difficult bottom topography it was not considered practical to determine the trawl stations in advance. Therefore, in order to cover the area systematically it was divided into squares of one degree longitude and half a degree latitude with each square being subdivided into four parts. The intention was if possible to obtain at least two hauls in each sub-square. The coverage of the area is shown in Fig. 36. The weather conditions were extremely unfavourable and restricted the planned operations considerably.

4.2. RESULTS

Although it was known before that this area would be extremely difficult for bottom trawling nevertheless several locations were identified as being suitable. A total of 49 hauls were taken of which 14 (i.e. 29%) were faulty. Generally the catch was rather poor. On the other hand the variety of species was fairly wide amounting to 90 fish species (Table 9).

Good catches of blue ling (*Molva dypterygia dypterygia*) were obtained at one location, and of deep sea redfish (*Sebastes mentella*) at another. Several other species are worth mentioning of having potential for exploitation e.g. black scabbard fish (*Aphanopus carbo*) and several dogfishes (Table 10). The roundnose grenadier (*Coryphaenoides rupestris*) was fairly common but the catches consisted mostly of small juveniles. Large specimens were rather scarce. Orange roughy *(Hoplostethus atlanticus)* were recorded at several locations but only as single specimens. Although the greater silver smelt (*Argentina silus*) and the North Atlantic codling (*Lepidion eques*) were frequently observed they are usually more common on the slope region off south and west Iceland. The same is true for the smooth-head (*Alepocephalus bairdii*). Halibut (*Hippoglossus hippoglossus*) in spawning condition was observed at about 61° 00'N 27° 33'W, in 800 to 900 m depth, and in temperatures of 4.8 to 5.5 deg.C. Greenland sharks (*Somniosus microcephalus*) were caught which are usually encountered in colder waters such as north off Iceland and East and West Greenland.

4.3. COMMENTS

In general, the deep water region of the Reykjanes Ridge is very difficult for bottom trawling. Unavoidably, much time is spent in searching for spots where trawling might be possible. These do not, however, necessarily coincide with areas where fish are abundant at the same time. It should be pointed out that besides obtaining valuable information on species, species composition, biology etc. the importance of combined research and survey cruises such as this one are to a great extent in gathering knowledge about trawlable locations ,as well as about unsuitable areas, which is important information for the development of new fisheries.

4.4. NOTES ON THE SCATTERING LAYERS

Deep scattering layers are very common in the off shelf areas south and west of

Iceland. In the Irminger Sea, these scattering layers are more or less continuous throughout the whole region mostly in depths of 400/500 m to 700/800 m. The components of the main layer show a great variety of organisms. Very common are the following species or groups: *Myctophids*, viperfish (*Chauliodus sloani*), Boa dragonfish (*Stomias boa ferox*), Goitre black smelt (*Bathylagus euryops*), shortnosed snipe eel (*Serrivomer beani*), *Euphausids*, small squids and *medusae*. The density is quite variable. Frequently the layers are most dense at the edges of their distribution along the continental slopes. A part of the scattering layers undertakes a major diurnal migration of some hundreds of metres. With the onset of darkness this part of the scattering layer migrates to the upper water layers and even close to the surface but descends again by daylight. This migrating part is to a great extent composed of small *Myctophids* and *Euphausids*.

During the annual 0-group survey in the Irminger Sea carried out in August 1993, the relative density was recorded by means of a Simrad EK500 echo sounder (Fig. 37).

All the components of this layer are not known, nor are the regional differences in the composition. However, this layer represents a huge biomass.

TABLE 9. No. of recorded species by families, March 1993

		NO. OF SPECIES
CYCLOSTOMATA	**(Marsipobranchii)**	
Hyperotreta	*(Myxiniformes)*	1
SELACHII	**(Condrichthyes)**	
Pleurotremata		10
Hypotremata		4
Chimaerea	*(Chimaeriformes)*	3
Osteichthyes		
Isospondyli	*(Clupeiformes)*	22
Iniomi	*(Scopeliformes)*	9
Apodes	*(Anguilliformes)*	3
Heteromi	*(Halosauri & Notacanthiformes)*	1
Anacanthini	*(Gadiformes)*	13
Allotriognathi	*(Lampridiformes)*	1
Berycomorphi	*(Beryciformes)*	4
Percomorphi	*(Perciformes)*	7
Scleroparei	*(Scorpaeformes)*	7
Heterosomata	*(Pleuronectiformes)*	2
Pediculati	*(Lophiiformes)*	3
	TOTAL	**90**

6. Concluding remarks

The fish species discussed in this paper are only a small part of the fish fauna inhabiting the area in question. It is considered to be the starting point for a great deal of work which is still left to be done in this field of research.

It is obvious from this study that several deep sea fish species spawn in Icelandic waters, particularly in the Reykjanes Ridge area, and there are also nursery grounds for these species as well. Presumably this is also the case for many of the species not dealt with in this paper. The Reykjanes Ridge area is thus, of great scientific interest in this respect. It is also of vital importance to increase our knowledge on the life-cycle and magnitude of fish stocks which might be of potential interest for the fisheries in the near future. Due regard must be given to reasonable management measures. It will not be an easy task since the area is extremely difficult to investigate. However, the authors hope that there will be a continuation in this work.

TABLE 10. Main species by catch (kg). March 1993

SPECIES	NO. OF STAT.	TOTAL CATCH KG
Roundnose gr.	34	1725
Blue ling	31	52224
Black scabbard fish	17	776
Deep sea redfish	29	3376
Smooth-head	22	1042
Total	133	59143
		% 90.56
Dogfish, sharks:		
Greater lantern shark	30	489
Birdbeak dogfish	12	492
Portuguese shark	10	276
Lantern shark	2	1
Black dogfish	29	525
Leafscale gulper shark	4	56
Longnose velvet dogfish	11	367
Deep sea cat shark	19	87
Total	117	2292
		% 3.5
Others		3876
		% 5.93
Gr. Total	250	65311
		% 100

Acknowledgements

The authors wish to express their thanks to the staff of the Marine Research Institute, Iceland, for the assistance in preparing this paper. They are especially indebted to: Helga Óladóttir (Directorate of Fisheries), Garðar Jóhannesson, Höskuldur Björnsson, Jóhanna Erlingsdóttir, Kristín Jóhannsdóttir and Sigurdur Gunnarsson. All are staff members of the Marine Research Institute of Iceland.

180

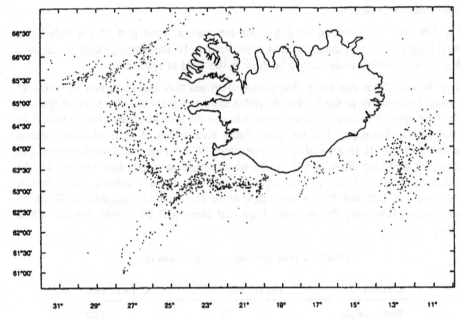

Fig. 1. Distribution of tows from which the samples have been derived (all species)

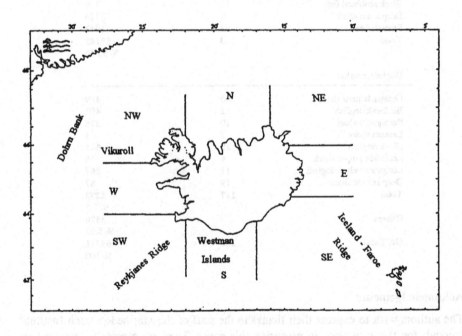

Fig. 2. Area division and local names used in the paper

Fig. 3 to Fig. 16: Distribution and relative abundance of 14 species discussed in the paper

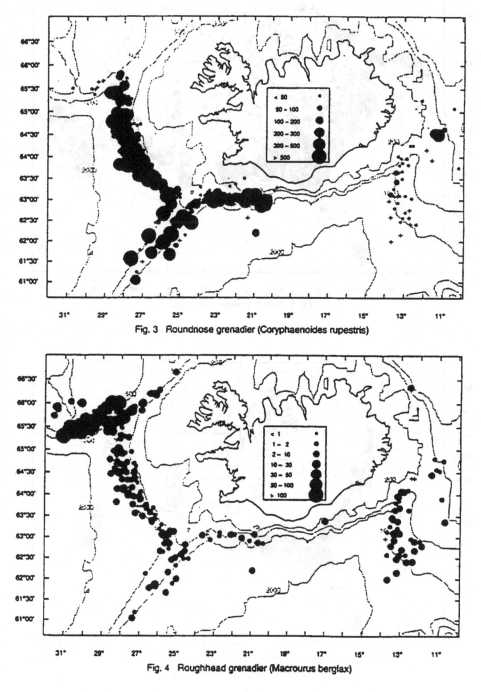

Fig. 3   Roundnose grenadier (Coryphaenoides rupestris)

Fig. 4   Roughhead grenadier (Macrourus berglax)

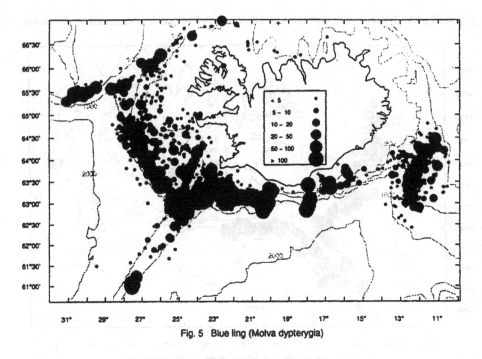

Fig. 5   Blue ling (Molva dypterygia)

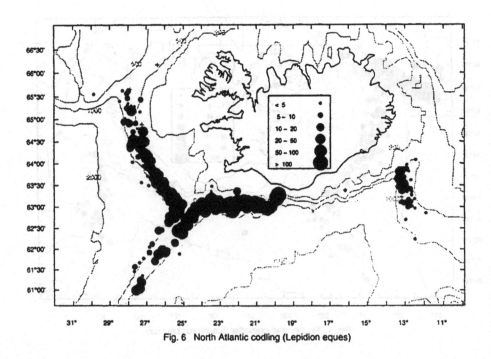

Fig. 6   North Atlantic codling (Lepidion eques)

183

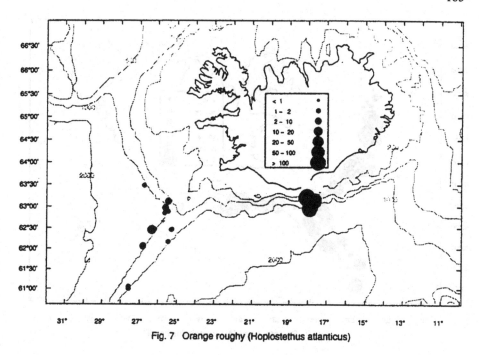

Fig. 7 Orange roughy (Hoplostethus atlanticus)

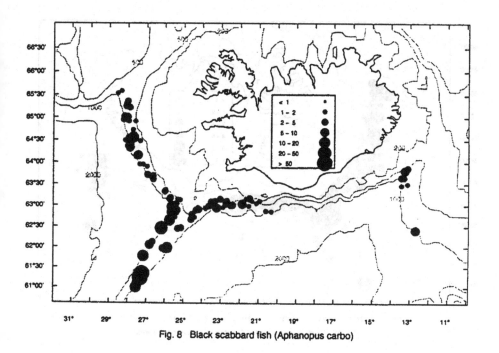

Fig. 8 Black scabbard fish (Aphanopus carbo)

184

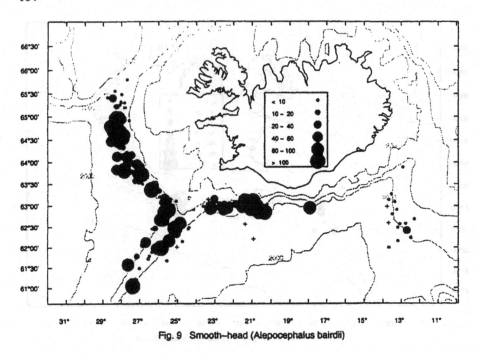

Fig. 9 Smooth–head (Alepocephalus bairdii)

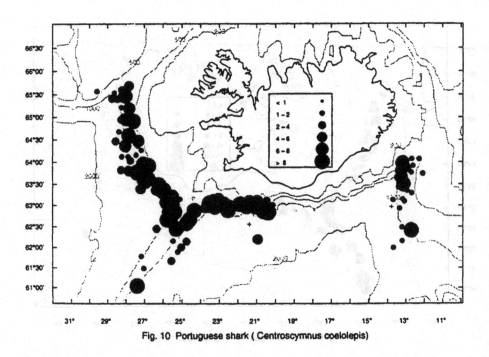

Fig. 10 Portuguese shark (Centroscymnus coelolepis)

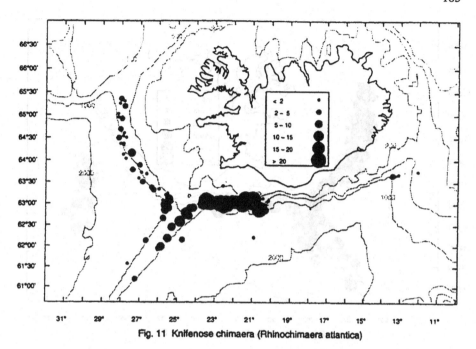

Fig. 11 Knifenose chimaera (Rhinochimaera atlantica)

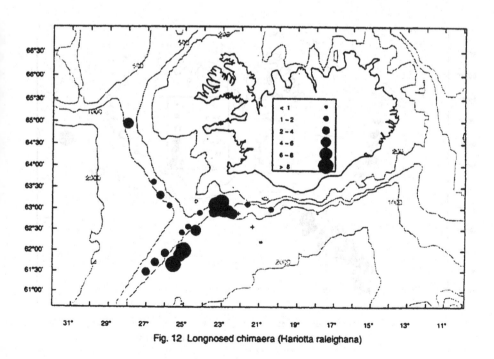

Fig. 12 Longnosed chimaera (Hariotta raleighana)

186

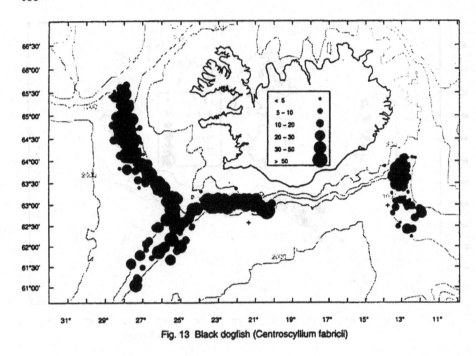

Fig. 13 Black dogfish (Centroscyllium fabricii)

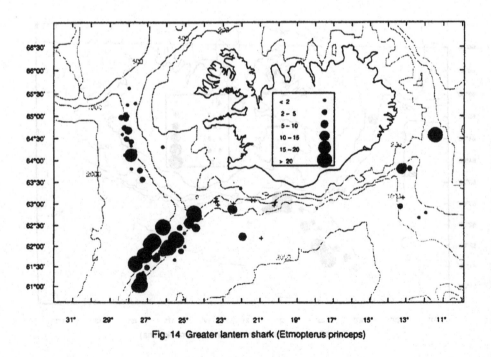

Fig. 14 Greater lantern shark (Etmopterus princeps)

187

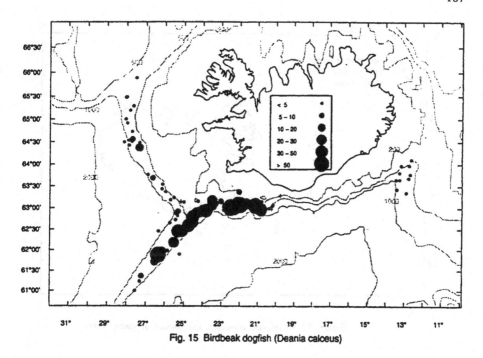

Fig. 15 Birdbeak dogfish (Deania calceus)

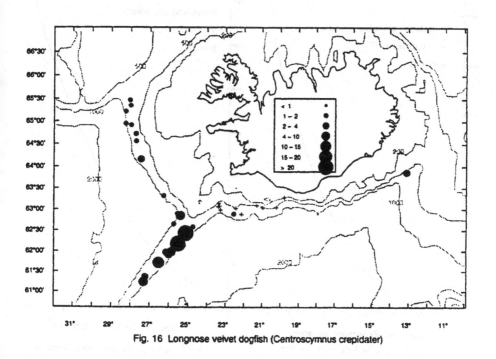

Fig. 16 Longnose velvet dogfish (Centroscymnus crepidater)

188

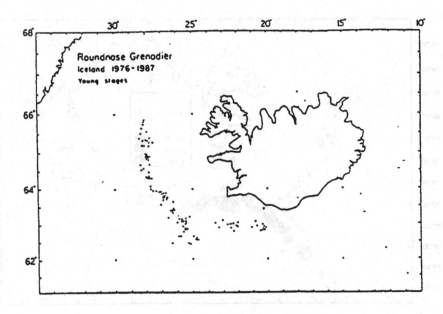

Figure. 17. Roundnose grenadier. Distribution of young stages

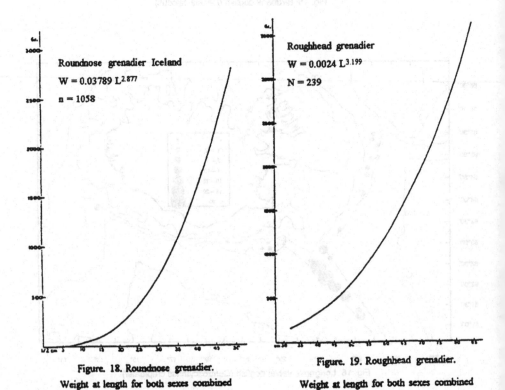

Roundnose grenadier Iceland

$W = 0.03789 \ L^{2.877}$

$n = 1058$

Roughhead grenadier

$W = 0.0024 \ L^{3.199}$

$N = 239$

Figure. 18. Roundnose grenadier.
Weight at length for both sexes combined

Figure. 19. Roughhead grenadier.
Weight at length for both sexes combined

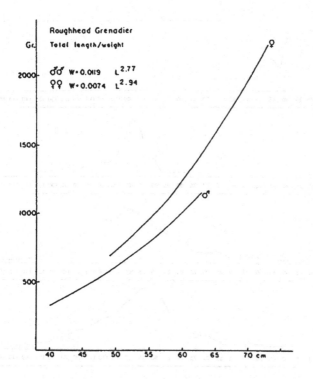

Figure. 20. Roughhead grenadier.
Weight at length for males and females separately

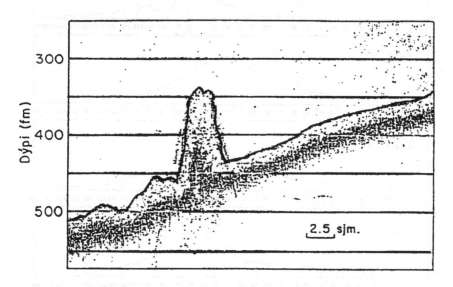

Figure 21. Underwater hill, a spawning location for blue ling south of the Westman Islands

190

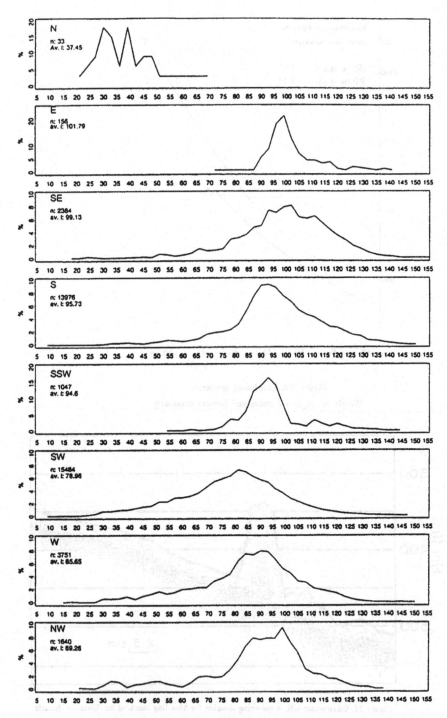

Figure. 23. Blue ling. Length distribution by areas

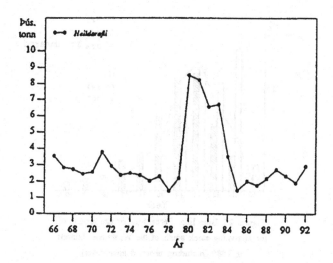

Figure. 22 . Blue ling. Total catch at Iceland (From Fjölrit Nr. 34, Hafranns.)

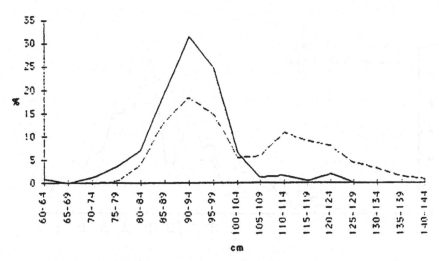

Figure. 24. Blue ling (*Molva dypterygia*). Length distribution at the spawning location on the Reykjanes Ridge

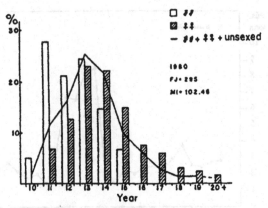

Figure. 25. Blue ling. Percentage of age groups of
the spawning stock south of the Westmann Islands
in 1980 (including unsexed specimens)

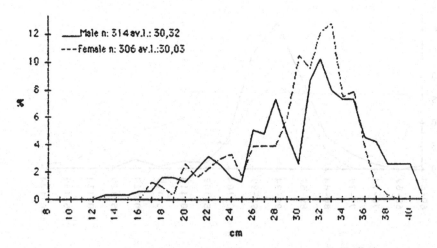

Figure. 26. North Atlantic codling (*Lepidion eques*). Length distribution

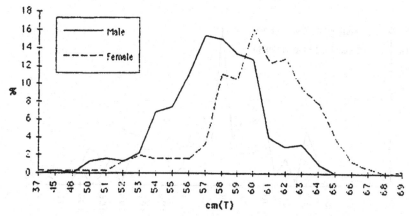

Figure. 27. Orange roughy (*Hoplostethus atlanticus*) by sex
Male n = 306 Av. length: 57.52 cm   Female n = 304 Av. length: 60.17 cm.

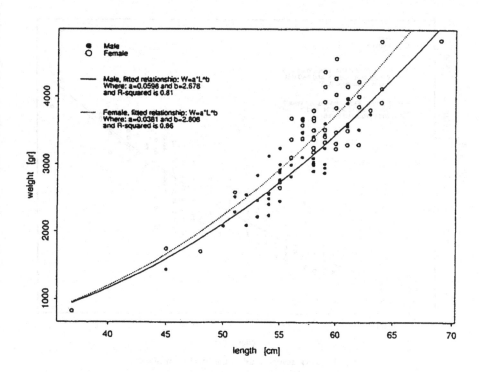

Fig. 28. Orange roughy. Length/weight relationship

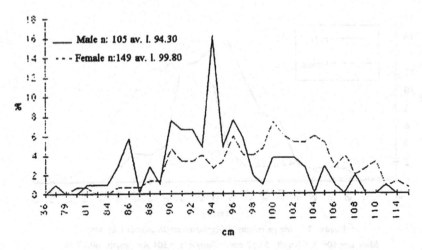

Fig. 29. Black scabbard fish (*Aphanopus carbo*). Length distribution

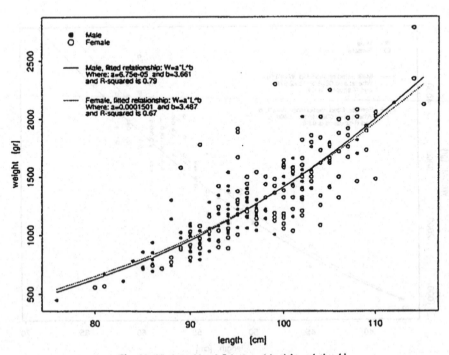

Fig. 30. Black scabbard fish. Length/weight relationship

195

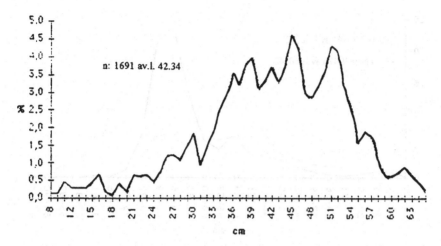

Fig. 31. Smooth-head (*Alepocephalus bairdii*) Length distribution

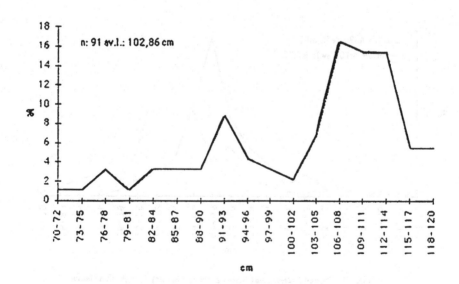

Fig. 32. Portuguese shark (*Centroscymnus coelolepsis*). Length distribution

196

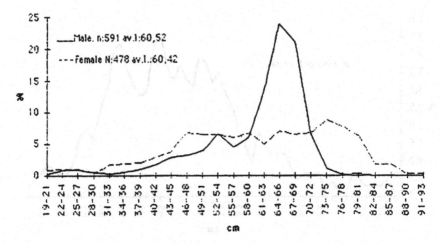

Fig. 33. Black dogfish (*Centroscyllium fabricii*). Length distribution

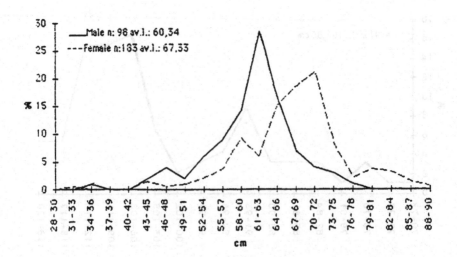

Fig. 34. Greater lantern shark (*Etmopterus princeps*). Length distribution

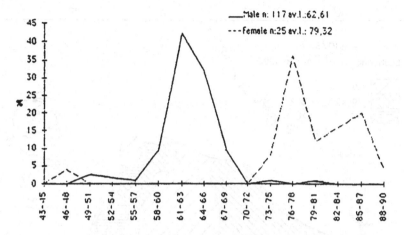

Fig. 35. Longnose velvet dogfish (*Centroscymnus crepidater*). Length distribution

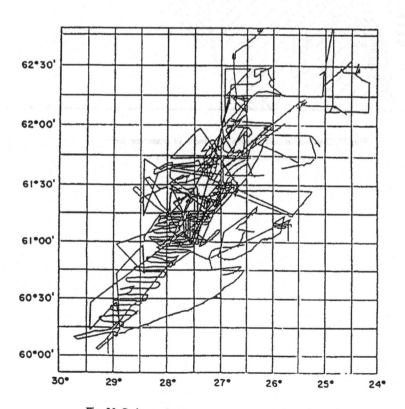

Fig. 36. Cruise tracks. Survey Reykjanes Ridge, in March 1993

198

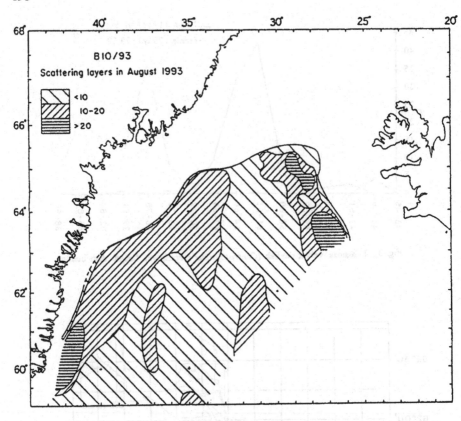

Fig. 37. Distribution and relative abundance of scattering layers
according to acoustic values ($m^2 \, nm^2$)

References

Anon, 1993: State of Marine Stocks in Icelandic waters 1992/93. Prospects for the Quota Year 1993/94. Hafrannsóknast. Fjölrit nr. 34, 2nd edition.

Jónsson, G. 1992: Íslenskir fiskar. Fjölva útg., Reykjavík.

Magnússon, J.V., 1977: Some notes on the spawning habits of *Macrouridae* at Iceland. ICES, C.M. 1977/F:49.

Magnússon, J.V., 1978: On the distribution and spawning grounds of the roughhead grenadier (*Macrourus berglax Lacépède*) west of Iceland. ICES. C.M. 1978/G:36.

Magnússon, J.V., 1982: Age, growth and weight of blue ling (*Molva dypterygia*) in Icelandic waters. ICES, C.M. 1982/G:22.

Magnússon, J.V., 1987: Grenadier Fish in Icelandic Waters. NAFO SCR Doc. 87/8, Ser. no. N1391.

Malmberg, Sv.A., 1978: Háfadjúp - Snæfellsnes 1971-1975. II Selvogsbanki. Hafrannsóknir 15. hefti 1978.

Pechenik, L.N. and F.M. Troyanovskii, 1970: Trawling Resources on the North Atlantic Continental Slope. Transl. from Russian by the Israel Program for Scientific Translations, Jerusalem 1971.

Thomas, R., 1980: Fischereibiologische Untersuchungen am Blauleng (*Molva dypterygia dypterygia*), (Pennant 1784 nach O.F. Müller 1776) im Seegebiet der Shetland - und Färöerinseln. Univ. Hamburg, Dipl. Arbeit, 109 pages.

Vilhelmsdóttir, V., 1982: Blálanga á Íslandsmiðum Ægir, 75. árg., 8. tbl.

Vilhelmsdóttir, V., 1992: Vannýttir djúpsjávarfiskar. Sjómannadasgbl. Neskaupst. 1992, 15. árg.

Vilhelmsdóttir, V. and J. Magnússon, 1992: Búrfiskur. Ægir, 1. tbl. 1992.

DEEP WATER RESOURCES IN FAROESE WATERS TO THE SOUTH, SOUTHWEST AND WEST OF THE FAROES - A PRELIMINARY ACCOUNT.

JÁKUP REINERT
Fiskirannsóknarstovan
Nóatún, P.O. Box 3051
FR-110 Tórshavn
Faroe Islands

ABSTRACT

Data from groundfish trawl surveys and exploratory longline fisheries have been used to illustrate the species composition and the relative abundance of the demersal fish assemblage in Faroese waters to the south, southwest and west of the Faroes. The hydrography of this part of the North Atlantic is very interesting because of the interfacing of several different water masses.

Based on the same data sources and data from exploratory trawl fisheries and commercial catch statistics, the resources of roundnose grenadier (Coryphaenoides rupestris), and scabbard fish (Aphanopus carbo),were analysed regarding their distribution, size and weight composition and catch rates. The composition of the catches in the trawls and longlines were quite different. In the trawl the dominant species were Sebastes mentella and Argentina silus. In the longline catches the former species was totally absent, whilst the incidence of the latter was very low. On the longlines the dominant species was Brosme brosme which was also present in the trawl but at much a reduced frequency

Following a request from the steering committee of this Workshop, a very brief description is given of the Faroese deep water tangle net fishery for monkfish (Lophius sp.), and of the Faroese pot fishery for the deep water red crab (Chaecon affinis). The deep water tangle net fishery is well established at present, whereas the pot fishery still is in the pioneer phase.

A. G. Hopper (ed.), Deep-Water Fisheries of the North Atlantic Oceanic Slope, 201–225.
© *1995 Kluwer Academic Publishers.*

1. Introduction

Several nations over the years, have carried out investigations and commercial fishing in the deep waters around the Faroes. Most of the work in the first half of this century was presented in the book "Marine and Freshwater Fishes" by Joensen and Tåning (1970). More recently, the former USSR, UK, Norway and the Faroes have investigated the deep water fish fauna, which are now commercially exploited mainly by Norway with longlines, France with trawls and the Faroes with both longlines and trawls.

This paper on deep water resources to the south, southwest and west of the Faroe Islands will only deal with data from recent Faroese activities, and some comercial data from French trawlers operating within the 200 mile EEZ. It is a part of a more comprehensive work on the deep water resources in the Faroe area, and, since it is based only on a part of the data sources available, it should be regarded as a preliminary account.

There are three purposes to this paper:

a) To use data from groundfish surveys and exploratory longline fisheries to illustrate the species composition and the relative abundance of the demersal fish assemblage in the area defined in section 2 below. Although it is certainly not a complete account of all fish species in the area, it should, however, give a reasonable overview of the assemblage.

b) To give a more detailed account regarding distribution, length composition, length-weight relationship and abundance of the roundnose grenadier (*Coryphaenoides rupestris*), and of the blackscabbard fish (*Aphanopus carbo*)

c) Acting upon a request from the steering committee of this workshop, to give a short description of the recently introduced deep water commercial fisheries for monkfish,(*Lophius sp).* and the deep water red crab (*Chaecon affinis* formerly *Geryon affinis*).

2 . Marine environment

The area of investigation in this paper is confined to the north by the 63°N latitude and the east by 6°W longitude, and is mainly restricted to demersal fishes below the 500 m depth contour line. This area is shown in Figure. 1, together with the topography of the seabed.

The water masses in these depth regions are dominated by Atlantic water deriving from the North Atlantic Current (MNAW), but to the north and east colder and less saline water is found deriving firstly from overflow of Deep Norwegian Sea Water (DNSW) through the Faroe Bank Channel (FBC) and over the Wyville-Thomsen Ridge (WTR), and secondly of water of the East Iceland Current (NI/AI) over the Iceland-Faroe Ridge (IFR). These are shown in Figure 2. The fourth water mass also shown in Figure. 2 is the North Atlantic Water (NAW) and is of minor importance for the purpose of this paper.

3 . Methodology and equipment

3.1. SPECIES COMPOSITION AND RELATIVE ABUNDANCE

3.1.1. *Faroese groundfish trawl surveys 1992-93.*

Faroese groundfish surveys, have been carried out annually in February to March since 1982 covering the Faroe Plateau, the Iceland-Faroe Ridge, the banks to the southwest and the Wyville-Thomsen Ridge down to about 500 m depth. In 1992 these surveys were revised, and the area to the west and southwest (strata 18-23 in Figure. 3) was expanded towards deeper water. At the same time all the squares (grid pattern) to be sampled were fixed from year to year.

For the present study only stations deeper than 500 m in the above defined area of investigation have been included (Figure. 1). The depth distribution of the stations is from 500 m to 900 m.

The Faroese research vessel Magnus Heinason has been used each year for the surveys since they commenced in 1982. It is a stern trawler of 136 ft. (41.5 m.) length and with an engine of 1800 HP. The gear used for the surveys was a 116 ft. (35.3 m.) bottom trawl with 40 mm stretched meshsize in the codend. The trawl was towed for 60 minutes at a speed of 3 knots. Trawling was only undertaken during daylight.

3.1.2. *Exploratory deepwater longline fishery.*

Between May and July 1988 the Fisheries Laboratory of the Faroes chartered a commercial longliner, m/s Hans Erik (an autoliner of 138 ft. (42.0 m.) length, 426 GRT and 1000 HP), to carry out exploratory deepwater longline fishing in national and international waters. The present account deals only with the results with codline (hook size = No.6) in national waters. The four fishing areas are shown in Figure. 4.

In order to cover as large an area as possible, the number of hooks per set was limited to between 1000 and 3000 approximately (Table 3). However, the number of hooks differs considerable between sets, and this has made it difficult to achieve a precise comparison between sets and areas. As only the number of species have been considered in this study, it has not been possible to standardize the sets to some unit effort. However, when plotting the number of species against number of hooks per set (Figure. 5), no clear relationship can be seen, so for the present purpose the sets seem to be comparable.

The depth distribution of sets varied between 500 m and 1350 m (Table 3).

3.2. ROUNDNOSE GRENADIER AND BLACK SCABBARD FISH INVESTIGA-TIONS.

In addition to the data sources mentioned in 3.1 the following material was used to illustrate distribution, length composition, length-weight relationship and abundance of roundnose grenadier and black scabbard fish..

3.2.1. *Exploratory deep-water trawl fishing.*

In January and February 1990 the Fisheries Laboratory of the Faroes carried out exploratory deep water trawl fishing with the r/v Magnus Heinason and a chartered commercial stern trawler, m/s Phoenix (150 ft. (45. m.) length, 699 GRT and 2400 HP). In both cases, trawling was performed day and night. The main target species were grenadier, black scabbard fish and deepwater sharks, especially *Centrophorus squamosus* and *Centroscymnus coelolepis*. The fishing areas are shown in Figure. 6. In the directed fishery for black scabbard fish, a bottom trawl with 80 mm stretched mesh size in the belly and codend was used. In all other cases the gear was a bottom trawl with 135 mm stretched mesh size in the codend.

3.2.2. *Commercial catch stastics.*
Logbook statistics from French trawlers for 1991 to 1993 for roundnose grenadier and black scabbard fish have been obtained from the Faroese Coastal Guard Service, and information on Faroese landings derives from the Faroese Statistical Office. These data have been used to show possible seasonal and general trends in the availability of these two species. Most of the French and Faroese deep water trawl activities must be characteristized as mixed deep water fisheries with varying target species. The availability of other more valuable species such as blue ling and redfish will be a decisive factor in determining the extent of a directed fishery for grenadier and black scabbard fish.

3.3. DEEP-WATER FISHERIES FOR MONKFISH AND DEEP WATER RED CRAB

3.3.1. *Tangle net fishery for monkfish.*
In 1991 and 1992 an exploratory tangle net fishery for monkfish was carried out in waters deeper than 250 m with two vessels, m/s Anita (109 ft.(30.5 m.) length, 265 GRT, 400 HP) and m/s Móanes (114 ft. (34.7 m.) length, 250 GRT and 560 HP). The gear used was monofilament (0.6 mm) tangle nets. The dimensions of each net were 50 m in length and 13 meshes in height; the mesh sizes varied between 28 and 36 cm stretched meshes.

Since 1993 the tangle net fishery has been managed as a licensed fishery, without a TAC. 4 vessels currently are liscensed. The fishing is restricted to a small area on the east side of the islands and to a larger area to the southwest of the islands (Figure. 7). The fishery must only take place in depths greater than 250 m.

The data used in this paper derives from the exploratory fishery with the above mentioned vessel; the m/s Móanes. In the treatment of catch and effort data, each month was divided into three periods of 10 days.

3.3.2. *Pot fishery for deep-water red crab.*
An experimental fishery with pots for the deep-water red crab commenced in 1993. This crab is frequently found as a by-catch in the tangle net fishery, and in order to investigate the possibilities regarding this resource, two preliminary licenses were allocated.

Different types of pots have been used. Some were small and conical or cylindrical with one tunnel, and others were larger 6x3x1 ft. and 7x7x3 ft. square pots, with 2 or 3 tunnels and with or without triggers. Several species have been used as bait, for example redfish, blue ling, ling and tusk.

The data used in this paper has been derived from an exploratory fishery in the period July to August 1993 with m/s Gudmundur (118 ft (36.0 m.) length, 262 GRT and 690 HP). The fishing area is shown in Figure. 8.

4. Results

4.1. SPECIES COMPOSITION AND RELATIVE ABUNDANCE

In the trawl surveys 1992-93 a total of 49 species were recorded (Table 1). From the frequencies of occurrence, calculated as the percentage of stations with catches of the species, it can by seen, that *Sebastes mentella* were by far the most common occuring on almost 80% of the stations followed, in second place, by *Argentina silus* which occured on 62% of the stations. Nine species occured on 25 to 50% of the stations, 16 species on 10 to 24% of the stations and 22 species were only recorded on one or two of the 29 stations. The recordings of *Ammodytes tobianus* and *Squalus acanthias* at these depths most likely occured from capture during shooting and hauling of the trawls.

In terms of relative abundance, the two most common species have reversed order, with the *A. silus* being the most abundant both by weight and in numbers followed by *S. mentella* in second place. Two other species account for about 5% of the weight (*Chimaera monstrosa* and *Aphanopus carbo*) and three species account for about 3% of the numbers (*Micromesistius poutassou*, *C. monstrosa* and *A. carbo*). No other species account for more than 2% of the total abundance.

In the exploratory longline catches a total of 35 species was recorded (Table 2). *Brosme brosme* was by far the most common of the species and occured on 83% of the stations. 8 species were recorded on 50 to 70% of the stations, 4 on 25 to 49% of the stations, 9 on 10 to 24% of the stations and 13 species occured on less than 10% of the stations.

A comparison of the four fishing areas in Figure 4 showed, that the number of species differ very little in areas 1, 2 and 3 (Tables 2 and 3). Area 4 has only 7 species, but only 2 sets were made there. The horizontal distribution (% occurrence) of the species in areas 1, 2 and 3 also showed, that the same species dominated in each of the areas, but with some deviating patterns which could be related to the different water masses, the stations are exposed to (Figure. 2). An attempt to analyze the effect of depth on the number of species (Figure. 9), revealed no clear relationship, although the least number of species was recorded both on the shallowest and on the deepest sets. The small number of species in the most shallow depths could be explained by the location of these stations lying within the cold water masses of the area.

Not surprisingly, the number of species and the species compositions in the trawl and longline catches were quite different. Much of this difference is because of the different catching processes of the two gears and the behaviour of different fish

206

TABLE 1. Species composition in the Faroese groundfish trawl surveys

No.	SPECIES	S'TONS	%	CATCH/kg.	FISH No.	REL/Wt	REL/No's
1	*Sebastes mentella*	23	79	5064.88	4102	31.83	21.75
2	*Argentina silus*	18	62	7757.75	11477	48.76	60.85
3	*Micromesistius poutassou*	14	48	95.091	699	0.60	3.71
4	*Chimaera monstrosa*	13	45	822.289	676	5.17	3.58
5	*Molva dipterygia*	12	41	252.426	81	1.59	0.43
6	*Brosme brosme*	12	41	49.76	25	0.31	0.13
7	*Reinhardtius hippoglossoides*	11	38	132.11	45	083	0.24
8	*Lepidion eques*	10	34	29.514	293	0.19	1.55
9	*Etmopterus spinax*	9	31	12.366	59	0.08	0.31
10	*Lophius piscatorius*	9	31	67.19	11	0.42	0.06
11	*Raja radiata*	8	28	147.825	128	0.93	0.68
12	*Raja fyllae*	7	24	7.35	16	0.05	0.08
13	*Sebastes marinus*	7	24	45.37	22	0.29	0.12
14	*Hippoglossoides platessoides*	7	24	5.495	17	0.03	0.09
15	*Lycodes esmarki*	7	24	19.5	35	0.12	0.19
16	*Coryphaenoides rupestris*	6	21	192.184	159	1.21	0.84
17	*Pollachius virens*	6	21	31.46	10	0.20	0.05
18	*Aphanopus carbo*	6	21	700.61	498	4.40	2.64
19	*Sebastes viviparus*	6	21	9.975	35	0.06	0.19
20	*Epigonus telescopus*	6	21	86.239	219	0.54	1.16
21	*Phycis blennoides*	4	14	16.13	9	0.10	0.05
22	*Helicolenus dactylopterus*	4	14	8.962	19	0.06	0.10
23	*Lepidorhombus whiffiagonus*	4	14	3.674	15	0.02	0.08
24	*Raja batis*	4	14	47.4	5	0.30	0.03
25	*Mora moro*	4	14	21.625	11	0.14	0.06
26	*Glyptocephalus cynoglossus*	4	14	2.373	10	0.01	0.05
27	*Deania calceus*	3	10	73.318	3	0.46	0.16
28	*Etmopterus princeps*	2	7	0.16	2	0.00	0.01
29	*Trachyrhynchus murrayi*	2	7	0.762	4	0.00	0.02
30	*Molva molva*	2	7	12.285	4	0.08	0.02
31	*Galeus melastomus*	2	7	7.28	13	0.05	0.07
32	*Bathyraja spinicauda*	2	7	22.13	6	0.14	0.03
33	*Centroscymnus crepidater*	2	7	57.742	39	0.36	0.21
34	*Centrophorus squamosus*	2	7	60.98	4	0.38	0.02
35	*Apristurus laurussonii*	2	7	1.785	2	0.01	0.01
36	*Xenodermichthys socialis*	1	3	0.025	1	0.00	0.01
37	*Anarhichas denticulatus*	1	3	13.4	1	0.08	0.01
38	*Eutrigia gurnardus*	1	3	8.029	11	0.05	0.06
39	*Macrourus berglax*	1	3	0.518	3	0.00	0.02
40	*Melanogrammus aeglefinus*	1	3	0.087	3	0.00	0.02
41	*Hippoglossus hippoglossus*	1	3	3.595	1	0.02	0.01
42	*Notacanthus chemnitzii*	1	3	2.755	1	0.02	0.01
43	*Gadiculus argenteusthori*	1	3	1.019	38	0.01	0.02
44	*Centroscymnus coelolepsis*	1	3	13	2	0.08	0.01
45	*Hydrolagus mirabilis*	1	3	0.06	6	0.00	0.03
46	*Onogadus argentatus*	1	3	0.03	1	0.00	0.01
47	*Ammodytes tobianus*	1	3	0.483	10	0.00	0.05
48	*Squalus acanthias*	1	3	1.83	1	0.01	0.01
49	*Halagyreus johnsonii*	1	3	0.01	1	0.00	0.01

Note: S'TONS= Positive stations, %= Occurence, FISH No.= Total individuals, REL= Relative abundance

TABLE 2. Species composition deeper than 500 m. in the exploratory long-line fishery- m/s Hans Erik in May-June 1988

No.	SPECIES	ALL AREAS		AREA 1	AREA 2	AREA 3	AREA 4
		Pos'veset	%	%	%	%	%
1	Brosme brosme	35	83	81	82	100	0
2	Molva dipterygia	29	69	50	82	77	100
3	Deania calceus	28	67	63	64	85	0
4	Centroscymnus crepidater	27	64	63	55	85	0
5	Mora mora	25	60	50	82	62	0
6	Centroscyllium fabricii	24	57	38	64	69	100
7	Centroscymnus coelolepsis	22	52	44	73	46	50
8	Lepidion eques	21	50	38	64	46	100
9	Galeus melastomus	21	50	38	55	69	0
10	Chimaera monstrosa	20	48	44	45	62	0
11	Centrophorus squamosus	19	45	50	27	54	50
12	Etmopterus spinax	17	40	31	45	54	0
13	Phycis blennoides	15	36	25	45	46	0
14	Raja nidarosiensis	9	21	13	45	15	0
15	Macrourus berglax	8	19	19	9	15	100
16	Pseudotriakis microdon	6	14	25	9	8	0
17	Molva molva	5	12	25	0	8	0
18	Hipoglossus hippoglossus	5	12	13	9	15	0
19	Raja bigelowi	4	10	0	9	23	0
20	Bathyraja spinicauda	4	10	13	0	0	100
21	Reinhardtius hippoglossoides	4	10	13	9	8	0
22	Raja fyllae	4	10	6	0	23	0
23	Raja circularis	3	7	0	0	23	0
24	Lophius piscatorius	2	5	0	9	8	0
25	Apristurus laurussonii	2	5	0	18	0	0
26	Synaphobranchus kaupi	2	5	6	9	0	0
27	Helicolenus dactyloptrus	1	2	0	0	8	0
28	Argentina silus	1	2	0	0	8	0
29	Somniosus microcephalus	1	2	0	0	8	0
30	Raja fullonica	1	2	0	0	8	0
31	Etmopterus princeps	1	2	0	0	8	0
32	Trachyrhynchus murrayii	1	2	0	9	0	0
33	Sebastes marinus	1	2	0	9	0	0
34	Cottidae sp.	1	2	6	0	0	0
35	Lycodes esmarki	1	2	6	0	0	0

Notes: Pos've set= Positive sets, %= Percentage of sets where species occured,
AREA 1 -16 sets, AREA 2-11 sets, AREA 3-13 sets, AREA 4-2 sets. See Figure 4 for the geographical location of the sets

TABLE 3. Exploratory long line fishery - m/s Hans Erik. May to July 1988

SET No.	DEPTH m.	No. of HOOKS	No. of SPECIES
AREA 1	Wyville - Thomson Ridge, Ymer Ridge, South of Bill Bailey Bank		
1	502	1390	4
2	723	1640	3
3	733	1390	5
4	887	1640	7
5	680	2830	10
6	1048	1350	10
7	735	1395	7
8	653	1773	7
9	818	1370	10
10	723	1370	6
11	504	1100	1
12	996	1180	8
13	673	1220	11
14	995	1200	12
15	654	900	10
16	1119	900	10
AREA 2	Lousey Bank		
1	944	1250	9
2	704	2750	13
3	1012	1350	9
4	706	1350	8
5	978	1350	10
6	710	1350	10
7	1265	1350	6
8	1016	1840	12
9	704	1650	14
10	1333	1840	5
11	979	1840	11
AREA 3	North of Bill Bailey Bank		
1	731	2700	12
2	926	2800	10
3	784	2760	18
4	1098	1350	8
5	717	2800	6
6	741	2545	13
7	745	2480	12
8	509	2480	9
9	1056	2425	15
10	734	1240	8
11	921	1110	6
12	912	1350	9
13	741	900	
AREA 4	Iceland - Faroe Ridge		
1	965	1610	7
2	1005	1610	5

species to trawls and longlines, and of the greater depths of some of the longline sets. The lack of the small species in the longline catches is particularly obvious. Typical is the total lack in the longline catches of *S. mentella*, which was the most common species in the trawl catches, and also of the very common "trawl-species" like *M. poutassou* and *Coryphaenoides rupestris*. The larger *Pleurotremata* species were more dominant in the longline catches and certainly this would have been even more apparent, if the line and hooks had been sized and baited for these species. Many hooks were lost because the line was damaged. A systematic list of the species is given in Table 4 with information on which gears they were caught.

4.2. ROUNDNOSE GRENADIER AND BLACK SCABBARDFISH

4.2.1. *The roundnose grenadier,(Coryphaenoides rupestris).*

In Figure. 10 the CPUE in the exploratory trawl fishery is plotted against depth. By far the highest catch rates were obtained on the deepest stations (below 1100 m) whilst in shallow water of less than 750 m the catch rates of roundnose grenadier were very poor. Figure. 11 shows, that there was no clear relationship between the catch rates and the time of the day during which the exploratory fishery was performed, although the highest catch rates were obtained in the evening.

In order to detect a possible seasonality in the availability of grenadier, the monthly catches by French trawlers have been plotted for the years 1991-93 (Figure. 12), and monthly Faroese landings in 1993 are shown in Fig. 13. The monthly catches were very variable throughout the year and between years. There were also differences between French and Faroese catches in the same year. One possible explanation for this is that grenadier and scabbard fish are not the main target for the commercial fleets, so the availability of other resources and other fishing areas will certainly influence the catch figures. Nevertheless the French catch figures indicate a decline in 1993 compared to 1991-92; again the reason could be solely a reduced activity of these vessels in Faroese waters.

The length distributions of roundnose grenadier in the groundfish surveys of 1992 and 1993 are shown in Figure 14. There was a shift in the length composition towards smaller sizes in 1993 compared to 1992. In 1992 pre-anal lengths of 15 to 20 cm dominated the catches, whereas lengths of 12 to 19 cm dominated the 1993 catches. The relationship between pre-anal length and weight in the groundfish surveys is given in Figure 15.

4.2.2 *The black scabbardfish.*

The plot of CPUE in the exploratory trawl fishery versus fishing depth (Figure 16) seems to indicate highest catch rates at about 700 to 900 m depths. No clear relationship seems to exist between catch rates and the time of the day (Figure 17).

No clear seasonality can be detected from the monthly Faroese and French catch figures in Figure 13 and Figure 18, respectively, but the French catches have been declining since 1991.

The length distributions in the groundfish surveys 1992 and 1993 are shown in

Figure 19. Lengths of 90 to 105 cm dominated the catches in both years. Figure 20 shows the length-weight relationship in the surveys.

5. Faroese deep-water fisheries for monkfish and redfish .

5.1. THE TANGLE NET FISHERY FOR MONKFISH

Figures. 21 to 23 show the development in the exploratory tangle net fishery by the m/s Móanes for monkfish from November 1991 to December 1992. The catch level in 1992 was generally higher than in 1992, but most of the reasons for this can be attributed to the increase in effort (Figure. 20). Firstly, the fishing time was increased, and gradually also the number of nets (Figure 22). However, this increase in effort was not accompanied by a corresponding increase in catch rate (Figure 21). On the contrary, the CPUE decreased after some months of fishing and seems to have stabilized at a much lower level than at the start of the fishery.

The fishing depths have varied between 300 m and almost 900 m (Figure 24), but the best catch rates have been obtained in depths less than 700 m. There did not seem to be any tendency towards a declining CPUE at the shallow end of the depth range.

The fishing times have varied between 30 and 175 hours without any clear trend in CPUE (Figure 25). However, fishing times between 60 and 160 hours seemed to give the best results.

It should be mentioned, that by-catches of other species were very small. For example, the total landings by m/s Anita in 1992 amounted to almost 220 tonnes. 95% of the catch was monkfish, 2% blue ling, 1% skate sp., 1% porbeagle, 0.2% redfish and 0.8% others.

5.2. POT FISHERIES FOR DEEP-WATER RED CRAB

The results of the exploratory fishery for red crab are shown in Figures. 26-28. The number of pots was very small, but kept consistent, with about 3 large and 15 small pots per set, but the catch rates were promising. This exploratory period was too short to show any seasonal variation of catches, and the plots of catch versus depth (Figure 26) and of catch versus fishing time (Figure 27) give no clear signals. Small by-catches of tusk were taken (Figure 28). Problems remain however, in the marketing of the crabs, and fishery on a commercial scale has not yet started.

References

Hansen, B., A. Kristiansen and J. Reinert 1990. Cod and haddock in Faroese waters and possible climatic influences on them. ICES C.M. 1990/G:33. 23 pp. Mimeo.

Joensen, J.S and Å.V. Tåning 1970. Marine and Freshwater Fishes. 1 th. ed, Vald. Pedersens Bogtrykkeri, København. 241 pp.

TABLE 4. A systematic list of the fish species composition in the groundfish trawl survey catches in March-April 1992-93 and the exploratory longline catches in May-July 1988, respectively, in Faroese waters deeper than 500 m to the south west (hatched area in Fig. 1)

SELACHII

Euselachii

Pleurotremata

S cyliorhinidae

1	*Apristurus laurussonii* (Saemundsson)	Trawl and longline
2.	*Galeus melastomus* (Rafinesque)	Trawl and longline

Pseudotriakidae

3.	*Pseudotriakis microdon* (Capello)	Longline

Squalidae

4.	*Centrophorus squamosus* (Bonaterre)	Trawl and longline
5.	*Centroscyllium fabricii* (Reinhardt)	Longline
6.	*Centroscymnuscoelolepis* (Bocage & Capello)	Trawl and longline
7.	*Centroscymnus crepidater* (Bocage & Capello)	Trawl and longline
8.	*Deania calceus* (Lowe)	Trawl and longline
9.	*Etmopterus princeps* (Collett)	Trawl and longline
10.	*Etmopterus spinax* (Linnaeus)	Trawl and longline
11.	*Somniosus microcephalus* (Bloch & Schneider)	Longline
12.	*Squalus acanthias* (Linnaeus)	Trawl

Hypotremata

Rajidae

13.	*Bathyraja spinicauda*(Jensen)	Trawl and longline
14.	*Raja radiata* (Donovan)	Trawl
15.	*Raja batis* (Linnaeus)	Trawl
16.	*Raja nidarosiensis* (Storm)	Longline
17.	*Raja circularis* (Couch)	Longline
18.	*Raja fullonica* (Linnaeus)	Longline
19.	*Raja bigelowi* (Stehmann)	Longline
20.	*Raja fyllae* (Lütken)	Trawl and longline

Holocephali

Chimaerea

Chimeridae

21.	*Chimaera monstrosa* (Linnaeus)	Trawl and longline
22.	*Hydrolagus mirabilis* (Collett)	Trawl

OSTEICHTHYES

Isospondyli

Alepocephalidae

23.	*Xenodermichthys copei* (Gill)	Trawl

Argentinidae

24.	*Argentina silus* (Ascanius)	Trawl and longline

Apodes

Synaphobranchida

25.	*Synaphobranchus kaupi* (Johnson)	Longline

Heteromi

Notacanthidae

26.	*Notacanthus chemnitzii* (Bloch)	Trawl

Anacanthini

Macrouridae

27.	*Coryphaenoides rupestris* (Gunnerus)	Trawl
28.	*Macrourus berglax* (Lacepède)	Trawl and longline
29.	*Trachyrhynchus murrayi* (Günther)	Trawl and longline

Gadidae

30.	*Gadiculus argenteus thori* (Schmidt)	Trawl
31.	*Melanogrammus aeglefinus* (Linnaeus)	Trawl
32.	*Micromesistius poutassou* (Risso)	Trawl
33.	*Pollachius virens* (Linnaeus)	Trawl
34.	*Brosme brosme* (Ascanius)	Trawl and longline
35.	*Molva dipterygia* (Pennant)	Trawl and longline
36.	*Molva molva* (Linnaeus)	Trawl and longline
37.	*Onogadus argentatus* (Reinhardt)	Trawl
38.	*Phycis blennoides* (Brünnich)	Trawl and longline

Moridae

39.	*Halargyreus johnsonii* (Günther)	Trawl
40.	*Lepidion eques* (Günther)	Trawl and longline
41.	*Mora moro* (Risso)	Trawl and longline

Percomorphi

Apogonidae

42.	*Epigonus telescopus* (Risso)	Trawl

Ammodytidae

43.	*Ammodytes tobianus* (Linnaeus)	Trawl

Trichiuridae

44.	*Aphanopus carbo* (Lowe)	Trawl

Anarhichadidae

45.	*Anarhichas denticulatus* (Kröyer)	Trawl

Zoarcidae

46.	*Lycodes esmarki* (Collett)	Trawl and longline

Scleroparei

Scorpenidae

47.	*Helicolenus dact. dactylopterus* (Delaroche)	Trawl and longline
48.	*Sebastes marinus* (Linnaeus)	Trawl and longline
49.	*Sebastes mentella* (Travin)	Trawl
50.	*Sebastes viviparus* (Kröyer)	Trawl

Triglidae

51.	*Eutrigla gurnardus* (Linnaeus)	Trawl

Cottidae

52.	*Cottidae sp.*	Longline

Heterosomata

Scophthalmidae

53.	*Lepidorhombus whiffiagonis* (Walbaum)	Trawl

Pleuronectidae

54.	*Glyptocephalus cynoglossus* (Linnaeus)	Trawl
55.	*Hippoglossoides platessoides* (Fabricius)	Trawl
56.	*Hippoglossus hippoglossus* (Linnaeus)	Trawl and longline
57.	*Reinhardtius hippoglossoides* (Walbaum)	Trawl and longline

Pediculati

Lophiidae

58.	*Lophius piscatorius* (Linnaeus)	Trawl and longline

214

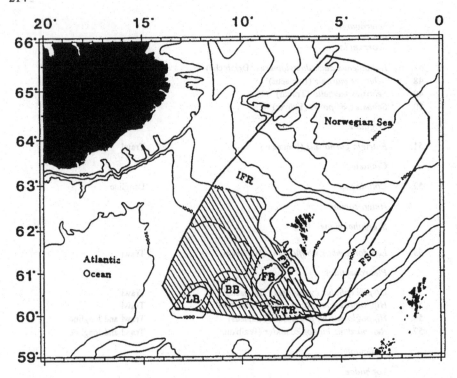

Figure 1. Topography of the waters around the Faroes and the 200 nm EEZ. The area dealt with in this report is hatched.

FB = Faroe Bank; BB = Bill Bailey Bank; LB = Lousy Bank; WTR = Wyville-Thomsen Ridge; IFR = Iceland-Faroe Ridge; FBC = Faroe Bank Channel; FSC = Faroe Shetland Channel.

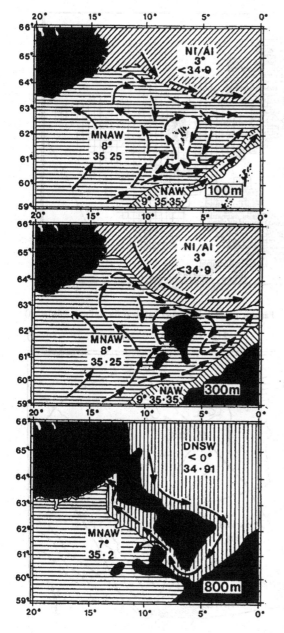

Figure 2. Distribution of main water masses and circulation patterns of waters around the Faroes at three different depths (adapted from Hansen et al., 1990).

MNAW = Modified North Atlantic Water;
DNSW = Deep Norwegian Sea Water;
NI/AI = North Icelandic/Arctic Intermediate Water;
NAW = North Atlantic Water.
Salinity and mean temperatures are also shown

216

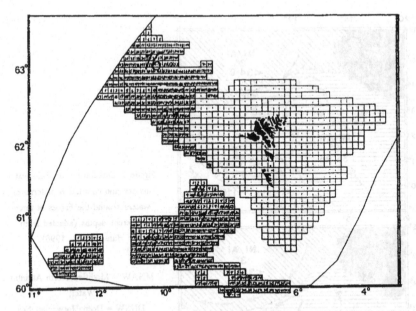

Figure 3. Stratification scheme for the Faroese groundfish surveys (see text for further explanation).

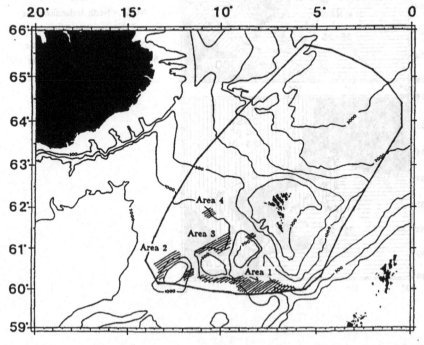

Figure 4. The four areas of investigation in the exploratory deep water longline fishery in May-July 1988.

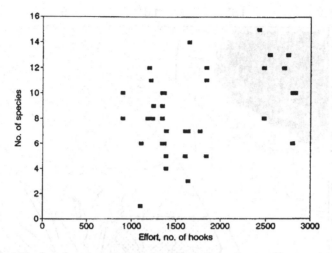

Figure 5. Number of species with increasing effort in the exploratory deep water longline fishery in May-July 1988.

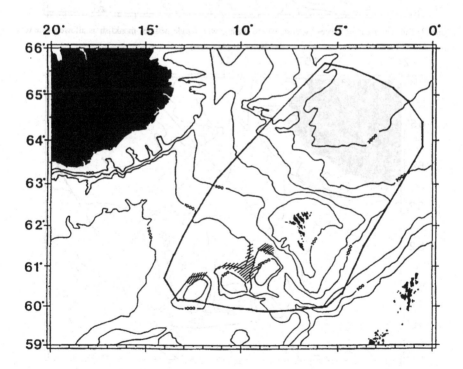

Figure 6. The areas of investigation in the exploratory deepwater trawl fishery in January-February 1990.

218

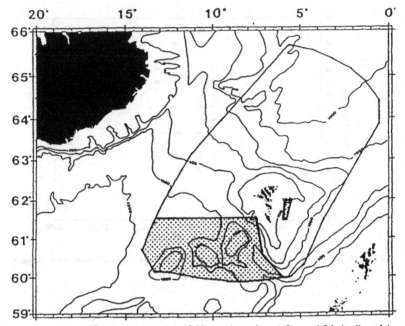

Figure 7. The two areas in Faroese waters, where fishing with tangle nets for monkfish is allowed (see text for further explanation).

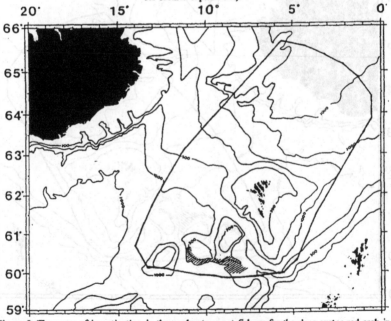

Figure 8. The areas of investigation in the exploratory pot fishery for the deep water red crab in July-August 1993. M/s Gudmundur.

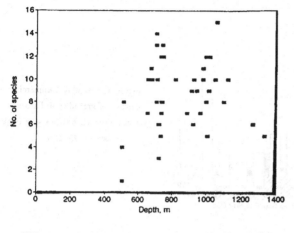

Figure 9. Number of species compared to depth in the exploratory deep water longline fishery in May-July 1988.

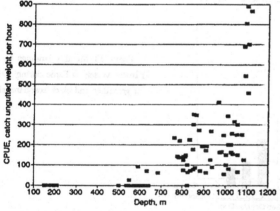

Figure 10. CPUE of grenadier compared to depth in the exploratory deep-water trawl fishery during January - February 1990

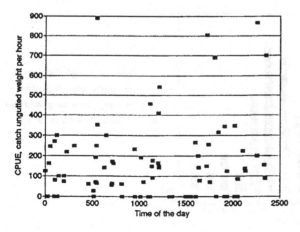

Figure 11.CPUE of grenadier compared with time of the day in the exploratory deepwater trawl fishery during January-February 1990.

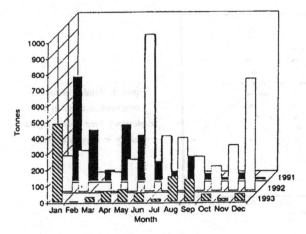

Figure 12. Monthly commercial catches of grenadier in Faroese waters by French trawlers -log book data for 1991- 93

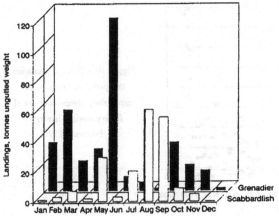

Figure 13. Monthly landings by Faroese vessels at Faroe during 1993 of grenadier and black scabbardfish

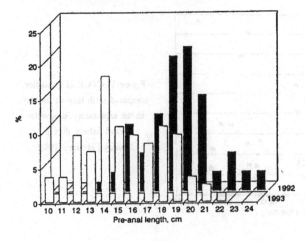

Figure 14. Length distribution (pre-anal length) of grenadier (*C.rupestris*) from the ground fish surveys - 1992 to 1993

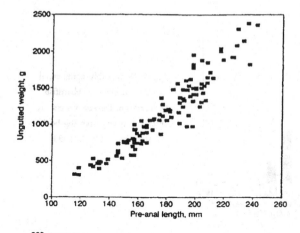

Figure 15. The relationship between pre-anal length and weight of grenadier (*C. rupestris*) from the groundfish surveys 1992 and 1993

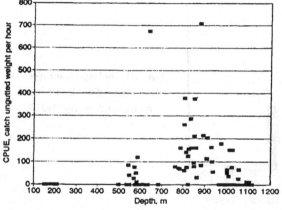

Figure 16. CPUE of black scabbardfish (*A. carbo*) with depth from the exploratory trawl fishery in January and February 1990

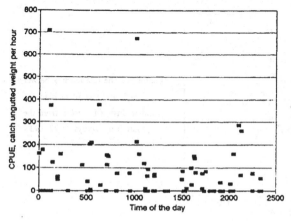

Figure 17. CPUE of black sacbbardfish (*A. carbo*) with time of day from the exploratory trawl fishery in January and February 1990

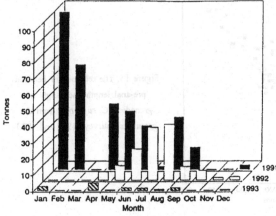

Figure 18. Monthly commercial
catches of black scabbardfish
(*A. carbo*) in Faroese waters by
French trawlers from log book
statistics - 1991 to 1993

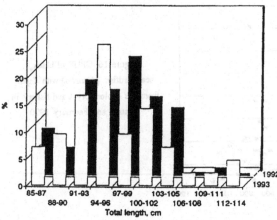

Figure 19. The length distributions of
black scabbardfish (*A. carbo*) from
the groundfish surveys 1992-93

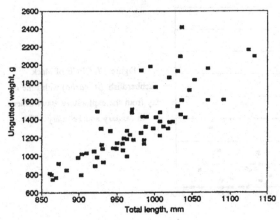

Figure 20. The relationship between
length and weight of black
scabbardfish (*A. carbo*) from the
groundfish surveys 1992 - 93

223

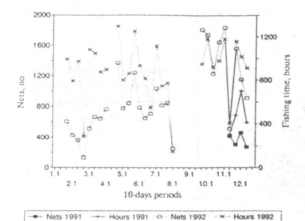

Figure 21. Fishing effort measured as No. of nets and fishing time from the exploratory fishery with tangle nets for monkfish (*Lophius sp.*) Nov 1991 to Dec 1992 - m.s. Móanes

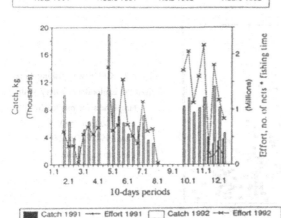

Figure 22. Catch and effort from the exploratory fishery with tangle nets for monkfish (*Lophius sp.*) Nov 1991 to Dec 1992 - m.s. Móanes

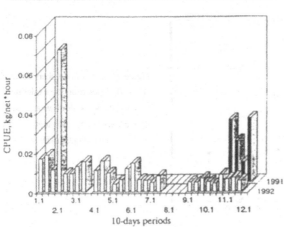

Figure 23. CPUE from the exploratory fishery with tangle nets for monkfish (*Lophius sp.*) Nov 1991 to Dec 1992 - m.s. Móanes

224

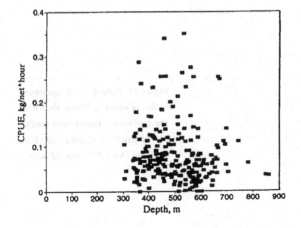

Figure 24. CPUE with depth from the exploratory fishery with tangle nets for monkfish (*Lophius sp.*) Nov 1991 to Dec 1992 - m.s. Móanes

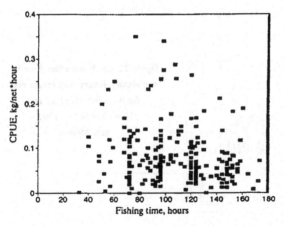

Figure 25. CPUE with fishing time from the exploratory fishery with tangle nets for monkfish (*Lophius sp.*) Nov 1991 to Dec 1992 - m.s. Móanes

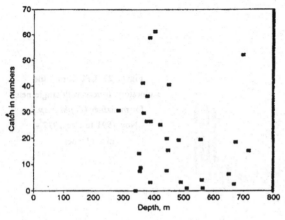

Figure 26. Catch in numbers for each set with depth from the exploratory pot fishery for deep-water crab (*G. affinis*), July and August 1993 -m.s. Gudmunder

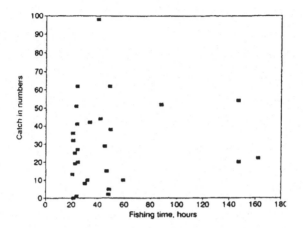

Figure 27. Catch in numbers for each set with fishing time in the exploratory pot fishery for the deep water red crab (G.affinis) in July-August 1993. m.s. Gudmundur.

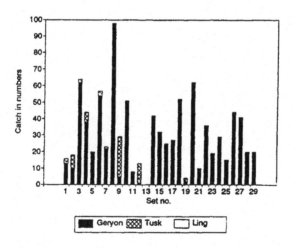

Figure 28. Catch in numbers for each set in the exploratory pot fishery for the deep water red crab (G. affinis) in July-August 1993. m.s. Gudmundur.

Figure 27. Catch (numbers) for each net with fishing time in the exploratory gill fishery for the deep-water red crab (*G. maritae*) in July-August 1993, near Guadeloupe.

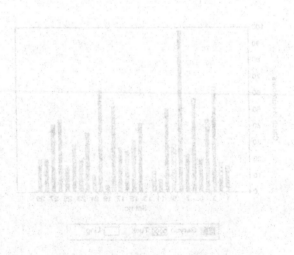

Figure 28. Catch in numbers for each set in the exploratory gill fishery for the deep-water red crab (*G. maritae*) in July-August 1993 near Guadeloupe.

COMPARISONS BETWEEN LONGLINING AND TRAWLING FOR DEEP-WATER SPECIES - SELECTIVITY,QUALITY AND CATCHABILITY - A REVIEW

NILS-ROAR HAREIDE
Møre og Romsdal Research Foundation
Box 5075
6021 Ålesund
Norway

ABSTRACT

A review of the literature on comparative fishing trials for deep water species using both trawl and longline is presented. Most attention is given to length and species selectivity. Other issues discussed are fish behaviour, energy consumption and quality of the products.

There are possibilities for both trawling and longlining in deep waters. The trawl is the most efficient gear on dense concentrations. However, trawling on dense concentrations may be harmful to the stocks. On scattered concentrations and on fish species with a large body size the relative catchability of longlines compared with trawls will be higher.

In some fisheries longlining will give the highest long term yield. For some fisheries the catchability is too low for longlines, and therefore trawling would be the best method for exploitation of these stocks.

A. G. Hopper (ed.), Deep-Water Fisheries of the North Atlantic Oceanic Slope, 227–234.
© *1995 Kluwer Academic Publishers.*

1. Introduction

As pressure on coastal fish stocks increases, more attention is being paid to deep-water species. In recent years new deep-water resources have been discovered and investigated, and some are already exploited. At present only a few deep-water species in the North Atlantic are subjected to quotas or other regulations. However, there is a growing realisation that the exploitation of these stocks has to be controlled or else they will be quickly reduced to levels from which they may take a long time to recover.

Many problems have to be dealt with in the development of fisheries in deep water. They include biological and technical issues and matters relating to product development and marketing. Once the resources are discovered and the utilization of the different deep-water stocks has commenced the problems connected with fisheries management, and the optimal biological, and economical yields will be important issues to investigate.

The type of fishing gear is of importance both for successful fishing and effective management. The two gears most often used in deep waters are the trawl and the longline. Gillnets are also used in depths down to at least 1800 m. in the North Atlantic.

Recently the question of whether trawling is an appropriate method of exploiting the deep water species has been raised, (Gordon and Swan, 1992).

In this paper the characteristics of the trawl and longline for fishing in deep water are compared and the advantages and disadvantages of both these gears are discussed.

2. Selection

The trawl and the longline are fundamentally different in the way they catch fish. The trawl herds the fish into the opening of the net while the longline lures the fish to the hooks by the smell of the bait. This fundamental difference in the way the two types of gear catch fish results in differences in selection for different species and differences in size selection within the same species when comparing catches from the two types of gear.

Some species are very seldom caught by longlines either because of their size or the characteristics of the bait. Other species and large specimens of certain species seem to be able to escape the trawl. Results from the Davis Strait (Jørgensen, this volume and Hareide, 1992) showed a marked difference in species composition between trawls and longlines. This was clearly seen for roundnose grenadier (*Coryphaenoides rupestris*), which was not caught at all by longlines, but was very abundant in the trawl catches. Other species such as smoothheads (*Alepocephalus* spp.) show more or less the same pattern (Jørgensen this volume). The opposite was observed for tusk (*Brosme brosme*), which was almost exclusively taken on longlines (Hareide, unpublished data from West Norwegian fjords).

The effect of species selection for different types of gear can be that some species may become heavily exploited while others are not likely to be exploited at all. This knowledge can be used as an effective measure in fisheries management.

According to Løkkeborg and Bjordal, (1992) and He, (1993) the swimming speed of a fish is proportional to its body size. The larger fish will therefore reach baited longlines more rapidly than the smaller fish. Furthermore, competition occurs among fish that are attracted to the bait. Behaviour studies show that large cod are able to frighten small cod away from baited hooks (Løkkeborg and Bjordal, (1992)), and hence the biggest fish are more likely to be caught.

The selectivity of the trawl is also influenced by the size of the fish because the fish have to maintain a high swimming speed and thus expend more energy to escape the gear. In these circumstances,the largest fish will have the best chance of survival.

The difference in length composition between the two gears was shown for Greenland halibut (*Reinhardtius hippoglossoides*) in the Davis Strait by Jørgensen, this volume. Nedreaas et al., (1993) presented data from their own investigations of Greenland halibut in the north east Atlantic, and their results are similar demonstrating that longlines tend to catch larger fish than the trawl (Figure 1).

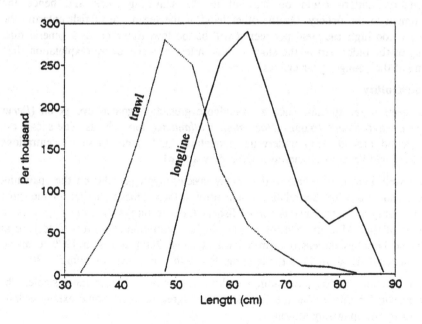

Figure 1. .Length distribution of Greenland halibut from trawl and longline catches in north east Atlantic.
(Nedreaas et al, (1993)).

Similarly, Klein, (1986) found the mean lengths to be 57.3 and 65.8 cm respectively, for the trawl and longline when comparing selectivity in the fishery for sablefish (*Anoplopoma fimbra*).

He (1993) has shown that size selection in the trawl is also dependent on trawling speed. When the speed is low the mean size of fish is smaller than when the trawling speed is high. The trawling speed is often less in deep water than in shallow water because of the extra power absorbed by the total trawling system when working with long lengths of warp and heavy gear. This will cause a bigger difference in size selection between the two gears as the fishery goes into deeper water.

The consequences of the differences in size selectivity is that the exploitation pattern for the two gears will be different. The yield per recruit (Y/R) and spawners per recruit will be different. The results from Davis Strait show that a higher biomass will remain in the sea if a given catch quota is taken with longliners than by trawlers, and the sustainable yield will be higher when using longlines compared with trawl. (Jørgensen, this volume).

Klein, (1986) also found that yield per recruit of sablefish is higher for longlines than for trawls, as long as fishing mortality (F) is lower than 0.7.

One effect of size selectivity is that there will be a greater proportion of mature fish in longline catches than in trawl catches. On heavily exploited stocks this exploitation pattern could be harmful to the spawning stock and hence the recruitment to the fishery. On the other hand, if the mean age of fish entering the fishery is too high, the yield per recruit will be too low. However, as a general rule, fishing of the older part of the stock results in less of a risk of overexplotation than fishing of the younger year classes.

3. Catchability

Some deep water species, such as roundnose grenadier, golden eye perch (*Beryx splendens*) and orange roughy (*Hoplostethus atlanticus*) form shoals. The shoals are often found around steep underwater mountains, and where these concentrations exist they will be most efficiently captured by a trawl.

It is a well known phenomena that many species aggregate during the spawning season, and as a result the relative catchability at these times is higher for the trawl than for longlines. The Icelandic trawl fishery for blue ling (*Molva dipterygia*) is an example of this. After discovering a spawning concentration southwest of Iceland in 1980, the landings increased from a level of about 2000 tonnes in 1979 to above 8000 tonnes in 1980. In 1985 the landings fell to the same level as before 1980.

During the same period the landings of ling (*Molva molva*) were more stable. The main reason for this is that the ling does not aggregate to the same extent as blue ling during the spawning season.

Shoaling species will often make daily vertical migrations. This behaviour pattern will favour use of the trawl, especially the pelagic or semipelagic trawl. Nevertheless

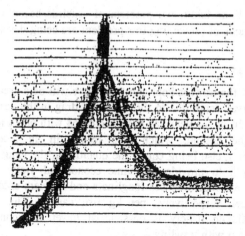

Figure 2. Shoal of golden eye perch (*Beryx splendens*) above an underwater mountain in the Mid-Atlantic Ridge Area.

trawling on shoaling fish can cause overexploitation of stocks, as has been shown for the orange roughy in New Zealand waters (Clark, this volume).

In areas where fish are found in heavy concentration the total catch per unit effort by longlines will be limited because the hooks quickly become occupied. The risk of over exploitation of fish stocks is therefore not as high as for the trawl.

4. Environmental effects

Bottom trawling with heavy gear will inevitably have an effect on bottom conditions and benthic communities. The long term impact of trawling in deep water is not yet known. By contrast longlines have very little impact on the sea bed or the benthic communities.

Both gears can easily be lost during fishing. This will be a form of marine pollution causing damage to other gear. Lost trawls will have a greater impact on commercial fishery activities than longlines because of their greater size and weight. Neither of the two gears however will cause problems by continuing to catch fish in what is known as "ghost fishing".

However both gears may cause stress to fish that are caught but not retained by the net or hooks. Fish that escape can be so severely injured that they die. Fish that have escaped from a trawl may be exhausted for at least 24 hours, and complete recovery from hooking can take several days. (Fernø, 1993)

Fish lost during the hauling of the longlines may have difficulties in returning to the bottom again because of expansion of the swim bladder. These fish may also be injured by the longline hooks.

232

With longlines there is an additional problem of seabirds becoming hooked when setting the lines. Methods of preventing this problem are being developed in Norway (Anon, 1993).

5. Energy consumption

If the towing speed of the trawl is kept constant, there will be a greater consumption of energy with increased depth due to the increased drag in the gear and the warps. Furthermore, the weight of the gear has to be increased with depth to ensure that the trawl remains on the bottom, which also results in an increase in energy consumption. Everything else being equal, the quantity or the value of fish then has to increase with depth to provide the same economical yield. Norwegian data (Bjørkum, (1992)) shows that deep sea longliners and trawlers used about 0.4 litres and 0.8 litres of fuel, respectively, per kg fish caught. This implies that an economical and profitable fishery is possible at lower fish concentrations with longlines by comparison with trawling.

Trawl fishing on shoaling fish such as spawning concentrations, will, however, only require short tows and in these circumstances a lower energy consumption could result in the trawl being the most efficient gear.

The Norwegian data on fuel consumption indicates that trawlers need more fuel per kg. fish than longliners. This is not always the case. If the fish concentrations are dense, large quantities of fish can be caught in short time. So when starting up "new" fisheries the fuel consumption will increase for both gears as the previously unexploited biomass is reduced. But probably the increase will be highest for trawlers.

6. Technical limitations

Deep water fisheries tend to be further offshore and therefore vessels will require a large freezing capacity and hold space. This together with extra space for longer and larger diameter warps, heavier gear, and a greater demand for energy calls for large trawlers with high engine power. Long, heavy lines and heavy anchors also requires the longliners to be of a certain minimum size. At the moment trawlers are able to trawl down to about 2000 m. and longliners are able to operate to about 3000 m. but fishing at such depths will not be efficient because there is insufficient technical knowledge.

On fishing grounds with a rough and or a steep bottom, trawling is difficult, and sometimes quite impossible. Under such circumstances the longline is the most effective gear to use.

7. Fish Quality

The quality of fish caught by hooks is generally better than the quality of fish caught in the trawl. One reason for this is that there is no compression from other fish on the longlines as often is the case in a trawl where the fish are forced into the codend. In longline fisheries most of the fish are still alive when brought on deck, and it is possible to cut the throat and bleed the fish immediately after it is caught resulting in a higher quality product. This higher quality often leads to better prices for fish caught by longlines.

Another criteria of quality is size. The bigger fish often fetch a higher price than the smaller fish. The average price per kg. of Greenland halibut landed in the Sunnmøre and Romsdal Area in 1993 was 23% higher for longline caught fish compared with trawl caught fish due mainly to the size of the fish.

Klein, (1986) found that longliners received a three times higher price than trawlers in the fishery for sablefish. This price difference was because of the size composition of the landings.

If there is usually a higher price paid for the bigger fish, longlines will give an even higher long term profitability than trawling. However, for species with high natural mortality this will not be true.

8. Conclusions

There is a difference in species selection between the two gears. Trawls tend to catch smaller fish than longlines. For most species the exploitation by longlines will give a higher Y/R and a lower risk of over exploitation for long lived fish such as those in deep water. These differences in the pattern of exploitation of the two gears could be used in management of the fish resources.

The trawl has high catchability on dense concentrations. Trawling on spawning concentrations and schooling species can overexploit deep water resources. Longlines will not have this effect.

None of the two gears have severe environmental impacts but trawling demands a higher consumption of energy per kg. fish caught.

Longlines result in better fish quality than trawled fish.

Although both trawls and longlines have been widely used in commercial fisheries for decades there has been very little investigation into comparing the effects of these gears. The fishermen choose their gear from several points of view such as fuel economy and capital cost, traditional preferences, quota opportunities, licences or other personal factors. Their choice is in this way based on their own terms, or it might be based on terms laid down by regulations. There is also a need for more comparative research into the advantages and disadvantages of other fishing methods such as gillneting and fish traps.

Little work seems to have been done on economic risk and uncertainty in fisheries. A total evaluation of the two gears should include economic theory and models for sustainable biological and economical yield rather than short term assessment of profit or loss.

234

References

Anon (1993) New "seabird scarer" solves age-old problem. Fishing Boat World, 5(1): 31.

Bjørkum I. (1992) Banklinefisket i et miljøperspektiv- Energiforbruk og forurensing. Paper presented on symposium "Banklineflåten mot år 2001". Ålesund 5.-6. October 1992.

Engås, A., Løkkeborg, S. and Ona E. (in press) Submitted to J. Northw. Atl. Fish. Sci.

Fernø, A. (1993) Advances in understanding of basic behaviour: consequences for fish capture studies. ICES mar. Sci. Symp., 196: 5-11.

Gordon, J.D.M. and Swan, S. C. (1993) The Scottish Association for Marine Science. Report 1992/10 to the Commission of the European Community DG XIV/C/1, 260 p.

Hareide, N. (1992) Forsøksfiske med linefartøyet M/S Skarheim ved Vest-Grønland. Mai - Juni 1992. Report from Møreforsking, Ålesund 27 p. (Partly translated by Canadian Translation of Fisheries and Aquatic Sciences No. 5605. Experimental fishery with the longline vessel M/V Skarheim at West Greenland May - June 1992.

He, P. (1993) Swimming speeds of marine fish in relation to fishing gears. ICES mar. Sci. Symp., 196: 183-189.

Jørgensen, O.A. and Boje, J. (1991) A comparison of the selectivity in trawl and long-line fishery for Greenland halibut. NAFO SCR Doc. 92/55.

Jørgensen, O.A. (this volume) A comparison of deep water trawl and long-line research fishery in the Davis Strait. Paper for the workshop on Deep Water Fisheries of the North Atlantic Oceanic Slope. UK, Hull 1.-4. March 1994.

Klein, S.J. (1986) Selectivity of Trawl, Trap, Longline and Set-net Gears to Sablefish, (Anoplopoma fimbria). Alaska Fisheries Center, NWAFC Processed Report 86-06: 84 p.

Løkkeborg, S. and Bjordal, Å. (1992) Species and size selectivity in longline fishing: a review. Fisheries Research, 13: 311-322.

Nedreaas, K., Soldal, A.V. and Bjordal, Å. (1993). Performance and biological implications of a multi-gear fishery for Greenland halibut (Reinhardtius hippoglossoides). NAFO -Symposium Dartmouth, NS, Canada, 13-15 September 1993, 29 p.

A COMPARISON OF DEEP WATER TRAWL AND LONG-LINE RESEARCH FISHING IN THE DAVIS STRAIT

O.A. JØRGENSEN
Greenland Fisheries Research
Tagensvej 135 1.
DK-2200 Copenhagen N,
Denmark

ABSTRACT

In August 1991 a long-line and a bottom trawl comparative survey, covering depths between 950 and 1450 m, was conducted off west Greenland. Great differences in catch composition and length frequency of Greenland halibut (Reinhardtius hippoglossoides) and roughhead grenadier (Macrourus berglax) were observed in the two types of gear. Calculation of relative selection (RS) of Greenland halibut showed that long-lines were up to 30 times more effective in catching large fish, and fishing with long-lines by comparison to the trawl allows a greater maximum sustainable yield (MSY). The RS for roughhead grenadier showed that the long-lines were about 18 times more effective for this species up to 71 cm total length but there is probably a drop in the RS for the very large fish. These results tend to confirm previous work in this field suggesting that for Greenland halibut in particular the long-line is probably the best method of fishing.

Roundnose grenadier were present in large numbers in the trawl but totally absent from the longline catches

A. G. Hopper (ed.), Deep-Water Fisheries of the North Atlantic Oceanic Slope, 235–250.
© *1995 Kluwer Academic Publishers.*

1. Introduction

Since 1987 bottom trawl surveys covering depths down to 1500 m have taken place annually off the west coast of Greenland in NAFO Subarea 1 principally for the purpose of estimation of trawlable biomass of Greenland halibut (*Reinhardtius hippoglossoides*). Over the same period surveys and commercial fisheries with long-lines and gill-nets have taken place in the fjords in north west Greenland (Riget and Boje, 1989). The length distribution obtained from the two areas has shown marked differences. The average sizes of fish taken in the fjords are considerably larger than fish taken off shore. In order to investigate the possibilities for an off shore long-line fishery for Greenland halibut, experimental fishing trials were carried out by a commercial long-liner in NAFO Division 1D in August, 1991. Simultaneously the same area was covered by the deep water trawl survey which allowed comparisons of the catch composition and selectivity in the two types of gear. The main species for which catch comparisons have been possible are the Greenland halibut and the roughhead grenadier (*Macrourus berglax*). Figure 1 shows the survey area.

2. Material and methods

2.1. LONG LINING

The long-line survey was conducted in NAFO Division 1D (Fig. 1) in the period 5th to 22nd August 1991 by the Faroese long-liner Varsol (593 GRT). The vessel was equipped with a Mustad Autoliner system, and fishing was carried out with 7.5 mm polypropylene long-lines with 75 cm ganglions spaced 1.8 m and mounted with 'EZ-baiter' circle hooks and baited with squid. A total of 23 settings covering depths between 970 and 1427 m were made. The number of hooks per set varied from 500 to 3925 and in total 57,598 hooks were used. Fishing was carried out during daylight and darkness and the average fishing time was about 5 hours. All catches of Greenland halibut from the 23 settings and the catches of roughhead grenadier (*Macrourus berglax*) from 5 line settings (15,063 hooks) were weighed and measured. From 3 of the 5 settings all fish were weighed (8,725 hooks).

2.2 TRAWLING

Only those trawl hauls carried out in the same area and depth range covered by the long-line survey are included in the analysis. These hauls were carried out in the period 5th to 8th August, and a total of 26 hauls were made. The trawl survey was carried out by the Japanese Research Vessel Shinkai Maru (Yano and Jørgensen, 1992). The trawling time was 30 min and trawling speed was 3.5 knots. The mesh size was 140 mm with a 30 mm mesh liner in the cod-end. The wing spread was approximately 45 m. Trawling was carried out in daylight only (for further information about vessel and gear see Yamada et. al., 1988a). In the calculation of relative selection (RS) between the long-line and the trawl, the catch per unit effort (CPUE) for the trawl is given per 0.15 km^2 swept area corresponding to an average haul.

TABLE I. Catches in kg. in 26 trawl hauls and on long lines in Davis Strait August 1991.

SPECIES	TRAWL		LONG- LINE		
	Number	Weight	Number	Weight	no*
Alepocephalus agassizzi	792	419.6			
Alepocephalus sp.a	33	10.6			
Anarhichas denticulatus	7	67.5	12	144.3	3
Anopologaster cornuta	5	0.4			
Antimora rostata	529	247.9	359	431.0	3
Bathylagus euryops	-	56.6			
Bathyraja spinicauda	3	19.7	5	50.4	3
Borostomias antarctica	29	2.8			
Centroscyllium fabricii	28	55.5	156	318.1	3
Chalinura brevibarbis	9	0.2			
Chauliodus sloani	9	0.6			
Chiasmodon bolangeri	17	0.6			
Clycothone microdon	+	0.0			
Coryphaenoides guntheri	1512	89.2			
Coryphaenoides rupestris	13567	3743.2			
Cottunculus microps	6	0.2	1		23
Cottunculus thomsonii	27	23.6			
Gigantactis sp.	1	0.6			
Gonostoma bathyphila	45	0.8			
Hippoglossus hippoglossus			1		23
Hippoglossoides platessoides	3	0.2			
Hydrolagus affinis	7	72.2	2	19.0	3
Lycodes vahli	12	2.3			
Lycodonus flagillicauda	77	1.6			
Macrourus berglax	620	302.3	1215	1024.0	5
Malacosteus niger	2	0.1			
Molva dypterygia	1	0.5			
Myctophidae	-	43.6			
Myxine glutinosa	11	1.0			
Notacanthus chemnitzii	80	104.5			
Oneirodes eschrichti	4	1.6			
Onogadus argentatus	3	0.0			
Onogadus ensis	54	10.4	4	2.8	3
Paraliparis copei	66	0.8			
Paralipidae	10	1.1			
Polyacantohonotus rissoanus	75	8.2			
Poromitra crassiceps	2	0.0			
Raja bathyphila	5	8.9			
Raja fyllae	12	8.9			
Raja hyperborea	1	9.0			
Raja lintea			27	287.0	3
Raja radiata			5	42.4	3
Raja spinacidermis	2	1.3			
Reinhardtius hippoglossoi	3478	5089.7	3207	15413.0	23
Rhodichtlys regina	1	0.0			
Saccopharynx ampullaceus	1	0.3			
Sagamichthys sachkenbech	27	1.6			

TABLE 1. (continued)

SPECIES	TRAWL		LONG-LINE		
	Number	Weight	Number	Weight	No*.
Scopelosaurus lepidus	374	58.6			
Sebastes mentella	24	14.3			
Serrivormer beani	36	4.2			
Somniosus microcephalus	2	612.0	2		23
Stomias boa ferox	6	0.1			
Synaphobranchus kaupi	787	91.1	9	1.7	3
Trachyrynchus murrayi	35	14.7			
Decapoda	47	6.8			
Lithodes maja	10	16.8	1		23
Natantia		29.3			
Octopoda		629.0			
Pandalus borealis		0.4			

No* refers to the number of settings on which the data are based

In the long-line survey the length of roughhead grenadier was measured as total length (TL), whilst in the trawl survey it was measured as snout to anal fin length (AFL). In order to make the length frequencies from the two surveys comparable the conversion factor TL=5.744 + AFL x 2.273 (data from research trawl catches in NAFO Div. 1C and 1D, unpublished) was used to transform AFL to TL.

Age-length keys, based on data from long-line and trawl surveys in NAFO Divisions 1C and 1D in the period 1987-1989, were used to calculate age distribution of Greenland halibut. The mean weight at age values used in the yield per recruit analysis come from the same set of data.

3. Results

A total of 51 fish or fish groups, of which most species are found widely spread in the North Atlantic, were represented in the trawl catches. Greenland halibut and roundnose grenadier (*Coryphaenoides rupestris*) were by far the most frequent and they were taken in all hauls together with roughhead grenadier and antimora (*Antimora rostrata*) (Table I). Other common species were *Alepocephalus spp.*, *Scopelosaurus lepidus* and *Synaphobranchus kaupi*. Less common were a number of skates and grenadier species, except those already mentioned. A number of pelagic and bathypelagic species such as *Bathylagus euryops* and *Myctophidae* among others were also captured in the trawl.

In total 15 different species were taken on the long-lines. Catches were totally dominated by Greenland halibut, but also present were antimora, black dogfish (*Centrocyllium fabricii*) and especially roughhead grenadier (*Macrourus berglax*) (Table 1)

For the fish that were present in significant numbers in both the trawl and on long-lines the mean weight by species was always greatest on the long-lines. This fact was most pronounced for Greenland halibut where the mean weight was more than three times greater on long-lines compared to the trawl (4.806 kg compared to 1.463 kg) and about twice as great for roughhead grenadier (0.843 kg compared to 0.488 kg) and antimora (1.201 kg compared to 0.469 kg). This is reflected in the length composition graphs shown for Greenland halibut and roughhead grenadier in Figs. 2 and 3, respectively. In the trawl survey the length of Greenland halibut ranged from 27 cm to 112 cm with a unimodal distribution showing a distinct mode at 48 cm. On the long-lines the length ranged from 42 cm to 120 cm and the length distribution differed markedly from the length distribution in the trawl catches. The long-line catches of Greenland halibut consisted, mainly, of fish between 50 cm and 95 cm with modes around 63 and 85 cm, the first mode being dominated by males whilst the second consisted exclusively of females.

The length distribution of roughhead grenadier in the trawl survey showed three modes about 20, 35 and 45 cm, respectively, whilst in the long-line survey, the distribution was bimodal with a pronounced mode about 40 cm and a smaller one around 55 cm.

The relative selection (RS) of long-lines as compared to the trawl is derived by comparing the CPUE by length group of the two gears according to a formulae proposed by Hovgård and Riget (1992), i.e :

$$\text{Relative selection(RS)} = \frac{\text{Longline CPUE (no. of fish caught/1000 hooks}}{\text{Trawl CPUE (no. of fish caught/0.15 km}^2 \text{ swept)}}$$

RS for Greenland halibut was estimated for two depth strata: 950 to 1200 m and 1200 to 1450 m and calculated in 6-cm length groups (Table 2). The length distribution in the two depth strata follow the total length distribution closely. There was a tendency, however, towards larger fish being slightly more abundant in the depth stratum 1200 to 1450 m. Catches outside the size range 42 to 101 cm in the shallow stratum, and 42 to 95 cm in the deep stratum were not included in the calculations due to too few observations being available.

CPUE values are usually subject to multiplicative errors and the RS values were therefore log-transformed. The In. RS increased with length to about 70 cm showing that only Greenland halibut at this size and above are fully recruited to the long-line as compared to the trawl (Fig. 4). From Fig. 4 and Table 2 it is seen that the relative selection is larger in the deep stratum. This is due to a combination of a decrease in CPUE for the trawl and an increase in CPUE on the long-line in this stratum as compared to the more shallow stratum.

The relative selection (RS) in a one-side ANOVA, i.e.

In. (RS) = (effect due to length) + noise

where the lengths are the 6-cm groups given in Table 2. The statistics and the estimates from the ANOVA (Anon, 1985) are given in Table 3.

240

240

TABLE 2. CPUE of Greenland halibut for each trawled 0.15 km^2 and for each 1000 hooks and relative selection (RS) by depth strata.

length.grp cm.	DEPTH STRATUM 980 - 1200 m 15 hauls, 20824 hooks			DEPTH STRATUM 1200 - 1450 m 11 hauls, 36774 hooks		
	Trawl CPUE	Long-line CPUE	RS	Trawl CPUE	Long-line CPUE	RS
24-29	0.0714					
30-35	0.7218					
36-41	6.7146			3.5208		
42-47	47.7216	0.0721	0.0015	16.8877	0.6012	0.0356
48-53	71.0105	3.4876	0.0491	34.1146	2.8735	0.0842
54-59	26.7540	4.5634	0.1706	14.5207	5.3184	0.3663
60-65	8.3978	4.9680	0.5916	5.5004	6.2834	1.1424
66-71	3.0460	4.1356	1.3577	1.3867	6.0188	4.3404
72-77	1.9253	3.9736	2.0639	1.3780	5.9862	4.3441
78-83	2.4588	5.1022	2.0751	0.8167	8.4168	10.3059
84-89	1.4848	5.8458	3.9371	1.0085	9.5632	9.4826
90-95	1.4230	3.6925	2.5949	0.5628	7.2653	12.9092
96-101	0.3344	3.1838	9.5209	0.4503	3.2463	7.2092
102-107	0.2690	0.8653	3.2167	0.1800	1.0356	5.7533
108-113	0.1344	0.7377	5.4888	0.1366		
114-119		0.0337				
120-125		0.0337				

TABLE 3. ANOVA of RS for the Greenland halibut length group.

STATISTICS

Source	Df	SS	MS	F	R^2
Model	9	94.46	10.50	9.76	0.907
Error	9	9.68	1.08		

ESTIMATES

Length group	Estimate	Retransformed Estimate
42-47	-4.919	0.013
48-53	-2.824	0.102
54-59	-1.387	0.428
60-65	-0.196	1.407
66-71	0.886	4.152
72-77	1.096	5.123
78-83	1.531	7.914
84-89	1.810	10.461
90-95	1.755	9.901
96-101	2.253	16.292

[1]Retransformed estimate = exp(ln estimate) x exp (MS error/2)

When transforming the RS back to an arithmetic scale the RS have been corrected by exponential (MS error/2) i.e. 1.712 (Table 3). From this the following equations for translating long-line catches (per 1000 hooks) to trawl catches (per 0.15 km² swept area) have been obtained for different size groups of Greenland halibut:

TABLE 4. Equations for the translation of long-line catches to trawl catches and the relative efficiency

SIZE	EQUATION	REL. EFFICIENCY of long-line to trawl
42-47	Trawl catch = 76.92 x long-line catch	0.02
48-53	Trawl catch = 9.80 x long-line catch	0.18
54-59	Trawl catch = 2.33 x long-line catch	0.76
60-65	Trawl catch = 0.71 x long-line catch	2.50
66-71	Trawl catch = 0.24 x long-line catch	7.41
72-77	Trawl catch = 0.20 x long-line catch	8.89
78-83	Trawl catch = 0.13 x long-line catch	13.68
84-89	Trawl catch = 0.10 x long-line catch	17.78
90-95	Trawl catch = 0.10 x long-line catch	17.78
96-101	Trawl catch = 0.06 x long-line catch	29.63

The equations relate somewhat incompatible units i.e. a trawl path of 3.2 km compared to a long-line setting of 1.8 km. Instead the catches are transformed according to Dickson (1986) to obtain comparable path lengths (a trawl path of 1 km to a long line on 1 km). These are shown in the right hand column of the above table as relative efficiency.

This implies that the efficiency of a long-line for Greenland halibut is 0.02 of that of a trawl in the size group 42-47 cm, while it is about 30 times more efficient for catching fish in the size group 96-101 cm.

Due to too few observations for roughhead grenadier the RS is estimated for the total depth range and calculated by 6-cm groups (Table 6). Although data is rather noisy it seems that ln. RS increased to about 40 cm showing that only roughhead grenadier at this size and above are fully recruited to the long-line as compared to the trawl (Fig. 5).

The following equations in Table 5 for translating long-line catches (per 1000 hooks) to trawl catches (per 0.15 km² swept area) have been obtained for different size groups of roughhead grenadier. As before the equations are transformed to comparable path length (ie.a trawl path of 1 km to a long line on 1 km) and the relative efficiency is shown in the right hand column

TABLE 5. Equations for the translation of long-line catches to trawl catches and the relative efficiency

SIZE	EQUATION	REL. EFFICIENCY of long-line to trawl
12-17	Trawl catch = 5.69 X long-line catch	0.31
18-23	-	-
24-29	Trawl catch = 1.41 X long-line catch	1.26
30-35	Trawl catch = 0.76 X long-line catch	2.34
36-41	Trawl catch = 0.16 X long-line catch	11.11
42-47	Trawl catch = 0.25 X long-line catch	7.11
48-53	Trawl catch = 0.25 X long-line catch	7.11
54-59	Trawl catch = 0.15 X long-line catch	11.85
60-65	Trawl catch = 0.12 X long-line catch	14.81
66-71	Trawl catch = 0.10 X long-line catch	17.78
72-77	Trawl catch = 0.38 X long-line catch	4.68
78-83	Trawl catch = 0.31 X long-line catch	5.73

This implies that the relative efficiency of a long-line is 0.31 of that of a trawl for roughhead grenadier in the size group 12-18 cm, whilst the long-line is about 18 times more efficient for catching fish in the size group 66-71 cm. The efficiency seems to drop off in larger size groups on the long-lines, although the results are based on a few observations only.

A yield per recruit analysis was performed on catch data for Greenland halibut using M=0.2. The relative selection at age in the different gears is given in Table 7 and is derived from the age distribution shown in Fig 6. The relative selection pattern used for the trawl is based on the fact that the same unimodal length distribution, with a mode around 47 to 49 cm. has been observed in all research surveys which have been conducted annually since 1987, and the growth of Greenland halibut that would have been expected in that period has not been observed (Yamada et al. 1988b; Yatsu and Jørgensen, 1989; Jørgensen and Akimoto, 1990; Jørgensen and Akimoto, 1991; Yano and Jørgensen, 1992; Satani et al., 1993; Ogawa and Jørgensen, 1994) indicating that fish above a certain size are able to escape from the trawl. The selection pattern used for the long-lines is based on the age distribution shown in Figure 6. from which it reasonable to assume that Greenland halibut is fully recruited to the long-lines at age 9

From Figure 7 it is seen that F_{max} is obtained at a lower level of $F_{0.28}$ per year and gives a higher yield 0.62 kg per recruit for the long-line compared to a F_{max} of 1.15 per year and yield per recruit of 0.45 kg for the trawl. Further it seems that at all levels of F the yield per recruit is significantly higher for the long-lines compared to the trawl. This means that a fishery using long-lines would permit a sustainable annual catch 1.9 times as great as the sustainable catch if the trawl only was used (ie. 1.15/0.62). The yield per recruit calculation is based on length frequencies obtained from tows with a research trawl fitted with a 30 mm. mesh liner in the cod

TABLE 6. CPUE for each 0.15 km² trawled,and for each 1000 hooks and the Relative Selection (RS) for roughhead grenadier.

26 hauls, 15063 hooks

LENGTH GROUP	TRAWL CPUE	LONGLINE CPUE	RELATIVE SELECTION
6-11	0.0746		
12-17	0.3089	0.0543	0.1758
18-23	1.9521		
24-29	1.4438	1.0243	0.7094
30-35	5.3562	7.0768	1.3212
36-41	4.6258	28.8584	6.2386
42-47	5.1030	20.7698	4.0701
48-53	2.2233	8.8938	4.0003
54-59	0.9098	5.9577	6.5484
60-65	0.5703	4.6932	8.2294
66-71	0.1964	1.9293	9.8233
72-77	0.1544	0.4041	2.6172
78-83	0.0405	0.1327	3.2765

TABLE 7.Relative selection and mean weights (kg.) used in the calculations of yield per. recruit.

		RELATIVE SELECTIVITY	
AGE	WEIGHT	TRAWL	LONG-LINE
5	.340	0.1	0.001
6	.580	0.5	0.05
7	.780	0.75	0.8
8	1.050	1.0	0.9
9	1.350	0.6	1.0
10	1.900	0.2	1.0
11	2.580	0.05	1.0
12	3.570	0.01	1.0
13	5.000	0.01	1.0
14	5.790	0.01	1.0
15	6.790	0.01	1.0
16	8.860	0.01	1.0
17	9.830	0.01	1.0
18	10.650	0.01	1.0

end. The length frequencies observed in the commercial fishery using a mesh size of 140 mm. are, however, almost identical with those from the research trawl (Jørgensen and Boje, 1993; Boje et al., 1994) and therefore it is reasonable to apply the same result to the commercial fishery.

4. Discussion

The species distribution in the catches of the two gears was significantly different. From Table 1 it seems that the trawl to some extent at least, takes a representative part of the total fish biomass, whilst the long-line takes only the larger specimens of carnivorous fish species which are capable of taking a certain size of prey. Fish species living on small sized prey are taken in the trawl only. This applies to the most abundant species in the trawl catches, the roundnose grenadier, and other common species such as *Alepocephalus spp.* and *Coryphaenoides guntheri*. Of the 15 species taken on long-line three were not represented at all in the trawl catches: Atlantic halibut (*Hippoglossus hippoglossus*), *Raja lintea* and *R. radiata*. However, these species have been observed in previous year's trawl catches in the same area, (data from trawl surveys conducted annually in the area since 1987, unpublished) and the difference should not be interpreted as a clear cut differences between the selectivity of the two gears.

Generally fish taken on long-lines are larger than fish taken in the trawl. This was especially true for Greenland halibut,and only to a slightly less extent for roughhead grenadier (Figures 2 and 3). The same difference in size selection of the two types of gear was observed for Greenland halibut in the Barents Sea by Nederaas et al. (1993), and for Atlantic cod (*Gadus morhua*) by Sætersdal (1963). The length distribution for Greenland halibut in the off shore long-line catches closely resembled the in shore long-line catches (Riget and Boje op. cit.).

Further, Hovgård and Riget (1992) made calculations to compare the relative selection (RS) of the trawl and long-line for cod at West Greenland and obtained comparable figures in good accordance with the findings of these investigations, as did Chumakov and Soshin (1991) who made the same experiment on Greenland halibut in NAFO Subarea 0. However, whilst their calculations of CPUE in the trawl were in the same size range, the efficiency of the Russian long-lines was about an order of a magnitude lower than observed in this study. This was probably due to less efficient hooks (Chumakov and Soshin op. cit.).

The observed selection pattern may be caused by at least two factors (or a combination of both):- 1) Large fish are able to avoid the trawl or 2) the long-lines . attract large fish from a vast area. The very stable stock structure seen from one year to the next without any sign of growth in the population, as mentioned previously implies that the large sizes of Greenland halibut, to a great extent, are able to escape the trawl.

The decrease in CPUE of Greenland halibut for the trawl coinciding with an increase in CPUE for the long-line observed in almost every length group, when going from the lowest depth stratum to the deepest (Table 2), is not immediately understandable.

It is probably caused by a decrease in trawl efficiency with depth combined with higher abundance of Greenland halibut as indicated by higher CPUE for long-lines, since there is no reason to believe that the long-line efficiency should increase with depth.

For both Greenland halibut and roughhead grenadier of the very large sizes there is a drop off in the selectivity of the long-line (Tables 2 and 6) but the few observations available make it difficult to draw any firm conclusions.

The age-length keys used are from 1987-1989, as no data for 1991 are yet available, and they might not reflect the actual growth in the stock. However, Boje and Jørgensen (1991) showed that there were no significant differences in growth of 5-11 years old Greenland halibut at a number of places in the north west Atlantic. One should therefore not expect great variation within years in the same area.

The yield per recruit analysis shows that the best exploitation of the Greenland halibut stock will be achieved by using long-lines, all other things being equal. However, the analysis is based on data from 1991 and since then the fishery has increased (Anon. 1994), and the results based on the 1991 data might not be applicable on the current stock structure, although the trend is most likely to be the same.

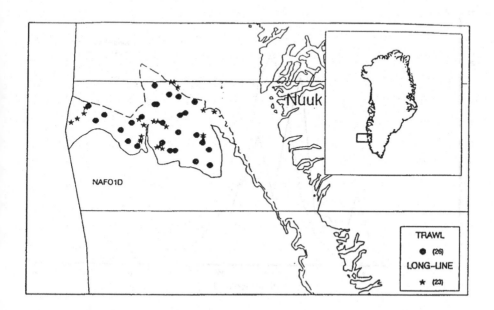

Figure. 1. Distribution of trawl and long-line stations and indication of 1000 m and 1500 m depth contour line.

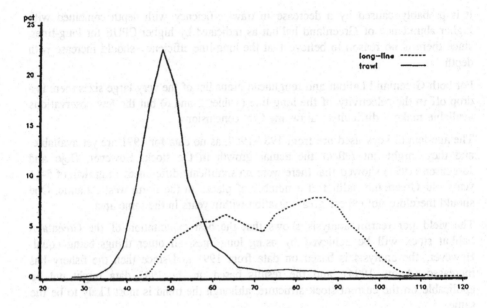

Figure. 2. Greenland halibut-length distribution (in 3-cm groups) by gear.

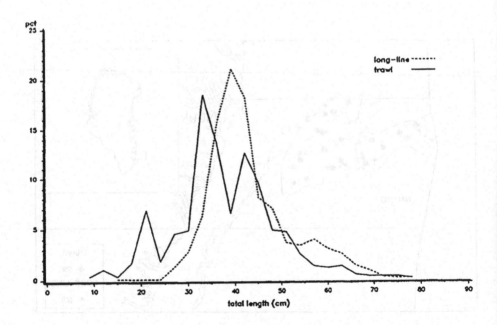

Figure. 3 Roughhead grenadier-length distribution (in 3-cm groups) by gear.

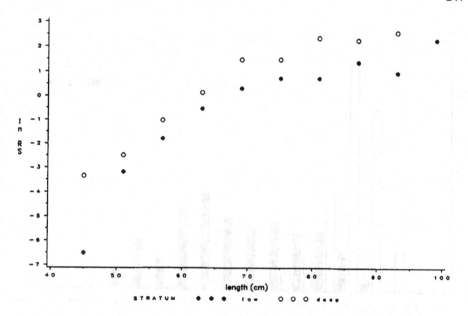

Figure. 4. Relative selection(RS) of long-line to trawl of size of Greenland halibut by depth stratum.

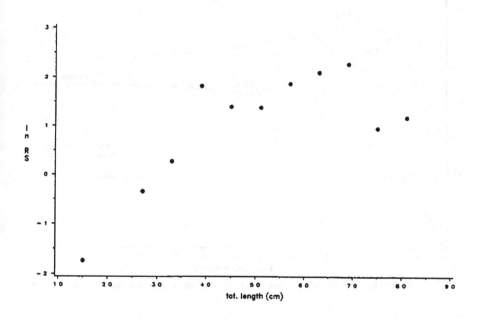

Figure. 5 Relative selection (RS) of long-line to trawl of size of roughhead grenadier.

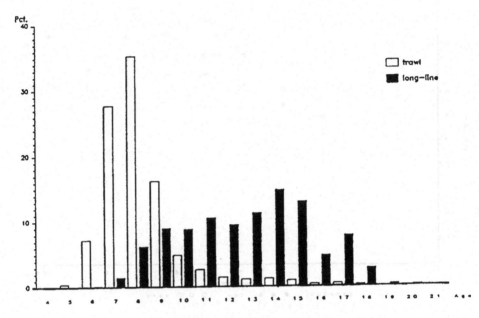

Figure 6. Greenland halibut-age distribution by gear.

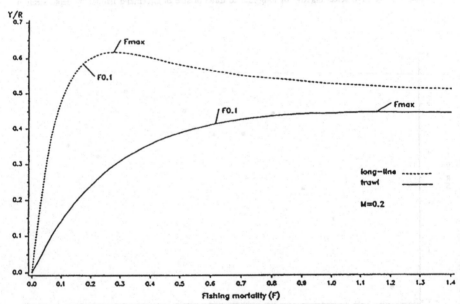

Figure 7. Greenland halibut-yield per recruit (kg.) by gear, based on the relative selectivity given in
TABLE 7

References

Anon., 1985. SAS User's Guide; Statistics, Version 5 Edition. Cary,NC., USA: SAS Institute Inc. 956pp.

Anon., 1994. Report of the Scientific Council, 2-16 June Meeting. NAFO SCS Doc. 93/17.

Boje, J. and Jørgensen O., 1991. Growth of Greenland halibut in the Northwest Atlantic. ICES CM 1991/40.

Boje, J., Jørgensen O. and Bech G., 1994. An assessment of the Greenland halibut stock component in NAFO Subareas 0 and 1. NAFO SCR Doc., 94/59

Chumakov, A.K. and Soshin S.M., 1991. Results of Stratified Random Bottom Trawl and Long-line Surveys on Greenland halibut in NAFO Div. 0B in 1990, NAFO SCR Doc. 91/66.

Dickson, W., 1986. Cod gillnet effectiveness. Report from the Institute of Fishery Technology Research, Bergen, Norway.

Hovgård, H. and Riget F.F., 1992. Comparison of long-line and trawl selectivity in cod surveys off West Greenland. Fish.Res.,13:323-333.

Jørgensen, O. and Akimoto K., 1990. Results of a Stratified Random Bottom Trawl Survey in NAFO Subarea 1 in 1989. NAFO SCR Doc. 90/39.

Jørgensen,O. and Akimoto K., 1991. Results of two Trawl Surveys in NAFO Subarea 1 in 1990. NAFO SCR Doc. 91/50.

Jørgensen,O. and Boje, J., 1993. An assessment of the Greenland halibut stock component in NAFO Subareas 0 and 1. NAFO SCR.Doc., 93/80

Nederaas, K., A.V. Soldal A.V. and Bjordal Å., 1993. Performance and biological implications of multi-gear fishery for Greenland halibut (Reinhardtius hippoglossides). NAFO-Symposium on Gear Selectivity/Technical Interaction in Mixed Species Fisheries. Dartmouth, NS, Canada, 13-15 September, 1993.

Ogawa, M., Yokawa, K. and Jørgensen O., 1994. Results of a stratified random bottom trawl survey off West Greenland in 1993. NAFO SCR Doc. 94/31

Riget, F. and Boje J., 1989. Fishery and some biological aspects of Greenland halibut in West Greenland Waters. NAFO Sci. Coun. Studies 13 pp. 41-52.

Satani, M., Yawahara S. and Jørgensen O., 1993. Results of two stratified random bottom trawl surveys off West Greenland in 1992. NAFO SCR Doc. 93/58.

Sætersdal, G. 1963. Selectivity of long lines. In: The Selectivity of Fishing Gear. Int. Comm. Northwest Atl. Fish., Spec. Publ., Dartmouth, NS, Canada, pp. 189-192.

Yamada, H., Okada K. and Jørgensen O., 1988a. West Greenland ground fish biomasses estimated from a stratified-random trawl survey in 1987. NAFO SCR Doc., 88/31.

Yamada, H., Okada K. and Jørgensen O., 1988b. Distribution, Abundance and Size Composition Estimated from a Stratified-Random Trawl Survey off West Greenland in 1987. NAFO SCR Doc. 88/34

250

Yano, K. and Jørgensen O., 1992. Results of two trawl surveys at West Greenland in 1992. NAFO SCR Doc., 92/48.

Yatsu, A. and Jørgensen O., 1989. Distribution, Abundance, Size, Age, Gonad Index and Stomach Contents of Greenland halibut off West Greenland in September/ October 1988. NAFO SCR Doc. 89/31.

EXPERIENCE WITH MANAGEMENT OF ORANGE ROUGHY (Hoplostethus atlanticus) IN NEW ZEALAND WATERS, AND THE EFFECTS OF COMMERCIAL FISHING ON STOCKS OVER THE PERIOD 1980- 1993

MALCOLM CLARK
MAF Fisheries, Greta Point,
PO Box 297, Wellington,
New Zealand.

ABSTRACT

Orange roughy (Hoplostethus atlanticus) occurs throughout New Zealand waters at depths of 700m to 1500m. Commercial fishing dates from 1980 with the discovery of large concentrations of spawning fish on the Chatham Rise. Since then, further spawning and feeding grounds have been identified, and fishing occurs in 8 separate regions of the New Zealand EEZ. Catches increased rapidly, and throughout much of the 1980's the total reported catch was 40-50 000 tonnes per year. This made orange roughy New Zealand's single most valuable fish species.

Management of the fishery has been based on a system of regional Total Allowable Catches (TACs), with the quota allocated amongst fishermen, rather than on an open competitive basis. TACs were initially set using very limited research data. It has since been recognised that orange roughy are slow growing and long lived, with low fecundity and low productivity. Combined with improved estimates of relative abundance from time series of surveys, it is clear that for several of the stocks the initial TACs were set at levels considerably higher than would be sustainable in the long term. Sustainable yields are around 1 to 2% of virgin biomass, with the "optimal' stock size at about 30% of virgin.

Several stocks have been overexploited, and reduced to levels of 15-20% of virgin biomass. Associated with such a decline in biomass on the north Chatham Rise and

A. G. Hopper (ed.), Deep-Water Fisheries of the North Atlantic Oceanic Slope, 251–266.

the Challenger Plateau were contractions in the areas of high density, and the apparent fishing-out of some aggregations. However, no marked biological changes have been observed. The size structure of the populations has not altered, and the location and timing of spawning has remained constant between years.

It is likely that orange roughy populations take a long time to recover from heavy fishing, yet they can be reduced quickly to low levels by commercial operations. The New Zealand experience shows that careful and controlled development of the fishery is required from the outset. Research should occur in advance of substantial commercial fishing effort, so that baseline data on distribution, abundance and biology are collected. This would avoid a number of management problems later on.

1. Introduction

The last few years have seen increasing fishing activity for deepwater fish resources of the North Atlantic, with particular interest in orange roughy *(Hoplostethus atlanticus)*. While it is a relatively new commercial species in this region, orange roughy has been the target of commercial fishing around New Zealand since the late 1970s, where it has become an important and valuable fishery. This paper is an overview of the New Zealand experience with orange roughy, focusing on the salient features of the commercial fishery, its research, and management, and the lessons learned over the past 10 to 15 years which could be relevant for any developing deep water fisheries in the North Atlantic.

2. The Fishery

Orange roughy is a deepwater species, generally occurring on the continental slope at depths between 700 m and 1500 m. It occurs all around New Zealand at these depths, but forms localised aggregations (either for spawning or feeding) in a number of specific areas which are commercially fished (Figure 1). There are 7 major fishing grounds inside the Exclusive Economic Zone (EEZ), with additional fishing areas outside the EEZ on the Lord Howe Rise. The fisheries operate on spawning fish during winter (June-August) in the areas of the Challenger Plateau, Cook Canyon, Ritchie Banks, and north eastern Chatham Rise, as well as on non-spawning (generally spring/summer) aggregations off the Wairarapa and Kaikoura coasts, and regions of the Puysegur Bank and southern Chatham Rise.

Orange roughy have been known from New Zealand waters since the early 1960s, but there was no known commercial fishery until 1979 (Robertson and Grimes, 1983 and Robertson, 1991) when large concentrations of spawning orange roughy were

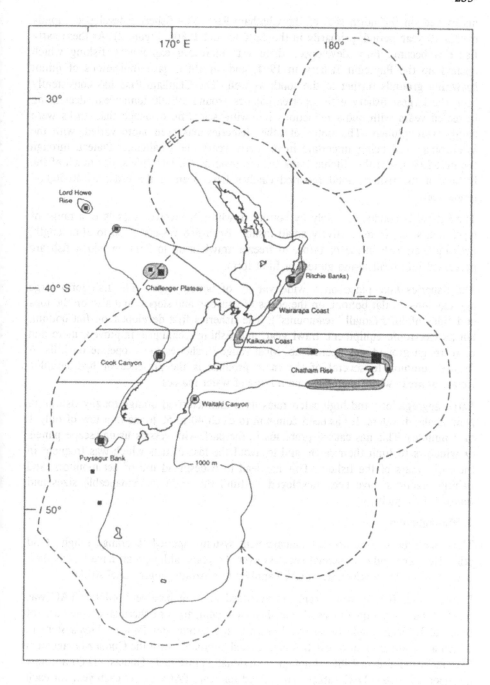

Figure 1. The New Zealand EEZ, showing location of major fishing grounds for orange roughy (stippled), main spawning locations (black squares), and Quota Management Area boundaries.

discovered on the north side of the Chatham Rise. The fisheries developed rapidly over a 10 year period, primarily in the early to mid 1980s (Figure 2). As these early fisheries became fully developed, there was increased exploratory fishing which opened up the Puysegur fishery in 1991, and, in 1993, gave indications of minor spawning grounds further to the south as well. The Chatham Rise has consistently been the largest fishery with reported catches around 30 000 tonne/year, decreasing in recent years with quota reductions following scientific evidence that stocks were being overexploited. The status of other fisheries has been more varied, with the Wairarapa coast being important in the early years, the Challenger Plateau through the mid-1980s, and the Ritchie Bank fishery towards the late 1980s. For much of the history of the fishery, total reported catches have been of the order of 40-50,000 tonne/year.

The fishery is conducted solely by bottom trawling. It involves vessels in a range of types and sizes, from relatively small inshore fresh-fish vessels of 20 to 30m length, packing their fish in ice or brine, to freezer trawlers up to 95m on which fish are processed into headed-and-gutted, or fillet form.

The fisheries take place on a wide variety of bottom types. The fish form large aggregations on flat bottom, on the edges of canyons and slopes, and also on the tops and sides of hills (small "seamounts"). The fisheries first developed on flat bottom, but as electronic equipment, trawl gear, and fishing techniques improved, more and more rough ground was opened up. Most roughy fisheries today operate on hills. A feature common to several of the main grounds is the presence of hydrographic fronts, or areas where there is some mixing of water masses.

Large aggregations and high catch rates have characterised orange roughy fishing in New Zealand waters. It has been common to catch 40 to 50 tonne in a tow of only 1 or 2 minutes. This has caused problems in the past, with vessels using escape panels or windows to limit their catch, and to avoid the loss of nets which was frequent in the early years of the fishery. This has lead to widespread use of net monitors, and fishing practices have been developed to limit the catch to manageable sizes and improve fish quality.

3. Management

There have been a number of management systems applied to orange roughy (and other New Zealand commercial species) over the years, although all have had as their basis a Total Allowable Catch (TAC) applied to separate geographical stocks.

Before 1982, fishing was largely unrestricted and competitive until the TAC was reached, but then a quota was allocated to each company or fisherman on the basis of their catch history under what was known as the 'Deepwater Policy'. This system in which a certain amount of catch was allocated continued with the Quota Management System introduced in 1986, although 'Individual Transferable Quotas' (ITQ) are now a proportion of the TAC, rather than a fixed tonnage. TACs are set each year for each orange roughy stock. The New Zealand EEZ is divided into 7 main management areas for this purpose (see Figure 1), although informal subdivisions occur in cases

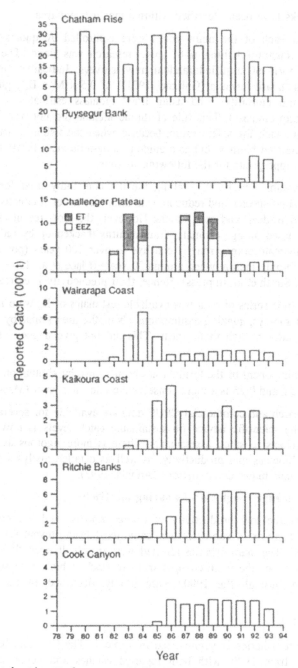

Figure 2. Reported catches (t) of orange roughy from the main areas by fishing year, 1979 to 1992. (fishing year = 1 October-30 September; 1978 = 1977/78 fishing year; Challenger Plateau, EEZ = inside EEZ, ET = outside EEZ; Note different scales between areas)

where separate stocks have been identified within a management area.

Initially catches for each of the main stocks were controlled by quotas set at an arbitrary level to constrain fishing until some research was done. The research generally took the form of stratified random trawl surveys. Indicative yields were estimated using a standard formula (Gulland, 1971) of $0.5 \times M \times B_0$, guessing M (instantaneous natural mortality), and using trawl biomass estimates as absolute tonnages (B_0 = virgin biomass). This rule of thumb method did not work too well. This did not matter much for a few years, because when the fishery started it was fishing down accumulated biomass but as a guide to long-term yield levels for orange roughy, it was an inappropriate for the following reasons:

1) M was assumed to be 0.1, taking the 0.2 commonly used for northern hemisphere shelf species and reducing it to allow for lower productivity which was expected in deep and cold waters. However, the most recent estimate of M is 0.045, based on age estimates from otoliths (backed up by radio-isotope work) that indicate orange roughy can live to over 100 years (possibly up to 120-130), and only mature when around 30 years (Mace et al, 1990; Fenton et al, 1991 and Smith et al, in press) Hence, their productivity is extremely low.

2) Before a time series of data was available, estimates of B_0 were made from single trawl surveys, making assumptions about the area swept by the trawl, and vulnerability of fish to the gear. This did not give a good estimate of biomass.

3) The 0.5 component of the formula has been shown by simulation modelling to be too high, and 0.25 is a more conservative and safer level (Mace, 1988).

Time series of research and commercial CPUE data are available for several years in some areas, and the scientific advice on sustainable catch levels is now based on much more comprehensive stock assessments. There is more rigorous use of much improved data on biomass and productivity, as well as detailed analyses of the risk associated with various management options (Annala, 1993).

4. Changes in the orange roughy stocks during the 1980s

The recognition of the low productivity of orange roughy, and improvements in survey design and analysis of survey results, came too late to prevent several stocks being overexploited. The main fisheries affected were the Challenger Plateau and the Chatham Rise. Some of the main changes in these stocks which were subjected to heavy exploitation through the 1980s are briefly described in the following paragraphs.

4.1 THE FISHERIES

The history of both fisheries is summarised in Table 1. The Chatham Rise fishery developed rapidly from 1979, with both reported catches and quota at levels of around 30 000 tonne to 1990, after which there were reductions in catch following decreases in the TAC.(Francis et al, 1993) The Challenger Plateau was, and is, a much smaller fishery. It developed from 1982, with quota increases to 1988. Catches

TABLE 1. Summary of orange roughy reported catches and quota levels (tonnes) for Chatham rise and Challenger Plateau

(Notes- in years prior to 1982 and 1984 for the respective grounds when area-specific TACs were introduced the catch was limited by total EEZ restrictions; the figures are based on an October-September fishing year; 1980 = 1979/80; TAC = quota available to the fishery)

	Chatham Rise		Challenger Plateau			
				Inside EEZ	Outside EEZ	
Year	TAC	Catch	TAC	Catch	Catch	Total
1979		11 800				
1980		31 100				
1981		28 200		1	32	33
1982	23 000	24 888		3 539	709	4 248
1983	23 000	15 434		4 535	7 304	11 839
1984	30 000	24 818	4 950	6 332	3 195	9 527
1985	30 000	29 340	4 950	5 043	74	5 117
1986	29 865	30 075	6 190	7 711	42	7 753
1987	33 000	30 689	10 000	10 555	937	11 492
1988	33 000	24 214	12 000	10 086	2 095	12 181
1989	33 300	32 785	8 200	6 791	3 450	10 241
1990	32 787	31 669	2 500	3 709	600	4 309
1991	23 787	21 521	1 900	1 340	17	1 357
1992	18 787	17 500	1 900	1 911	0	1 911

TABLE 2: Mean size (cm, standard length) of orange roughy measured during research surveys of Chatham Rise and Challenger Plateau spawning grounds

	Chatham Rise		Challenger Plateau	
Year	Male	Female	Male	Female
1984	34.10	35.33	31.47	32.78
1985	33.99	35.56	32.15	33.59
1986	34.01	35.30	32.05	33.39
1987	34.62	35.91	31.82	33.14
1988	33.89	35.28	31.33	33.00
1989	34.09	35.60	31.48	33.45
1990	33.65	35.06	31.44	33.62
1991				
1992	33.80	36.09		

258

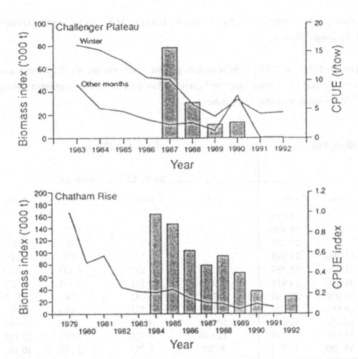

Figure 3. Abundance indices from trawl survey (Biomass, stippled bars) and catch-effort (CPUE, line) analyses for Challenger Plateau and Chatham Rise orange roughy. (Indices are 'mid-season' = July that year)

generally exceeded the TAC (more than on the Chatham Rise) because it was a straddle stock fishery, with some fishing outside the EEZ which until 1990 was not regulated (Clark, 1991 and Clark, 1991). Signs of over-exploitation appeared earlier than on the Chatham Rise, and after 1988 there were relatively substantial reductions in allocated quotas (Clark, 1992).

Data on the stocks were gathered during extensive research programmes in both areas involving regular trawl surveys beginning in 1983-84. Commercial catch and effort data have also been extensively analysed. These sources give detailed information on orange roughy abundance, distribution, and biology in the two areas, and an indication of the effects of fishing.

4.2 ABUNDANCE

Trawl survey and commercial catch per unit of effort indices showed marked changes over time (Figure 3). On both grounds there was a strong declining trend. These indices were used in stock reduction analyses, where the relative change in indices, the known commercial catch, and life history parameters were used in an age-structured population model to estimate virgin biomass, and assess the current state of the stocks (Clark, 1992 and Francis, 1992).

In Figure 4 general trends in the computer models of the Challenger and Chatham Rise stocks are shown (based on deterministic model results (Francis et al, 1993 and Clark, 1992). In both cases the stocks declined rapidly, and have been driven below their optimal size (B_{MSY} as estimated from computer simulations is about 30% of B_o), and close to or below that considered biologically safe (20% of B_o). Recent estimates of the size of the two stocks in 1992 suggested the Challenger was about 20% of virgin size, and the Chatham Rise about 15%.

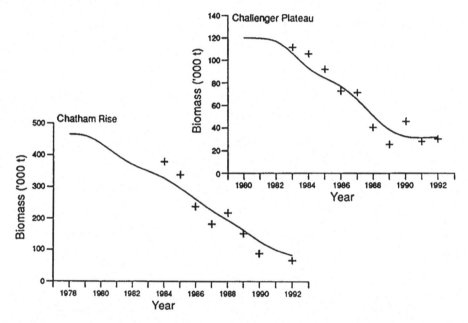

Figure 4. Computer model plots of population size history for Chatham Rise and Challenger Plateau orange roughy stocks (line), with scaled abundance indices from which virgin biomass was derived (crosses). (Year refers to July that year)

4.3 DISTRIBUTION

Associated with the decline in abundance have been marked contractions in distribution. On the Challenger Plateau (Figure 5), catch rates recorded during research trawl surveys over a wide area on flat bottom progressively shrank. Historically stable schools were fished out, and the fishery shifted more to hill fishing. In the period 1982-87 over 80% of the catch each year was taken from the flat area, but by 1989-90 this had dropped to 40-45%, with most of the catch coming from the hills (Clark and Tracey, 1994).

A similarly dramatic contraction in the area of high catch rates was evident on the Chatham Rise (Figure 6). There were large areas in which catch rates over 5 tonne/km were trawled in 1984, but by 1992 such catch rates occurred in one small

260

area.

The biomass, and distribution, of other species measured in the trawl surveys have generally remained steady or declined. The surveys were designed specifically for orange roughy, and are of limited value in assessing changes of other species, but on the Challenger Plateau there has been no indication of any replacement of the depleted stocks of orange roughy by any other fish species (Clark and Tracey, 1994).

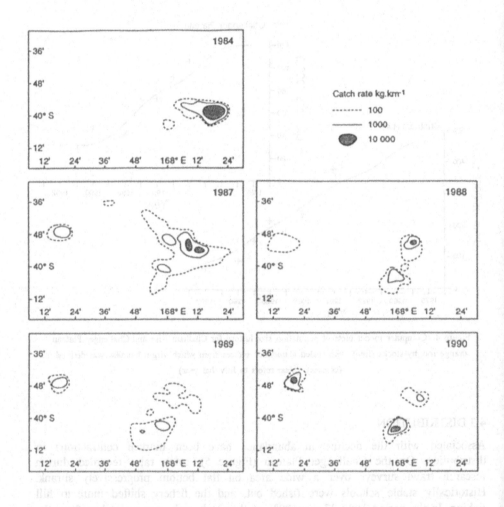

Figure 5: Contour plots of orange roughy catch rate from trawl surveys on the Challenger Plateau. (Trawl survey indices in the years shown (1984-90) were 143,000, 75,000, 29,000, 11,000, and 13,000)

4.4 BIOLOGICAL ASPECTS

4.4.1 Size structure

Length frequency distributions of all the exploited orange roughy stocks in NZ waters have remained relatively constant over time. On the Challenger Plateau, there has consistently been a strong unimodal distribution, with the peak at 32-33 cm standard length, while that of the Chatham Rise has its peak at 35-36 cm. The similarity between years can be seen by looking at the mean of the distributions (which have coefficients of variation of 1 to 2%) from the main spawning area on the Challenger Plateau and Chatham Rise (Table 2). There were no significant differences between years and no consistent trend in size over time.

4. 4. 2 Reproduction

The main areas of spawning appear to have remained the same. There is no evidence of the fish moving elsewhere to spawn, or forsaking some areas, in response to heavy fishing.

The timing of spawning has also been relatively consistent. Using an arbitrary date at which 20% of female fish were first recorded as spent during research surveys on the Challenger Plateau, this has ranged from 09 to 16 July, a period of only one week (Clark and Tracey 1994).

Fecundity did not change significantly over a period from 1987 to 1990 (Clark et al, 1994). Estimates of total fecundity each year were between 36,000 and 40,000 eggs per average-sized female. Fecundity for a number of species has been reported to increase as well as decrease with a decline in stock size, but available New Zealand orange roughy data do not show a trend.

In summary,the New Zealand studies show 3 major conclusions about changes to orange roughy stocks observed with heavy fishing:

1) Relatively rapid reduction in biomass

2) Progressive contraction in distribution

3) No apparent short-term changes in biological features (size structure, reproduction)

The first two features are quite clear. Nevertheless because there are no data on the size of the stocks prior to their exploitation, the level of any natural fluctuations in population size or distribution is unknown, and therefore, it is not certain how much was attributable directly to the fishery. However, the longevity of orange roughy suggests natural changes would occur relatively slowly, and so the rate and extent of decline suggests that fishing is most likely to have been the major factor in the observed depletion of the stocks.

The lack of change in size structure of the populations, given the extent of depletion, is of interest and of some concern for management. Generally with fished populations, there is a truncation of the length frequency distribution, and a reduction

262

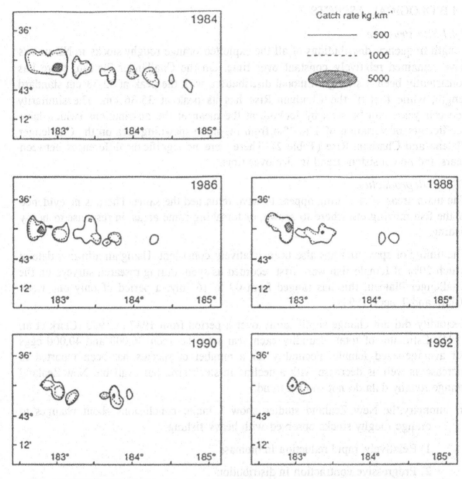

Figure 6: Contour plots of orange roughy catch rates from trawl surveys on the Chatham Rise. (Trawl survey indices in the years shown (1984 to 1992) were 164,000, 103,000, 95,000, 38,000, and 29,000)

in the mean size of fish, as larger and older fish are removed, and new recruits enter the population. However, few studies have described changes in a species as long-lived and slow growing as orange roughy. The length frequency distributions of spawning orange roughy consist largely of fish between 25 and 45 cm, which are probably all fully vulnerable to the trawl gear. Hence there could be relatively constant fishing mortality across all adult size groups, and consequently no major change in the proportion of larger fish. However, smaller fish should be recruiting to the population. Recruitment should be evident through a reduction in mean size, or a strengthening of the left hand limb of the length distribution. There has been no sign of strong recruitment, or even average levels expected from virgin recruitment (Francis et al, 1992); (fish recruit into the adult population at about 30 years, so current recruitment is still that of the virgin population). Indications are that there has been very low recruitment over the 10-15 year period of exploitation.

The stability of the size structure of the population, as well as the slow growth and high longevity of orange roughy, means that marked short-term changes in reproductive aspects such as size-at-maturity are unlikely, although in theory fecundity could change with food availability. The time period of 10-15 years is relatively short in the 100 year life of an orange roughy to see any functional response in biological features.

5. Lessons for fishery management

On the basis of the information presented in this paper, it is now possible to consider some of the implications of the New Zealand experience for effective management of orange roughy populations.

Orange roughy are slow growing and long-lived, with low productivity. This makes them highly susceptible to the effects of exploitation. Long-term sustainable yields are thought to be between 1 and 2% of virgin biomass, at a biomass level of 30% of virgin (Francis et al, 1993). Estimates of long-term yield will therefore appear extremely low against the high catch levels likely when a fishery begins. The schooling behaviour of orange roughy means that large catches can be taken in a short time, and, for several years, high catch rates may be maintained despite declining biomass. It could be very easy to 'overshoot' the 30% mark in only a few years.

The recovery of orange roughy from heavy fishing may be slow. Their fecundity is low, at between 30,000 and 50,000 eggs per female (Clark et al, 1994; and Pankhurst and Conroy, 1987) . There may also be long periods between strong year classes, as hypothesised for long-lived species in the northeast Pacific (Leaman and Beamish, 1984), and to an extent indicated by the orange roughy mean length data from the New Zealand fishery.

These are important considerations in management of orange roughy fisheries. Data over a comparatively long time period are required to provide a sound basis for management of a long-lived species such as orange roughy, and development of a fishery needs careful control from the outset. It is essential that research occurs in advance of major fishing, so that baseline data on distribution, abundance, and biology are collected to enable changes caused by fishing to be quickly detected.

Stock assessment of orange roughy, like most species, is not an easy task, but some of the species' particular characteristics make it more so. High longevity (over 100 years) means natural fluctuations can occur over long time periods, so it can be hard to isolate the impact of the fishery. Their depth distribution at 750-1500m makes some biological work very difficult (e.g., tagging for abundance/stock structure, acoustic target strength). Aggregation behaviour, with dense spawning or feeding schools causing trawl gear saturation, can limit the usefulness of CPUE, and constrain the use of trawl survey results as absolute estimates. Finally their distribution in a range of habitat types can require different methods or survey design to sample them in different areas.

The most commonly used techniques in New Zealand and Australia (trawl and acoustic surveys) provide relative abundance indices, and therefore require surveys over several years before true biomass can be estimated from stock reduction methods. Egg production estimates of spawning stock biomass shows promise in some areas, but until recently there has been a very incomplete knowledge of the early life history stages and egg development(Zeldis et al, in press). CPUE has not proven generally useful as an abundance measure. Although on the Challenger Plateau there was a relatively consistent reduction in CPUE (Clark, 1992), on the Chatham Rise and other grounds the decrease in mean catch rate was less consistent, but there has been a marked and progressive shortening of the period over which large catches were taken in winter (Doonan, 1991 and Field et al, 1993).

The desire to obtain information rapidly may lead to considering 'adaptive management', whereby a stock is fished hard enough to provide measurable changes in abundance indices over a short time. This was, in fact, one of the reasons the Challenger Plateau quota was raised from 6,000 tonne to 12,000 tonne in the mid-1980s (Robertson, 1986). This experience demonstrates some of the issues that should be considered. Good data were not available at the time on the abundance of the stock against which to measure any change. A comparative trawl survey series had not been developed, and commercial CPUE was thought unreliable. Thus there was little information on what sort of changes to expect, or how to measure them effectively. The trawl survey indices declined markedly between 1987 and 1988, but without previous comparable data it was not clear what the change meant. It declined again the following year, by which time the stock was at low levels. It would have been desirable to have had at least two-three years of data before fishing pressure increased. A second feature of adaptive management is the recognition that it can involve high risk to the stock, because changes in biomass can occur rapidly. There must be the ability in the fishery for it to be flexible and enable catch levels to be reduced rapidly. A further difficulty with orange roughy is that potential effects of any change in spawning stock size on recruitment won't be evident for 25 to 30 years when the recruits of that years spawning enter the commercial population (research surveys of pre-recruits have not proven very useful because of their long-term nature). Hence, until that time, the fishery will be removing only accumulated adult stock, at low levels of virgin stock recruitment.

With some of these thoughts in mind, if the fishery is carefully regulated in the first few years while the necessary data are collected, later management problems such as over-capitalisation, too many vessels and too much catching and processing capacity in the fishery, and the need to quickly and substantially lower catch levels, can be avoided. The New Zealand experience can hopefully be of benefit to others as North Atlantic deepwater fisheries develop.

References

Annala, J.H. (comp) (1993) Report from the Fishery Assessment Plenary, May 1993: stock assessments and yield estimates. 241 p. (available from MAF Fisheries, P O Box 297, Wellington, New Zealand).

Clark, M.R. (1991) Orange roughy fisheries in the mid-Tasman Sea. In. Abel, K.; Williams, M.; & Smith, P. (eds), Australian and New Zealand southern trawl fisheries conference. Australian Bureau of Rural Resources proceedings 10: 223-229.

Clark, M.R. (1991) Commercial catch statistics for the orange roughy fishery on the Challenger Plateau, 1980/90. New Zealand Fisheries technical report No. 27. 11 p.

Clark, M.R. (1992) Assessment of the Challenger Plateau (ORH 7A) orange roughy fishery for the 1992/93 fishing year. NZ Fisheries assessment research document 92/6. 13 p. (available from MAF Fisheries, P O Box 297, Wellington).

Clark, M.R.; & Tracey, D.M. (1994) Changes in a population of orange roughy *(Hoplostethus atlanticus)* with commercial exploitation on the Challenger Plateau, New Zealand. Fishery bulletin, U.S. 92(2): 236-253.

Clark, M.R., Fincham, D.J., and Tracey, D.M. (1994) Fecundity of orange roughy *(Hoplostethus atlanticus)* in New Zealand waters. New Zealand journal of marine and freshwater research 28: 193-200.

Doonan, I.J. (1991) Orange roughy fishery assessment, CPUE analysis-linear regression, NE Chatham Rise 1991. NZ Fisheries assessment research document 91/9. 48 p. (available from MAF Fisheries, PO Box 297, Wellington).

Fenton, G.E., Short, S.A., and Ritz, D.A. (1991) Age determination of orange roughy, *Hoplostethus atlanticus* (Pisces: Trachichthyidae), using 210Pb:226Ra disequilibria. Marine biology 109: 197-202.

Field, K.D., Francis, R.I.C.C., and Annala, J.H. (1993) Assessment of the Cape Runaway to Banks Peninsula (ORH 2A, 2B, and 3A) orange roughy fishery for the 1993/94 fishing year. NZ Fisheries assessment research document 93/8. 23 p. (available from MAF Fisheries, PO Box 297, Wellington).

Francis, R.I.C.C. (1992) Use of risk analysis to assess fishery management strategies: a case study using orange roughy *(Hoplostethus atlanticus)* on the Chatham Rise, New Zealand. Canadian journal of fisheries and aquatic sciences 49: 922-930.

Francis, R.I.C.C., Robertson, D.A., Clark, M.R., and Coburn, R.P. (1992) Assessment of the ORH 3B orange roughy fishery for the 1992/93 fishing year. NZ Fisheries assessment research document 92/4. 45 p. (available from MAF Fisheries, PO Box 297, Wellington).

Francis, R.I.C.C.; Robertson, D.A.; Clark, M.R.; Doonan, I.J.; Coburn, R.P.; & Zeldis, J.R. (1993) Assessment of the ORH 3B orange roughy fishery for the 1993/94 fishing year. NZ Fisheries assessment research document 93/7. 43 p. (available from MAF Fisheries, P O Box 297, Wellington).

Gulland, J.A. (comp,ed) (1971) The fish resources of the ocean, Fishing News (Books) Ltd, Surrey, England.

Leaman, B.M. and Beamish, R.J. (1984) Ecological and management implications of longevity in some Northeast Pacific groundfishes. International North Pacific Fisheries Commission bulletin 42: 85-97.

Mace, P.M. (1988) The relevance of MSY and other biological reference points to stock assessment in New Zealand. NZ Fisheries assessment research document 88/30. 41 p. (available from MAF Fisheries, PO Box 297, Wellington).

Mace, P.M., Fenaughty, J.M., Coburn, R.P., and Doonan, I.J. (1990) Growth and productivity of orange roughy (Hoplostethus atlanticus) on the north Chatham Rise. New Zealand journal of marine and freshwater research 24: 105-119.

Pankhurst, N.W. and Conroy, A.M. (1987) Size-fecundity relationships in the orange roughy, Hoplostethus lanticus. New Zealand journal of marine and freshwater research 21: 295-300.

Robertson, D.A., and Grimes, P.J. (1983) The New Zealand orange roughy fishery. In Taylor, J.L., and Baird, G.G (comps & eds). New Zealand finfish fisheries: the resources and their management. Trade Publications Ltd, Auckland, pp. 15-20.

Robertson, D.A. (1986) Orange roughy. In Baird, G.G. and McKoy, J.L. (comps & eds). Background papers for the Total Allowable Catch recommendations for the 1986/87 New Zealand fishing year, pp. 88-108. (Available from MAF Fisheries, PO Box 297, Wellington).

Robertson, D.A. (1991) The New Zealand orange roughy fishery: an overview. In. Abel, K.; Williams, M.; & Smith, P. (eds), Australian and New Zealand southern trawl fisheries conference. Australian Bureau of Rural Resources proceedings 10: 38-48.

Smith, D.C., Fenton, G.E., Robertson, S.G., and Short, S.A. (In press) Age determination and growth of orange roughy (Hoplostethus atlanticus): a comparison of increment counts with radiometric ageing. Canadian journal of fisheries and aquatic sciences

Zeldis, J.R.; Grimes, P.J.; and Ingerson, J.K.V. (In press) Ascent rates, vertical distribution, and a thermal history model of development of orange roughy (Hoplostethus atlanticus) eggs in the water column. Fishery bulletin, U.S..

AGE DETERMINATION OF DEEP-WATER FISHES: EXPERIENCES, STATUS AND CHALLENGES FOR THE FUTURE

O .A. BERGSTAD
Ministry of Fisheries,
Institute of Marine Research,
Flødevigen Marine Research ,
N-4817 His, Norway

ABSTRACT

A review is presented of the literature on aging of deep water fishes and of experiences from studies of Coryphaenoides rupestris, Argentina silus, Molva molva, Molva dipterygia, Brosme brosme, and Glyptocephalus cynoglossus of the Skagerrak.

It is suggested that the otoliths provide the most reliable age readings. The otoliths of many deep water species show growth zones which are similar to those interpreted as annulli in shallow water species. Indeed, in many deep water species the otolith zones are exceptionally distinct, particularly those deposited after the juvenile period. In some species, the number of growth zones and hence the age estimates become very high (100 - 150 yrs). Recent attempts to validate such readings for deep water fishes are reviewed.

In view of the significance of information on age for studies of growth, population structure, mortality, production, and hence the assessment of sustainable yield, the importance of age studies and in particular validation of the aging techniques is stressed.

Validation of age readings, intercalibration of readings and improvement of preparatory techniques for more species from different environments are challenges for the future.

A. G. Hopper (ed.), Deep-Water Fisheries of the North Atlantic Oceanic Slope, 267–283.
© *1995 Kluwer Academic Publishers.*

1. Introduction

The focus of deep water fish research has been gradually changing from exploration and description of species and distributions to analyses of life histories, population dynamics and ecology. If not always essential, information on age is very valuable for analyses of population processes such as growth, recruitment, mortality and maturation and hence for studies of life history strategies and ecology. Future management of deep water species may require analytical assessments which depend on age-structured data. Therefore, the development of accurate and practical aging techniques should have high priority in deep water fish studies. A number of reviews and symposia on fish aging illustrate the high level of activity and the priority given to this field of research (e.g. Bagenal 1974, Prince and Palos 1983, Prouzet and Souplet 1985, Summerfelt and Hall 1987, Casselman 1989, Smith 1992, Otolith Research Symposium, Carolina, USA 1993). Still, however, relatively few deep water species have been aged.

The final goal of all aging work should be methods which are accurate and precise and also practical, i.e. can be used to age large fish samples. Beamish and Mcfarlane (1983) in their extensive literature survey entitled "The forgotten requirement for validation" emphasised the need for validation of the age assessments, an essentially obvious point also stressed in more recent papers (e.g. Casselman 1989, Hancock 1992). However difficult and time-consuming, the gain in confidence and accuracy resulting from validation exercises is substantial.

The author holds the view that future efforts along three lines are important, particularly when considering deep water species:

1) Validation of aging techniques for more species and different environments.

2) International intercalibration exercises to improve precision and consistency of age assessments.

3) Improvement of processing techniques, particularly for otoliths.

This paper summarises recent developments and results of age determination studies on deep water fishes, and also experiences from the author's own studies of the Skagerrak deep water species (*Coryphaenoides rupestris, Argentina silus, Molva molva, Molva dipterygia, Brosme brosme,* and *Glyptocephalus cynoglossus*). It should not be considered an exhaustive literature review, the aim has rather been to emphasise main advances and suggest directions for future work.

2. Aging Techniques

The techniques which have been used for aging deep water fishes, essentially teleosts only, are basically the same as those applied for shallow water species (e.g. Casselman 1989 and references therein). This paper primarily focuses on the methods which are most common, and the author believes are the most promising for deep water fishes. These all involve the counting of growth zones which are assumed to be formed annually in hard parts such as otoliths or scales. Often different techniques have been applied for the same species or genus. The debate over whether to use scales or otoliths for aging *Sebastes* species (e.g. Nedreaas 1990) is an example of a particularly long-lasting controversy over what method should be adopted.

2.1 AGING BY OTOLITHS

The most common aging method is no doubt the counting of growth zones in otoliths. In cases where different hard parts from the same fish have been compared, otoliths have proven to be the preferred structures (e.g. Six and Horton 1977, Beamish and McFarlane 1987, Withell and Wankowski 1988). Otoliths appear in the inner ear during the embryonic phase of the fish and under normal circumstances otolith material is not resorbed, hence it is assumed to contain information on the entire growth history of the individual fish.

Otoliths have been used to age deep water species using a variety of techniques. In most fishes the sagitta, i.e. the largest otolith, can easily be extracted from the inner ear. There are several good descriptions of the techniques used for extracting, storing and preparing otoliths for examination (e.g. Chilton and Beamish 1982, Casselman 1989). The methods vary with the anatomy of the fish and the morphology of the otolith. The final aim is to expose annulli from the entire life span of the fish.

Some otoliths are thin and transparent and without further preparation growth zones are readily counted when the otolith is viewed under a dissecting microscope using either transmitted or reflected light. The slow-growth zones (sensu Pannella 1974), often referred to as the annulli, will appear transparent and fast-growth zones opaque. More robust otoliths are however common, and in these breaking or sectioning is required to expose the innermost growth zones.

The breaking technique is quick and used routinely for a number of common commercial species such as cod (*Gadus morhua*) (Rollefsen 1936; Williams and Bedford 1974) and other gadoids. The otolith is essentially broken into two halves, ideally through the nucleus, and the annulli are read from the exposed surfaces by a combination of sidewise illumination and shading. The so-called "break and burn" technique (Chugunova 1959; Christensen 1964) is used by many laboratories (e.g. Chilton and Beamish 1982, Hancock 1992). Charring the exposed surface in or near an alcohol flame accentuates the annulli and contrast is further enhanced by wetting the surface with oil. There are several modifications of this technique.

Sectioning, normally by a double-blade diamond saw, may be an alternative technique to breaking. The method tends to be time-consuming but modifications of the equipment for cutting large samples have reduced the processing time to the extent that sectioning is now used routinely in many laboratories. A number of sectioning techniques have been published (Rauck 1976, Chilton and Beamish 1982, Bedford 1983, McCurdy 1985, Augustine and Kenchington, 1987).

Some of the different techniques can be illustrated by examples from the author's own work on various species from the deep water basin of the Skagerrak. All these species are commercially harvested in that area and in other deep shelf or slope waters of the north Atlantic.

2.1.1 *Whole otoliths* (*Argentina silus* and *Glyptocephalus cynoglossus*)
The greater silver smelt, *Argentina silus* (Argentinidae) and the witch flounder
Glyptocephalus cynoglossus (Pleuronectidae) have otoliths which are morphologically
different but rather thin and transparent, hence aging is possible using whole otoliths
(Molander 1925, Thorsen 1979, Magnusson 1990, Bergstad 1993). For the smelt the
following procedure has been adopted in Norway. After extraction the otoliths are
stored in 70 % alcohol. Before viewing all tissues are cleaned off and the otoliths are
placed anti-sulcal (proximal) side up on black grooved plastic plates. When
submerged in a clear liquid (water or glycerol) or mounted in a clear mounting
medium (e.g. Eukitt) and illuminated by white reflected light the opaque fast-growth
zones appear as white relatively broad bands, whereas the narrower slow-growth
zones (annulli) appear dark. Most *Argentina silus* otoliths have clear patterns (Fig. 1)
and interpretation problems are minor compared with many other species. Counting is
easiest along an axis from the nucleus to the rostrum. The first 15 - 20 zones are
normally easy to distinguish but resolving the very narrow zones near the rostrum
becomes increasingly difficult with increasing age. In some individuals 30 - 40 zones
have been counted, but normally all fish 20 yrs old and older are pooled in a 20+
group.

The witch flounder (*Glyptocephalus cynoglossus)* otoliths are processed and viewed
in the same way as the smelt otoliths. This method was first adopted by Molander
(1925). Prolonged storage in alcohol may cause fading and the otoliths may
alternatively be stored dry. Dried otoliths may however need to be cleared or ground
on the convex side before viewing (Bowering 1976). The growth pattern of the witch
otolith is quite different from that of the smelt but similar to that of other
pleuronectid otoliths. Interpretation problems are greater for the witch than for
Argentina silus since often the growth zones are more diffuse.

2.1.2 *Thin sections. Coryphaenoides rupestris* and *Molva dipterygia.*
The roundnose grenadier Coryphaenoides ruspestris is the only Macrourid in the
North Sea and Skagerrak. It is only abundant in the southern and eastern parts of the
Norwegian Deep or Channel at depths greater than 280 m (Bergstad 1990). The
species has been aged by scales (Savvatimskiy 1971, Kosswig 1986, Eliassen 1986,
Danke 1987) and otoliths (Koch 1976, Bridger 1978, Gordon 1978, 1979, Bergstad
1990) Wilson (1982) compared ages estimated from scales with those from vertebrae
of a related species, *Coryphaenoides acrolepsis.*

In studies of the Skagerrak population otoliths are used (Fig. 2). Transverse sections,
0.1 - 0.3 mm thick, are produced by a double-blade diamond saw. The sections are
cleaned, dried and mounted on glass slides in a clear synthetic mounting medium (see
Bergstad 1990 for more details). A compound microscope is used for viewing and 40
or 100 X magnification is suitable. The first 10 - 12 growth zones appear similar to
those considered to represent slow-growth and fast-growth zones in gadoid otoliths
(e.g. Williams and Bedford 1974). After a certain size the otoliths appear to grow in
thickness, particularly on the proximal surface (sulcus side) rather than in length or
width along the antero-posterior or dorso-ventral axes, hence thin marginal growth

zones deposited on the proximal surface were included in the counts (Fig. 2b). A similar growth pattern has been described for a number of long-lived species (e.g. Beamish 1979, Kenchington and Augustine 1987, Fujiwara *et al.* 1988). The oldest grenadiers in the Skagerrak may be around 70 years old, and the trawl catches are dominated by 20 - 30 year old fish.

Much like the grenadier, otoliths from blue ling (*Molva dipterygia*) are rather opaque and robust and must also be sectioned to reveal growth zones. Several attempts were made to age blue ling (Magnusson 1982, Engås 1983, Ehrich and Reinsch 1985, Thomas 1987). In the Skagerrak and North Sea the method and interpretation of Engås (1983) was adopted. As for the grenadier, the otolith starts to grow in thickness after a certain size. Reducing the thickness by grinding the proximal and distal surfaces as suggested by Thomas (1987) will remove marginal growth zones in otoliths from old fish and probably lead to underestimated ages. As for the grenadier, rather thin outer growth zones on the proximal side are accepted as annulli, and these are often easy to resolve compared with the innermost broad growth zones. Between 70 and 80 % of the otoliths examined yield acceptable readings.

2.1.3 *Combination of methods Molva molva* and *Brosme brosme.*

In some species otoliths from specimens below a certain size can be viewed intact while sectioning is required for big fish (e.g. Augustine and Kenchington, 1987). This is the case for ling (Molva molva) and tusk (Brosme brosme), and the body length threshold appears to be around 90 and 70 cm for the two species. When examining whole otoliths, the ling otolith must be viewed submerged in glycerol *sulcus-side down*, either by reflected light against a black background or by bright transmitted light (see Molander 1956). The tusk otolith is best viewed *sulcus-side up* by reflected light. The otoliths from big specimens must be sectioned and processed as described for the grenadier and the blue ling.

With both species but especially the tusk, interpretation problems are serious and many big fish cannot be aged at all. In a tusk sample from the Skagerrak 20 - 30 % of the fish in the length range 40 - 70 cm were considered unreadable and only very few big fish can be aged with confidence. Indeed, age readings of all the three Lotinae will probably remain comparatively imprecise.

2.2 AGING BY SCALES

Many teleosts have large body scales and in salmonids and clupeids scales are used routinely for age determination. Checks, breaks or changes in spacing of the circuli may represent annual patterns. Scales were also used for some deep water fish, i.e. *Sebastes* sp. and Macrourids (Kosswig 1986, Savvatimskyi *et al.* 1977), but then zones revealed by transmitted polarised light were counted. Kosswig (1973), who refined the technique, showed that even in scales which do not show the conventional periodic compression of circuli normally accepted as annual growth marks, very distinct zonal patterns appeared when scales "impregnated" by silver nitrate were viewed by transmitted polarised light. The same patterns are also visible without the silver nitrate impregnation. Russian scientists have used similar techniques for scale reading and traditionally favoured scales for otoliths for many fish species. However, for redfish the Russians now use otoliths (Nedreaas, pers. commn, 1994).

At the start of the studies of *Coryphaenoides rupestris* in the Skagerrak scales prepared according to Kosswig (1973) were used for aging in addition to otoliths. A detailed discussion of comparisons of scale and otolith readings was given in an earlier paper (Bergstad 1990). Scale age appeared higher than otolith age at low ages and substantially lower at high ages, thus it did not seem possible to calculate one from the other. Wilson (1982) also questioned the validity of the polarised light method. Otoliths are now considered to produce the most reliable age estimates.

There are a number of reasons for not using scales for aging, particularly for long-lived species (e.g. Six and Horton 1977, Beamish and Chilton 1982, Casselman 1989). Scales probably stop growing when the somatic growth rate is reduced with age, hence there is a high risk of underestimating the age of fish above a certain size/age. The variation between replicated readings is high in many cases (Casselman 1989). There is also shedding and regeneration of scales. In many fishes scales are simply too small or absent or show no clear patterns at all.

3. The need for validation

So far this paper has focused on aging techniques, i.e. methods for revealing growth zones representing the entire growth history of the individual fish. The underlying assumption of all these methods is that the growth zones are indeed structures formed annually and thus that counting growth zones yield accurate estimates of age. Validation of aging methods is both difficult and time-consuming, nonetheless very important (Beamish and Mcfarlane 1983, Casselman 1989, Hancock 1992). Casselman (1989) reviewed the methods used for validating, but several of these are seldom if ever applicable for deep water fishes. None of the age readings for the Skagerrak species were strictly validated. (The witch may be the exception, see Powles and Kennedy 1967). The growth zones counted appear similar to those seen in species for which validation exercises have been made, but observing similarity constitutes no validation in the strictest sence.

Only for very few deep water species were validation of age readings attempted (Table 1). Various validation methods such as length frequency analyses and seasonal studies of the character of the edge of the otolith were applied. Length frequency analyses (e.g. Powles and Kennedy 1967, Lear and Pitt 1975, Mace *et al.* 1990) are only useful for young fish or short-lived species showing rapid growth and/or for species with exceptionally strong or weak yearclasses. Seasonal studies of the otolith edge to determine at what time of the year the annullus is deposited (e.g. Powles and Kennedy 1967, Lear and Pitt 1975) has proven valuable but are not useful for validating ages over the entire age range of long-lived species with narrow growth zones. Wilson (1988) used primary growth increments in otoliths to validate age readings for two abyssal Macrourids, and results were promising although dependent on the assumption that primary increments were indeed formed daily in these species.

The methods used in the most recent studies, particularly for long-lived species, were radiometric dating and mark-recapture experiments (Table 1). Bennett *et al.* (1982) were the first to show by radiometric dating that fishes could probably live to considerably greater ages than indicated by conventional counting of annulli on the surface of otoliths. The species studied was *Sebastes diploproa* for which a radiochemical assay of whole otoliths using Pb-210:Ra-226 ratios was conducted. The basis of this method is the incorporation of the calcium analog Ra-226 into the growing otolith. Ra-226 decays radioactively to Pb-210, and the ratio of the two isotopes becomes an index of the time elapsed since the Ra-226 was incorporated. The Pb-210 half-life is 22.5 years which makes it particularly suitable for age determination of long-lived fish species. For short-lived species Th-228:Ra-228 ratios have been used by Smith *et al.* (1991). Good descriptions of the methods, theory and assumptions of radiometric aging was provided by Smith *et al.* (1991) and Fenton and Short (1992).

Campana *et al.* (1990) refined the technique and analysed the cores rather than the entire otoliths of *Sebastes mentella.* By using cores the need to apply assumed models for the accretion of of Ra-22 through models of otolith mass growth rate were avoided, a problem recognised by Bennett *et al.* (1982). Radiometric ages corresponded well with age estimates from counts of annulli in broken and burnt otoliths according to the interpretation by Beamish (1979). The study strongly suggested that thin growth zones at the sulcus side of the otolith should be accepted as annulli, and indicated that aging by scales and conventional otolith aging would underestimate ages of old fish. Previous suggestions of high ages, i.e. 70 -75 years, of *Sebastes* appeared to be well founded.

Radiometric aging was later used for a few other species (Table 1), most notably the orange roughy *Hoplostethus atlanticus* from Australian waters (Fenton *et al.* 1991). Although based on whole otoliths rather than cores, the results indicated that the growth of the species was very slow as suggested by Mace *et al.* (1990), and that the oldest fish was as much as 149 yrs old. Short *et al.* (1993) showed for orange roughy that counting growth zones from longitudinal sections of otoliths probably yield age estimates which correspond better than the rather low surface counts with age estimates from radiometry. In the same paper equivalent results were presented for another long-lived deep water species, *Alocyttus verrucosus,* for which an age range of 7 - 95 years was found. Fenton *et al.* (1990) attempted to validate section ages of the hoki (*Macrouronus novaezelandiae*) by radiometry, but in this case the assumption of a constant rate of incorporation of Ra-228 throughout the life of the fish was violated. This was probably related to the ontogenetic shift in habitat and environment of this species, an observation which indicates that radiometric aging is most suitable for species spending their entire life cycle in the same environment. West and Gauldie,(1994) drew attention to this and other potential problems with the past and present use of radiometric aging.

A very different approach, i.e. oxytetracycline marking of the otoliths, was used for validating broken and burnt otolith ages of *Anoplopoma fimbria* (Beamish and

Chilton 1982, McFarlane and Beamish 1993). The result of this long-lasting and otherwise extraordinary marking experiment is still to be published. In the course of 13 years 1255 marked fish were recovered. Of these, 122 were at liberty for more than 5 years. Analysis of annulus formation outside the oxytetracycline mark, which was deposited immediately after marking, supports the interpretation that narrow zones in broken and burnt sections form annually. Sablefish can apparently become around 70 years old. In a recent paper Kastelle et al.(1994) used radiometric aging for sablefish and the results seem to support the continued use of broken and burnt sections.

4. Status and challenges

Many, but not all deep water species have otoliths which show growth patterns similar to those used for aging shallow water species (e.g. Morales-Nin 1990, Gauldie *et al.* 1991). Although the scales of many species show zonal patterns which can be interpreted as annual growth zones or checks, there are probably few deep water species which can be aged reliably by scales. When comparisons were made, otoliths always became the preferred structure.

For several deep water species good techniques have been developed which reveal growth zones in otoliths. In some species, here exemplified by *Molva molva*, *Brosme brosme* and *Glyptocephalus cynoglossus*, sectioning or breaking may not be necessary, at least not for small fish. For others, sectioning or breaking is unavoidable, e.g. *Coryphaenoides rupestris* and *Molva dipterygia*. The techniques for sectioning have now been refined to the extent that large samples can be processed at reasonable speeds.

A detailed discussion of otolith growth is beyond the scope of this paper. A number of very interesting microstructural studies of otoliths were made recently, at least partly with the aim to obtain a growth theory for otoliths which would help determine which of the many patterns seen in otoliths, i.e. alternate hyaline and opaque zones, checks and assumed daily increments, are true time-marks (e.g. Gauldie and Nelson 1990). Such a theory would make more traditional validation excercises unnecessary. Unfortunately this approach has thus far not been very fruitful (Paul 1992). Studies of the physiological mechanisms underlying the deposition of daily increments (Gauldie and Nelson 1990, Mugiya 1987) have provided arguments for the interpretation that classical hyaline and opaque zones reflect cycles of widely spaced and compressed daily increments related to seasonal cycles in somatic growth rate. At least in species which show seasonal growth cycles, hyaline and opaque zones appear to constitute time marks. The problem is that daily increments in most species can only be used for young stages, and are thus not very useful for aging or validation of older age-groups of long-lived species.

Some of the recently published radiometric aging experiments, particularly those using otolith core samples, and also the oxytetracycline study by McFarlane and Beamish (1993), seem to offer strong support for continued use of hyaline and opaque zones in otolith sections or broken and burnt otolith surfaces for aging. In species which have otoliths that after a certain size continue to grow in thickness (i.e. most of the growth is seen on the proximal surface), narrow marginal zones should also be counted.

There is however still good reason to stress the need for more basic otolith studies and validation of age reading techniques. Only for a few species were acceptable validation experiments conducted and then mostly for species from upper or middle slope waters. In the North Atlantic some 340 deep water fish species were recorded (Haedrich and Merrett 1988) and few of these were ever aged, only the *Sebastes* according to a validated procedure. Some 40 elasmobranchs occur in the area and ageing techniques, perhaps based on counting annual zones in vertebra, will have to be developed and validated. Clearly developing and validating ageing methods for fishes from a greater taxonomical and environmental range is a great challenge for the future.

Even when good techniques exist and ages have been validated, there is need for intercalibration of readings between readers and laboratories. This is the case because precision may well remain low while the accuracy is acceptable. In other words, while accuracy depends on validation exercises, consistency between readers can only be achieved by mutual agreement on interpretation, and practical experience. Exchanging material between workers is recommended since comparison of independent readings forms the only basis for estimating precision. Within international fora such as ICES (The International Council for the Exploration of the Sea) calibration exercises and workshops are held frequently on a number of species and prove very useful for achieving consistency among readers, countries and time periods. Within a project on long-line resources of the Northeastern Atlantic funded by the Nordic Council of Ministers, intercalibration exercises involving Icelandic, Faroese, French and Norwegian workers are underway, and this may result in recommendations for processing methods and interpretations for both ling, tusk, and blue ling.

The author has primarily focused on aging techniques involving counting of annual growth zones in otoliths or scales. Particularly for short-lived species, counting daily increments may be feasible. Some benthopelagic and mesopelagic species show very clear primary increments which are probably deposited daily (e.g. Wilson 1988, Gjøsæter 1987, Morales-Nin 1990, Linkowski *et al.* 1993), and the existence of daily increments is claimed to be almost universal (Pannella 1974). Otoliths of several demersal deep water fishes also show primary increments which may be daily (Gauldie *et al.* 1991 and references therein). Other aging methods involving cohorts rather than individuals, i.e. length frequency analyses or modal progression analyses, may prove useful for species with short reproductive seasons and fast growth.

Table 1. Deep water fish species for which age readings by otoliths have been validated.

Species	Habitat	Technique	Validation method	References
Glyptocephalus cynoglossus	Outer shelf	Whole otoliths	Edge character, length freq.anal.	Powles and Kennedy 1967
Reinhardtius hippoglossoides	Shelf & slope	Whole otoliths	Edge character, length freq.anal.	Lear and Pitt 1975
Coryphaenoides armatus variabilis	Abyssal	Transv. sections	Daily growth increments	Wilson 1988
Coryphaenoides yaquinae	Abyssal	Transv. sections	Daily growth increments	Wilson 1988
Anoplopoma fimbria	Deep shelf	Transv. sections	Tag-recapture,oxytetracycline	Mcfarlane & Beamish 1993
Anoplopoma fimbria	Deep shelf	Transv. sections	Radiometry	Kastelle et al. 1994
Sebastes diploproa	Outer shelf	Transv. sections	Radiometry	Bennett et al. 1982
Sebastes mentella	Outer shelf	Broken and burnt	Radiometry	Campana et al. 1989
Sebastes rufus	Outer shelf	Transv. sections	Radiometry	Watters et al. 1993
Sebastolobus altivelis	Slope	Transv. sections	Radiometry	Kline et al. 1993
Hoplostethus atlanticus	Slope	Whole otoliths	Length frequency analysis	Mace et al. 1990
Hoplostethus atlanticus	Slope	Whole otoliths	Radiometry	Fenton et al. 1991
Hoplostethus atlanticus	Slope	Transv. sections	Radiometry	Short et al. 1993
Allocyttus verrucosus	Slope	Transv. sections	Radiometry	Short et al. 1993

277

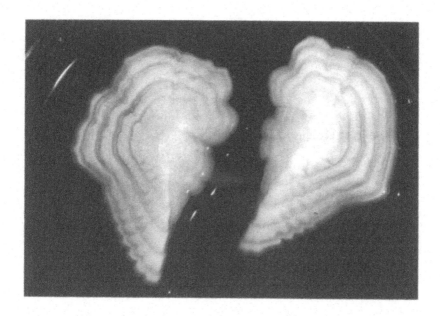

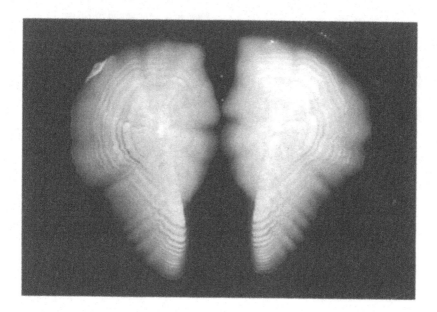

Figure 1. Otoliths (sagittae)from *Argentina silus*. Whole otoliths mounted in clear synthetic mounting medium (Eukitt) on black background.

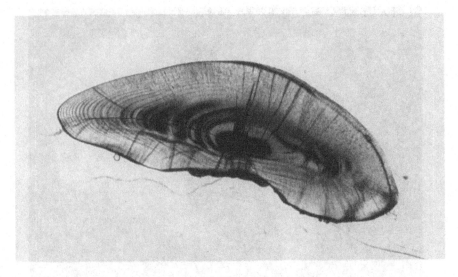

a.

b.

Figure 2. Transverse sections of otoliths (sagittae) from *Coryphaenoides rupestris*. a) whole section, b) detail from dorsal end of section of otolith from an old fish showing gradually thinner growth zones which were deposited on proximal (left) face.

References

Augustine, O. and Kenchington, T.J. (1987) A low-cost saw for sectioning otoliths. *J. Cons. int. Explor. Mer* 43, 296-298.

Bagenal, T.B. (1974) *Ageing of fish.* Unwin Brothers Ltd., London, UK.

Beamish, R. J. (1979) New information on the longevity of Pacific ocean perch (*Sebastes alutus*). *Journal of the Fisheries Research Board of Canada* 30, 607-616.

Beamish, R.J. and Chilton, D.E. (1982) Preliminary evaluation of a method to determine the age of sablefish (*Anoplopoma fimbria*). *Can. J. Fish. Aquat. Sci.* 39, 277-287.

Beamish, R.J. and McFarlane, G.A. (1983) The forgotten requirement for age validation in fisheries biology. *Transactions of the American Fisheries Society* 112, 735-743.

Beamish, R.J. and McFarlane, G.A. (1987) Current trends in age determination methodology, in R.C. Summerfelt and G.E. Hall (eds), *Age and growth of fish.* Iowa State University Press, Ames, Iowa, USA, pp. 15-42.

Bedford, B.C. (1983) A method for preparing sections of large numbers of otoliths embedded in black polyester resin. *J. Cons. int. Explor. Mer* 41, 4-12.

Bennett, J.T., Boehlert, G.W. and Turekian, K.K. (1982) Confirmation of longevity in *Sebastes diploproa* (Pisces:Scorpaenidae) from Pb-210/Ra-226 measurements in otoliths. *Marine Biology* 71, 209-215.

Bergstad, O.A. (1990) Distribution, population structure, growth and reproduction of the roundnose grenadier *Coryphaenoides rupestris* (Pisces: Macrouridae) in the deep waters of the Skagerrak. *Marine Biology* 107, 25-39.

Bergstad, O.A. (1993) Distribution, population structure growth, and reproduction of the greater silver smelt, *Argentina silus* (Pisces: Argentinidae), of the Skagerrak and the north-eastern North Sea. *ICES J. mar. Sci.* 50, 129-143.

Bowering, W.R. (1976) Distribution, age and growth, and sexual maturity of witch flounder (*Glyptocephalus cynoglossus*) in Newfoundland waters. *Journal of the Fisheries Research Board of Canada* 33, 1574-1584.

Bridger, J.P. (1978) New deep-water trawling grounds to the west of Britain. *Lab. Leafl. Fish. Lab. Lowestoft* 41, 1-40.

Campana, S.E., Zwanenburg, K.C.T. and Smith, J.N. (1990) Pb-210/Ra-226 determination of longevity in redfish. *Can. J. Fish. Aquat. Sci.* 47, 163-165.

Casselman, J.M. (1989) Determination of age and growth. Ch. 7, in A.H. Weatherley and H.S. Gill (eds) *The biology of fish growth.* Academic Press Ltd, London, pp. 209-242.

Chilton, D.E. and Beamish, R.J. (1982) Age determination methods for fishes studied by the groundfish program at the Pacific Biological Station. *Can.Spec. Publ. Fish. Aquat. Sci.* 60, 1-102.

Christensen, J.M. (1964) Burning of otoliths, a technique for age determination of soles and other fish. *J. Cons. Int. Explor. Mer* 29, 73-81.

Chugunova, N. I. (1959) Age and growth studies in fish. Izv. Akad. Nauk SSR. (Transl. from Russian by Israel Prog. Sci. Transl., Jerusalem, 1963).

Danke, L. (1987) Some particularities of roundnose grenadier (Coryphaenoides rupestris Gunn.) in the North Mid-Atlantic Ridge region. NAFO Scient. Counc. Res. Docums 87/78, 1-10.

Ehrich, S. and Reinsch, H.H. (1985) Bestandskündliche Untersuchungen am Blauleng (Molva dypterygia dypt.) in den Gewässern westlich von der Britische Inseln. Archiv für Fishereiwissenschaft 36, 97-113.

Eliassen, J.-E. (1986) Undersøkelser av utbredelse, forekomster og bestandsstruktur av skolest (Coryphaenoides rupestris Gunnerus) i Trøndelag. Fisken Hav. 1986, 1-19.

Engås, A. (1983) Betydning av ulike redskapsfaktorer i garnfisket etter blålange (Molva dipterygia Pennant 1784). Thesis, Dept. of Fisheries Biology, Univ. Bergen, Norway.

Fenton, G.E. and Short, S.A. (1992) Fish age validation by radiometric analysis of otoliths, in D.C. Smith (ed.), Age determination and Growth in Fish and Other Aquatic Animals. Aust. J. mar. Freshwat. Res. 43, 913-922.

Fenton, G.E., Ritz, D.A. and Short, S.A. (1990) Pb-210/Ra-226 disequilibria in Otoliths of blue grenadier, Macruronus novaezelandiae; problems associated with radiometric ageing. Aust. J. Mar. Freshwater Res. 41, 467-473.

Fenton, G.E., Short, S.A. and Ritz, D.A. (1991) Age determination of orange roughy, Hoplostethus atlanticus (Pisces: Trachicthyidae) using Pb-210/Ra-226 disequilibria. Marine Biology 109, 197-202.

Fujiwara, S. and Hankin, D.G. (1988) Aging discrepancy related to assymmetrical otolith growth for sablefish Anoplopoma fimbria in northern California. Jap. Soc. Sci. Fish. 54, 27-31.

Gauldie, R.W. and Nelson, D.G.A. (1990) Otolith growth in fishes. Comp. Biochem. Physiol. 97A, 119-135.

Gauldie, R.W., Coote, G., Mulligan, K.P., West, I.F. and Merrett, N.R. (1991) Otoliths of deep water fishes: structure, chemistry and chemically-coded life histories. Comp. Biochem. Physiol 100A, 1-31.

Gjøsæter, H. 1987. Primary growth increments in otoliths of six tropical myctophid species. Biological Oceanography 4, 359-382

Gordon, J.D.M. (1978) Some notes on the biology of the roundnose grenadier Coryphaenoides rupestris to the west of Scotland. ICES C.M. 1978/G:40, 1-12.

Gordon, J.D.M. (1979) Lifestyle and phenology in deep sea anacanthine teleosts. Symp. zool. Soc. Lond. 44, 327-359.

Haedrich, R.L. and Merrett, N.R. (1988) Summary atlas of deep-living demersal fishes in the North Atlantic basin. Journal of Natural History 22, 1325-1362.

Hancock, D.A. (1992) Australian society for fish biology - 1990 National workshop, in Smith, D.C. (ed) Age determination and growth in fish and other aquatic animals. Collected reprints from Aust. J. Mar. Freshwater Res. 43/5, pp. 1-6.

Kastelle,C.R., Kimura, D.K., Nevissi,A.E. and Gunderson,D.R. (1994). Using Pb-210/ Ra-226 disequilibria for sablefish, (Anoplopoma fimbia)age validation.Fishery Bulletin U.S. 92,292-301.

Kenchington, T.J. and Augustine, O. (1987) Age and growth of blue grenadier, *Macruronus novaezelandiae* (Hector), in south-eastern Australian waters. *Aust. J. mar. Freshwat. Res.* 38, 625-646.

Kline, D.E., Coale, K. and Cailliet, G.M. (1993) Radiometric age verification for two deep-sea rockfish (*Sebastolobs altivelis* and *Sebastolobus alascanus*). Paper read at the international symposium 'Fish otolith research and application', Hilton Head Island, South Carolina, USA. (unpublished)

Koch, H. (1976) A contribution on the methodics of age determination in roundnose grenadier (*Coryphaenoides rupestris* Gunn.). *ICNAF Res. Doc. 76/VI/* 28, 1-3.

Kosswig, K. (1973) Weitere Mitteilungen zur methodik der Altersbestimmung am Rotbarsches (*Sebastes marinus* L. und *S. mentella* Travin). *Ber. dt. wiss. Kommn. Meeresforsch.* 23, 84-89.

Kosswig, K. (1986) Investigations on grenadier fish (*Macrourus berglax* and *Coryphaenoides rupestris*) by the Federal Republic of Germany in 1986 and 1987. *Annls Biol., Copenh.* 40, 176-179.

Lear, W.H. and Pitt, T.K. (1975) Otolith age validation of Greenland halibut (*Reinhardtius hippoglossoides*). *Journal of the Fisheries Reserach Board of Canada* 32, 289-292.

Linkowski, T.B., Radtke, R.L. and Lenz, P.H. (1993) Otolith microstructure, age and growth of two species of *Cerastoscopelus* (Osteichthyes: Myctophidae) from the eastern North Atlantic. *J. Exp. Mar. Biol. Ecol.* 167, 237-260.

Mace, P.M., Fenaughty, J.M., Coburn, R.P. and Coburn., I.J (1990) Growth and productivity of orange roughy (*Hoplostethus atlanticus*) on the north Chatham Rise. *New Zealand Journal of Marine and Freshwater Research* 24, 105-119.

Magnusson, J.V. (1982) Age, growth and weight of blue ling (*Molva dipterygia*) in Icelandic waters. *ICES C.M. 1982/G:22*, 1-10.

Magnusson, J.V. (1990) Ageing and age composition of silver smelt (*Argentina silus* Asc.) in Icelandic waters. *ICES C.M. 1990/G:42*, 1-9.

McCurdy , W.J. (1985) A low-speed alternative method for cutting otolith sections. *J. Cons. int. Explor. Mer* 42, 186-187.

McFarlane, G.A. and Beamish, R.J. (1993) Validation of the otolith cross section method of age determination for sablefish (*Anoplopoma fimbria*) using oxytetracycline. Paper read at the international symposium 'Fish otolith research and application', Hilton Head Island, South Carolina, USA. (unpublished).

Molander, A. (1925) Observations on the witch (*Pleuronectes cynoglossus* L.) and its growth. *Conseil Permanent international pour L'exploration de la Mer, Publication de Circonstance* 85, 1-15.

Molander, A. (1956) Swedish investigations on ling (*Molva vulgaris* Fleming). *Institute of Marine Research, Lysekil. Series Biology, Report* 6, 1-36.

Morales-Nin, B. (1990) A first attempt at determining growth patterns of some Mediterranean deep-sea fishes. *Sci. Mar.* 54, 241-248.

Mugiya, Y. (1987) Phase difference between cacification and organic matrix formation in the diurnal growth of otoliths in the rainbow trout, *Salmo gairdneri*. *Fish. Bull. US* 85, 395-401.

Nedreaas, K.H. (1990) Age determination of Northeast Atlantic *Sebastes* species. *J. Cons. int. Explor. Mer* 47, 208-230.

Pannella, G. (1974) Otolith growth patterns: an aid in age determination in temperate and tropical fishes, in T. Bagenal (ed.) *Ageing of fish*. Unwin Brothers Ltd., London, UK, pp. 28-39.

Paul, L.J. (1992) Age and growth studies of New Zealand marine fishes, 1921-90: A review and bibliography. *Aust. J. Mar. Freshwater Res.* 43, 879-912.

Powles, P.M. and Kennedy, V.S. (1967) Age determination of Nova Scotia greysole, *Glyptocephalus cynoglossus* L., from otoliths. *Int. Comm. Northwest Atl. Fish. Res. Bull.* 4, 91-100.

Prince, L.D. and Palos, L.M. (eds) (1983) *Proceedings of the international workshop on age determination of oceanic pelagic fishes: tunas, billfishes and sharks*. United States National Marine Fisheries Service, NOAA Technical Report 8, Miami, Florida, USA.

Prouzet, P. and Souplet, A. (1985) Séminaire: Détermination de l'age et de la croissance à partir de pièces calcifiées chez les poissons. *Rev. Trav. Inst. Peches marit.* 49, 67-88.

Rauck, G. (1976) A technique of sawing thin slices out of otoliths. *Ber. dt. wiss. Kommn. Meeresforsch.* 24, 339-341.

Rollefsen, G. (1936) The otoliths of the cod. *Rep. Norw. Fish. and Mar. Inv.* 4, 1-14.

Savvatimskiy , P.I. (1971) Determination of age of grenadiers (order Macruriformes). *J. Ichthyol* (USSR) 11, 397-403.

Savvatimskiy, P.I., Koch, H. and Ernst, P. (1977) On comparison of methods on age determination of grenadiers (Macrouriformes, Pices) of the North Atlantic. J. Ichthyol (USSR) 17. 323-327.

Short, S.A., Smith, D.C., Fenton, G.E., Stewart, B.D., Robertson, S.G. and Ritz, D.A. (1993) Longevity in orange roughy (*Hoplostethus atlanticus*) demonstrated by two independent studies: Radiometry and longitudinal section counts. Paper read at the international symposium 'Fish otolith research and application', Hilton Head Island, South Carolina, USA. (Unpublished).

Six, L.D. and Horton, H.F. (1977) Analysis of age determination methods for yellowtail rockfish, canary rockfish, and black rockfish off Oregon. *Fish. Bull. US* 75, 405-414.

Smith, D.C. (ed.) (1992) *Age determination and growth in fish and other aquatic animals*. CSIRO Publications, Australia.

Smith, J.N., Nelson, R. and Campana, S.E. (1991) The use of Pb-210/Ra-226 and Th-228/Ra-228 dis-equalibria in the ageing of otoliths of marine fish, in P.J. Kershaw and D.S. Woodhead (eds) *Radionuclides in the study of marine processes*. Elsevier Applied Science, pp. 350-359.

Summerfelt, R. C. and Hall, G.E. (eds) (1987) *Age and growth of fish*. Iowa State University Press, Ames, Iowa, USA.

Thomas, R. (1987) Biological investigations on blue ling *Molva dipterygia* (Pennant 1784 after O.F. Müller 1776), in the areas of the Faroe Islands and to the west of the Shetland Islands. *Archiv für Fischereiwissenschaft* 38, 9-34.

Thorsen, T. (1979) Populasjonsparametrar hos vassild, *Argentina silus,* utanfor Møre-Trøndelag og i Skagerrak. Thesis, Dept. of fisheries biology, Univ. Bergen, Norway.

Watters, D.L., Kline, D., Coale, K. and Cailliet, G.M. (1993) Radiometric age confirmation of otoliths in the bank rockfish, *Sebastes rufus.* Paper read at the international symposium 'Fish otolith research and application', Hilton Head Island, South Carolina, USA. (Unpublished).

West, I.F. and Gauldie, R.W. (in press, 1994). Determination of fish age using Pb-210/Ra-226 disequilibrium methods. Can. J. Fish. Aquat. Sci.

Williams, T. and Bedford, B.C. (1974) The use of otoliths for age determination, in T. Bagenal (ed.) *Ageing of fish.* Unwin Brothers Ltd., London, UK, pp. 114-123.

Wilson, R.R. Jr. (1982) A comparison of ages estimated by polarized light method with ages estimated by vertebrae in females of *Coryphaenoides acrolepis* (Pisces: Macrouridae) *Deep-Sea Research* 29, 1373-1379.

Wilson, R.R. (1988) Analysis of growth zones and microstructure in otoliths of two macrourids from the North Pacific abyss. *Environ.Biol.Fish.* 21(4), 251-261.

Withell, A.F. and Wankowski, J.W. (1988) Estimates of age and growth of ocean perch, *Helicolenus percoides* Richardson, in south-eastern Australian waters. *Aust. J. Mar. Freshwater Res.* 39, 441-457.

SECTION 2   **Review papers**

SPANISH NORTH ATLANTIC DEEP-WATER FISHERIES

S. IGLESIAS and J. PAZ
Instituto Español de Oceanografía,
Apartado 1.552,
36280-Vigo,
Spain.

ABSTRACT

The Spanish fishing fleet catches deep-water species on both sides of the North Atlantic Ocean. In terms of catches, the most important commercial species is the Greenland halibut (Reinhardtius hippoglossoides) which is caught in the NAFO Divisions 3L, 3M and 3N. In this fishery most research has focused on the Greenland halibut and the paper contains some preliminary information on length distribution, spawning and feeding. The fishery first commenced in 1990 and the fishing effort increased from 9 vessels in 1990 to 33 vessels in 1993. Each vessel makes a voyage of 5 to 6 months. It is the practice to monitor catches, catch information and biological data by employing trained observers on the vessels.

Spanish vessels also fish for deep-water species in the eastern North Atlantic, and the paper describes the fishery for various species of deep-water sharks in depths from 400 m. to in excess of 1000 m. using longlines. These fisheries take place in ICES areas VII, VIIIc and IXa. Also in the eastern Atlantic there are Spanish fisheries for monkfish (Lophius piscatorius and L. budegassa) using gillnets, and a small fishery for Chaecon affinis,(formerly Geryon affinis) on the Banco de Galicia using longlines and traps

Spanish scientists and fishermen have also carried out surveys in deep-water since 1974 using trawl gear down to 2000 m. and more recently with longlines down to 3400 m.

A. G. Hopper (ed.), Deep-Water Fisheries of the North Atlantic Oceanic Slope, 287–295.
© *1995 Kluwer Academic Publishers.*

1. Introduction

The Spanish fishing fleet fishes for deep-water species on both sides of the North Atlantic Ocean. In terms of catches and value, the most important fishery is that of Greenland halibut (*Reinhardtius hippoglossoides*) in the western Atlantic, caught in the NAFO Divisions 3L, 3M and 3N. Deep water sharks (*Somniosus rostratus, Deania calceus, Centrophorus granulosus, Centroscymnus coelolepsis* and others), and monkfish (*Lophius spp.*)are the main species fished in the eastern Atlantic.

In recent years most research efforts have been focused on the Greenland halibut fishery, and this has also been monitored by trained scientific observers on board the commercial fishing vessels since the start of this fishery in 1990.

Deep-water fisheries in the eastern Atlantic have been largely as a result of the decline of the hake fishery along the edge of the continental shelf, and the need for the vessels to find alternatives.

2. Northwest Atlantic fisheries

2.1.FISHERY FOR GREENLAND HALIBUT (*Reindhartius hippoglosoides*)

The Spanish fishery for Greenland halibut has developed since 1990. This took place originally on the boundaries of NAFO Divisions 3L and 3M and later extended in 1992 to the north of Division 3N. These areas are shown in Figure. 1, (Junquera (1992)).

From the commencement of this fishery until 1994, the technology and the skills of the fishermen steadily improved and the fishing methods for the deeper water are now designed to operate as deep as 2000 m. The Spanish fleet is mainly composed of bottom trawlers, and the mean fishing depth is 900 m. Each vessel makes one trip of a duration of 5 to 6 months.

Fishing activity by the Spanish fleet has been monitored by observers on board the vessels. In 1990, 8 observers were employed, whilst in 1991, 1992 and 1993, there were 21, 30 and 12 respectively. The observers provide information on fishing areas, catches and fishing effort and also take biological samples from the catch.

Table 1 summarises the number of ships operating each year in the western Atlantic during the period 1990 to 1993, classified according to GRT.

TABLE 1 Spanish vessels operating in the western Atlantic 1990 to 1993

Tonnage group	1990	1991	1992	1993
< GRT 600	1(12%)	5(21%)	12(41%)	15(46%)
GRT 600-1000	4(44%)	7(37%)	9(31%)	10(30%)
>GRT 1000	4(44%)	8(42%)	8(27%)	8(24%)
Totals	9	20	29	33

The traditional the small Spanish freezer trawlers target flatfish, mainly American

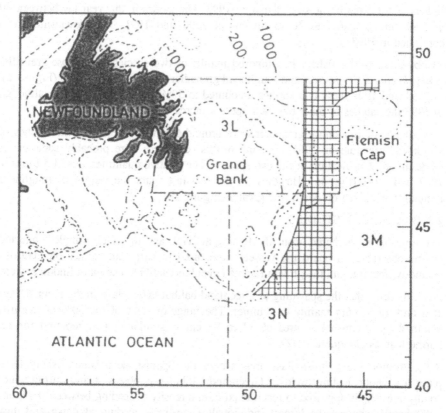

Figure 1. Fishing grounds used by Spanish trawlers in the western North Atlantic for Greenland halibut

plaice, and some also fished for Greenland halibut in 1992 and 1993. The activity of this fleet is not however included in this paper.

The annual catches of Greenland halibut by the Spanish vessels are about 30,000 tonne The main by-catch species are the grenadiers (*Macrouridae spp.*),(Figure. 2).

Following a recommendation of the NAFO Scientific Council in 1992, the standardization of the catch rate data was undertaken for the whole available time series of data from 1990 to 1993 from the commercial fishery (Cardenas et al.,(1993)). Catch rates were adjusted to a multiplicative model for CPUE and the results indicated that there were no significant differences in the year factor for 1991 and 1993,and a clear tendency was not observed, in spite of their annual variation. There was a clear seasonal trend in catch rate, with January being the month with the highest rates. In addition, catch rates declined slightly as the fishery moved into deeper water.

The length distribution of catches of Greenland halibut in the period 1990 to 1992 indicated a decline in the size of fish by 1992. The modes in this year lay between 40 and 50 cm compared to 50 to 60 cm in 1990 and 1991. This downward trend continued in 1993.

The by-catch in this fishery is composed mainly of two species: roundnose grenadier (*Coryphaenoides rupestris*) and roughhead grenadier (*Macrourus berglax*) (Figure. 2). In 1991 and 1992 these two species amounted to 50% of the total by-catch, although in 1993 the catches of other by-catch species increased (Figure. 3).

The catches of roundnose grenadier were around 6 times greater than the catches of roughhead grenadier in 1991, although in this year there were probably some errors made in the identification of the two species. The relative quantities were 1.5 times in 1992 and 1993 (Fig. 4a). However in the two last years the weight of roughhead grenadier processed on board was greater (Figure. 4b).

2.3. BIOLOGICAL STUDIES

The information gathered about the fishing activity, and the catch samples collected by the observers on board the vessels, were used to carry out studies on maturity, fecundity, feeding and the growth rate of Greenland halibut. The main findings were:

2.3.1. It seems that the spawning of Greenland halibut takes place in the Flemish Pass area and this occurs mainly in summer. The range of sizes of the 50% of maturity obtained for Flemish Pass area of 67 to 73 cm is similar to that reported for the northern areas (Junquera 1992).

2.3.2. Around the Flemish Pass area, where the Spanish deep water fishery takes place a 24 hour feeding cycle of Greenland halibut was observed and during which young individuals appeared to reach maximum intensity of feeding between 0600 and 1200 hours, whereas the largest individuals show peak feeding at dawn and dusk (Junquera (1993)).

2.3.3. The most important food prey items were identified as, Cephalopods (32%), decapod crustaceans (22%) and other fishes (39%), with cannibalism reaching 2%. It would seem that at depths of between 700 and 1200 m. squid play a similar roll in the diet of Greenland halibut to that of capelin on the continental shelf. In individuals over 60 cm an abrupt diet change appears to take place, and fish and waste products become dominant. The annual average of empty stomachs was 69% and the feeding intensity diminished throughout the year for specimens over 60 cm. The frequency of occurrence of the main food items bears more relation with size than to depth (Rodriguez-Marin et al.,(1993)).

At the present studies on growing, fecundity and feeding are continuing along with other work on by-catch species.

3. Northeast Atlantic fisheries

3.1. ICES SUB-AREA VII

A fishery for several species of deep water sharks commenced in 1991 in ICES Sub-area VII. A number of longliners which had traditionally fished for hake in this area, were no longer able to be profitable With the developing market for the livers of these species for the production of oils, these vessels began to fish for sharks in waters of depths greater than 1000 m.

In Galicia, in the northwest of Spain, the landings are made principally in the port of La Coruña. In the 1991 shark fishery, 43 trips were made by 10 vessels whilst in 1992 there were 79 trips from 12 vessels. In 1993, 10 ships made 38 trips. Almost all of the vessels involved operated under the flag of Ireland or United Kingdom and only one or two of the vessels could be strictly referred to as Spanish flag vessels.

The sharks captured are a mixture of the species *Somniosus rostratus, Deania calceus, Centrophorus granulosus, Centroscymnus coelolepis* and others. Their livers amount to one third to one fifth of the total body weight and of which approximately 70 to 80% of the liver weight can be converted to oil. The livers are the main justification for the capture of these sharks. On occasions only the liver is retained and the remainder of the fish is discarded.

In 1991 the weight of all deep-water sharks landed (skinned and gutted) in north Galicia was 180 tonnes whilst the corresponding quantity for 1992 was 340 tonnes, and in 1993 the catches were 234 tonnes of sharks and 29 tonnes of *Phycis spp*. The average catch rate in 1993 over the year was 5 tonnes/trip and no seasonal variation was observed.

3.2. ICES SUBAREA VIIIc-IXa

A fishery for deep-water sharks has also developed to a limited degree on the continental slope off Cantabria to the north and northeast of Spain (ICES Division VIIIc). Fishing for sharks occurs when the traditional target species, such as hake and red sea bream, are not available. The highest catches and prices for sharks occur in winter.

This fishery is carried out by vessels of 20 to 75 GRT which are licensed to gain access to this fishery. The larger vessels tend to target the *Mora moro* and *Phycis blennoides* when fishing for deep-water species but sharks are also caught. The gear usually consists of a single longline with about 4000 large hooks, and is fished at depths of 400 to 700 m.

In 1992 17 vessels from Spain's Asturian and Cantabrian ports were participating in this fishery and landed 340 tonnes of sharks composed of the species *Scyliorhynus canicula, Galeus melastomus, Centrophorus spp, Etmopterus spp, Delatias licha,* and *Deania calceus*. In 1993 10 vessels landed 452 tonnes of sharks.

In both of the above-mentioned fisheries, the current practice of skinning the shark carcases which are landed and/or retaining on board only the shark livers whilst discarding the rest of the fish makes it difficult and indeed impossible to obtain accurate statistics of landings or the catches by species.

3.3. GILLNET FISHING FOR MONKFISH (*Lophius spp*)

Two species of monkfish (*Lophius piscatorius and L. budegassa*) are caught by gillnets sometimes at depths greater than 400 m. off the Cantabrian and north east coasts of Spain. This fishery by comparison with other deep-water fisheries has been established for some years, and Spanish National Regulations of 1983 define the fishing areas, a minimum mesh size of 280 mm. and the maximum length of net in relation to the tonnage of the vessels.

The statistical data does not permit discrimination of the catch by depth.

3.4. DEEP-WATER CRAB (*Chaecon affinis*) ON THE BANCO DE GALICIA.

In 1980 and 1981 Spanish scientists from Instituto de Investigaciones Marinas (Vigo) carried out several surveys on the Banco de Galicia, 130 miles east Cape Finisterre.The gears used were long line and traps and the fishing depths were 600 to 1000 m. From these surveys it was concluded that there is a potential fishery of royal crab (*Chaecon affinis*, formerly *Geryon affinis*) for 5 vessels each to fish for 210 days per year with 150 traps per day.

At present there is only a small fishery for this species and 27 trips were carried out in 1993 and 11 tonnes of crab were landed.

3.5. OTHER POTENTIAL FISHERIES IN THE NORTH EAST ATLANTIC

Spanish scientists and fishermen have investigated possibilities for fisheries in deep water in a number of areas of the north east Atlantic.

3.5.1. Investigations to the north and northwest of Spain (ICES Subareas VIIIc and IXa) From 1974 to the present a series of demersal trawling surveys by the research vessel Cornide de Saavedra have been carried out covering depths between 30 and 650 m. in the ICES Subareas VIIIc-IXa (Fariña, (1985); Sanchez, (1991)). At the edge of the continental shelf and on muddy bottoms in depths greater than 200 m. the major species of potential commercial interest are *Galeus melastomus*, *Etmopterus spinax* (lantern sharks), *Beryx decadactylus*, *Phycis blennoides* (greater forkbeard), *Chimaera mostrosa*, and a number of grenadiers including *Malacocephalus laevis* and *Nezumia aequalis*.

3.5.2. Investigations off Asturias (ICES subarea VIIc)

Two experimental surveys were carried out, in 1988 and 1989. In 1988 traps and longlines were deployed at depths between 150 and 3400 m. (Alcazar,(1992)). In 1989, specialised trawling gear was deployed at depths between 1000 and 2000 m.

One part of the catch consisted of species already known as a by-catch in the traditional Asturian trawl fisheries including *Mora moro*, *Phycis blennoides*, *Hoplostethus atlanticus*, *H. mediterraneus* and various species of those sharks now commercially exploited for their livers. The rest of the species are those for which at present no market demand exists and these included *Alepocephalus spp.*, *Antimora rostrata*, *Lepidion eques* and *Bathysaurus ferox*.

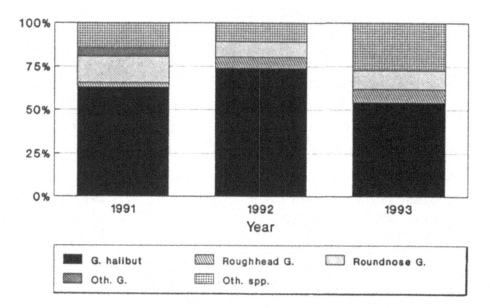

Figure 2. Main species in the catch composition of the Spanish Greenland halibut fishery. NAFO Divisions 3L & 3M, 1991 to 1993

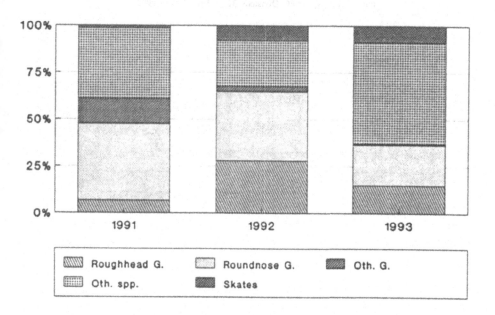

Figure 3. Proportion of the main by-catch species in the catch of the Spanish Greenland halibut fishery. NAFO Divisions 3L & 3M, 1991 to 1993

294

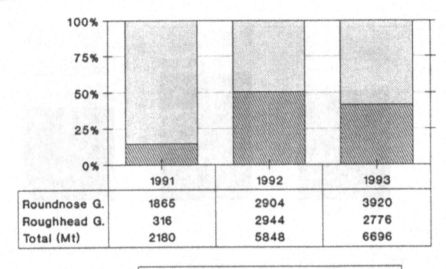

	1991	1992	1993
Roundnose G.	1865	2904	3920
Roughhead G.	316	2944	2776
Total (Mt)	2180	5848	6696

Roughhead G.   Roundnose G.

1991 six months

Figure 4a. Proportion and weight of roundnose and roughhead grenadiers caught in the Spanish Greenland halibut fishery, NAFO Divisions 3L & 3M, 1991 to 1993

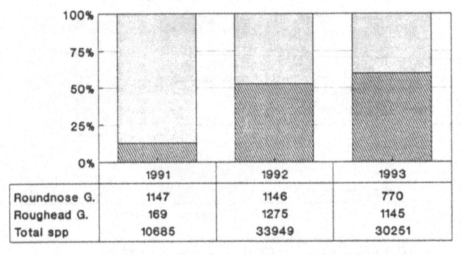

	1991	1992	1993
Roundnose G.	1147	1146	770
Roughead G.	169	1275	1145
Total spp	10685	33949	30251

Roughead G.   Roundnose G.

1991 Six months only

Figure 4b. Proportion and weight of roundnose and roughhead grenadiers processed on board in the Spanish Greenland halibut fishery, NAFO Divisions 3L & 3M, 1991 to 1993

References

Alcazar, J.L., Gonzalez, P., Fernandez, C.L., Garcia, L., Rucabado, J., Lloris, D. y Castellon, A. (1992) Prospecciones pesqueras en aguas profundas (Sector VIIIc-ICES), Recursos Pesqueros de Asturias no.6, Consejeria de Agricultura y Pesca del Principado de Asturias.

Cardenas, E. de, Junquera, S., Vazquez, A. (1993) Abundance indices of Greenland halibut in Deep water Fishing Zones of NAFO Divisions 3LMN. NAFO Sci. Counc. Res. Doc. 1993/61, Serial No. 2244.

Fariña, A.C., Pereiro, F.J. y Fernandez A. (1985) Peces de los fondos de arrastre de la plataforma continental de Galicia, Bol. Inst. Esp. Oceanog. 2(3), 89-98.

Junquera, S. and Zamarro, J. (1992) Sexual Maturity and Spawning of the Greenland Halibut (*Reinhardtius hippoglossoides*) From Flemish Pass Area. NAFO SCR Doc. 92/41, Serial No. N2092, 10p.

Junquera, S., Iglesias, S. and Cardenas, E. de, (1992) Spanish Fishery of Greenland halibut (*Reinhardtius hippoglossoides*) in 1990-91. NAFO, SCR Doc. 92/28, Serial No. N2075, 14p.

Junquera, S. (1993) Feeding cycles of Greenland halibut (*Reinhardtius hippoglossoides*) in the Flemish Pass area in relation to catch rates (1991-92). NAFO, Sci. Coun. Res. Doc. 1993/17, Serial No. N2194.

R-Marin, E., Punzon, A. and Paz, J. (1993) Greenland halibut (*Reinhardtius hippoglossoides*) feeding in Flemish Pass. NAFO Divs. 3LM. NAFO Sci. Counc. Res. Doc. 1993/18 Serial No. 2195, 9 pp.

Sanchez, F. (1991) Las comunidades de peces de la plataforma del Cantbrico, Tesis doctoral, Publ. espec. Inst. Esp. Oceanogr., No. 12.

THE IRISH EXPERIENCE OF DEEP-WATER FISHING IN THE NORTH EAST ATLANTIC

RICHARD MC CORMICK
Fishing Technology Executive
An Bord Iascaigh Mhara(BIM),
Dublin,
Ireland

ABSTRACT

Deepwater species fishing trials commenced in Ireland in 1988 assisted by the European Union Commission's Exploratory Fishing Voyage scheme. The objective was to develop a fishery for unexploited non-quota species and divert fishing effort from existing quota species which were under severe pressure. The first species targeted was blue whiting (Micromistius poutassou) quickly followed by argentines, also known as silver smelt (Argentina silus), and progressing to a wider range of species in even deeper water such as grenadiers (Coryphaenoides rupestris), black scabbard fish (Aphanopus carbo), Portuguese shark (Centroscymnus coelolepis) etc. This involved pelagic and demersal fishing techniques at a cost of considerable material damage as a result of working on unknown fishing grounds. Over the 3 year period 1988 to 1990 the catches of argentines suffered a rapid and alarming decline. The reasons for this are not known, although various possibilities are considered.

The technological development and food fish processing trials were carried out by An Bord Iascaigh Mhara (BIM), the state fisheries development agency, and parallel biological studies were undertaken by the Department of Marine's Fisheries Research Centre. This paper summarizes the progression from 1988 to date in the Irish context and points out, that whilst much has been learned, considerably more research and development needs to be undertaken to ensure an economically viable and biologically sustainable future for this emergent fishery.

A. G. Hopper (ed.), Deep-Water Fisheries of the North Atlantic Oceanic Slope, 297–306.
© *1995 Kluwer Academic Publishers.*

1. Introduction

BIM, the Irish Sea Fisheries Board is a semi-state organization charged with the responsibility of developing the Irish sea fishing industry. It provides assistance to the fisheries sector through research and development, grant aid, national and EU marine credit finance, training, aquaculture development, the supply of ice and also the provision of marketing facilities and services.

As Ireland has within its contiguous fishing zones 16% by area of all EU waters but only 4.4% of EU quotas, one of the principal functions of the Fishing Technology Section of BIM is to develop means whereby underutilized non-quota species can be exploited. This philosophy governs not only offshore non-quota species, but also non-quota fish and shellfish many of whom are caught within Irish limits by smaller coastal vessels. Down the years BIM has assisted the industry in developing sustainable fisheries for horse mackerel, squid, crab, whelks and shrimps to name but a few non-quota species. It was thus inevitable that as the size of Irish vessels increased from very humble beginnings in the early 1960's, involving 16 to 20 m. inshore vessels, to the large refrigerated seawater tank vessels of today measuring 27 to 71 m. in length, plus one 91 m. factory trawler recently destroyed in a fire, that the Irish fishermen's horizons were extended to encompass the untapped potential of the adjacent offshore waters including of course the Atlantic oceanic slope.

2. Deep-water trawling surveys, 1988 - 90

The Irish experience with deepwater trawling commenced in 1988, when two large refrigerated sea water (RSW) trawlers, the 71 m. Western Endeavour (4000 hp., Skipper John Bach) and the 65 m. Atlantic Challenge (3260 hp., Skipper Martin Howley), based in Killybegs on the north west coast of Ireland commenced exploratory fishing voyages, under the European Commission's Exploratory Fishing Voyage scheme. This was financed 20% by the EU, and between 10-20% by BIM with the balance being financed by the owners at their risk. Due to restricted quota opportunities of the more traditional species, these vessels had already turned to non-quota scad or horse mackerel (*Trachurus trachurus*) for which a food fish market had been developed by BIM. As this fishery is of short duration each year, and quality standards are very exacting, other non-quota possibilities had to be considered. In 1987 the Western Endeavour had, by chance, detected significant echo sounder marks in deep-water and a trial tow provided a small catch of argentines or silver smelt (*Argentina silus*), which was felt to be worth investigating further.

The 1988 series of exploratory fishing trips commenced in early April and was targeted on two species, blue whiting (*Micromistius poutassou*) and argentines principally in ICES area VIa. In 1989 and 1990 the same vessels conducted similar exploratory voyages extending the fishing zones to cover ICES areas VIb and large parts of area VII to the west and south west of Ireland with the following catches in tonnes being recorded by these two vessels.

Year	Blue Whiting	Argentines
1988	3,979	5,454
1989	2,289	6,103
1990	486	585

Tows of up to 400 tonnes of both species were caught in the early stages of this programme, and many valuable technical lessons were learned. In addition to the Western Endeavour and the Atlantic Challenge, other Irish vessels began fishing blue whiting and argentines commercially with mixed results. In the early stages fishmeal was the only outlet for these species, with the Western Endeavour on one occasion landing a world record of 2,278 tonnes of argentines into Esbjerg, Denmark.

The dramatic fall in catches in 1990 is still difficult to explain. Biological sampling undertaken by the Irish Department of Marine's, Fisheries Research Centre (FRC), which was responsible for the collection, analysis and interpretation of samples, indicated that 40% of argentines were in excess of 20 years old. Many were in spawning condition when caught, with surprisingly, no sign of immature fish. During 1988, conversations at sea with Norwegian fishermen fishing for blue whiting nearby, indicated that the few Norwegian skippers who had tried the argentine fishery, following observations of the success of the Irish vessels, had problems of adequate winch power and capacity. Thus there was little pressure on the stocks that year.

In 1989 it is believed that the Norwegians specifically targeted the argentine stock and in 1990 the Dutch were reputed to have a contract to sell 40,000 tonnes with evidently little success, thus mirroring the Irish experience. What then had happened to the stocks to result in such a dramatic decline in catches? Is it conceivable that they were fished out in two years? Though sterling biological analysis work was done by the FRC, which concluded that this was probably a virgin stock with an age structure which would suggest a susceptibility to overfishing, it is still true to say that little is really known to this day about this species. It should also be recalled that in the 3 years in question, unusually long periods of severe weather occurred of up to 3 months duration, which may or may not have influenced migration patterns.

However, whilst these developments were still unfolding, BIM which had always been more interested in developing a food fish market, particularly for argentines, had by 1989, initiated shore processing trials with a local processor at Killybegs, Killybegs Seafoods. These continued through 1990, augmented also by a separate onboard processing trial on the 91m. Killybegs based freezer trawler, MFV Veronica (6000 hp., Skipper Kevin Mc Hugh).

3. Technical observations on the 1988 - 90 trials

Blue whiting were observed to congregate in commercial quantities in depths between 275 and 500 m. with the most productive tows being made, in daylight

300

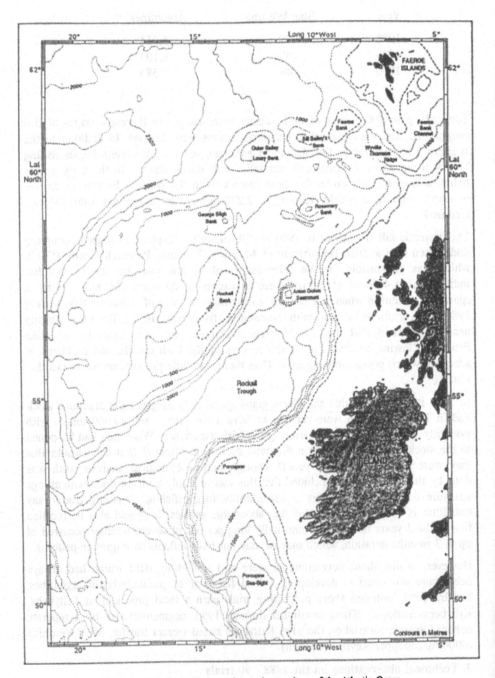

Figure 1. Fishing grounds on the north east slope of the Atlantic Ocean

only, whilst towing in a southerly direction, against the direction of migration. The morning tow seemed the most productive and it was noticeable that if this was not successful, the daily tonnage was usually poor. The species faithfully exhibited the well known diurnal vertical migration, and fishing was considerably better in calm, bright, sunny conditions. Blue whiting in April 1988 were found to be in poor condition and only fit for fishmeal and, as the peak of the fishery usually occurs earlier in the year, dedicated efforts were made to conduct night time echo sounder search patterns in 400 to 800 m. for argentines.

On the 9th May 1988 the first significant tow of 100 tonnes of argentine were brought aboard by the Western Endeavour. The most productive commercial quantities were found between 550 and 720 m. depth, west of St.Kilda on clean level ground, though echo sounder searches indicated marks persisting down to 880 m. depth. Pelagic trawls with headline heights varying between 50 and 75 m. and fitted with cable type headline transducers were used, The method was to fish the footropes tight down on the seabed. It was not considered prudent in the early stages to submit the cable electronics to the pressures expected beyond 720 m. Whereas pair pelagic trawling posed no difficulties for blue whiting, single boat pelagic trawling proved more efficient for argentines and these practices quickly became the norm.

The argentine fishery, on the other hand, proved considerably more hectic than blue whiting, with fishing taking place 24 h. per day and it was necessary to attempt to limit catches to around 200 to 300 tonnes to minimize gear damage. Catch rates of up to 700 tonnes per 24 h. were recorded. Most of the fishing effort was concentrated in 550 to 720 m. to ensure clean tows of argentines only. Argentines are caught in more shallow depths than this, but often intermixed with blue whiting. Argentines must be hauled with greater alacrity than blue whiting, as they exhibit a tendency to sink faster once the trawl has surfaced, due to burst air bladders and are somewhat difficult to bring aboard by the fish pump because of an excessively scaly skin.

One unique danger of this fishery manifested itself in the spectacle of a 4000 hp. vessel being pulled astern at 4 knots, with the main engines at full throttle as the codend, with 400 tonnes of argentines, commenced an uncontrolled ascent from the depths with a high risk of entanglement of the gear in the propeller. This problem was countered by turning approximately 20° to starboard for every 100 m. of warp hauled to ensure the gear would be clear of the vessel when the codend surfaced, usually like a submarine. A blowout joining round (safety panel) at the codend/net join also proved desirable, as it was better to haul the otter boards and the trawl independently of the codend. Despite these precautions, a 75 m. long codend worth £50,000, was lost by each of the vessels concerned.

Towards the end of May each year it was observed that argentines were migrating towards the north east from level to foul (very rough) ground with the attendant increased risk of gear damage, as the pelagic trawls were fitted with a conventional rope rounded footrope. In the 1988 season both vessels developed severe technical problems with the winches with the drum end plates expanding sideways up to 25 mm. per side and each had to proceed to Norway for repairs. In the case of one vessel, the winch drum plate had actually disengaged from the core along 75% of it's circumference. This damage is believed to be a consequence of a reduction in nominal warp diameter during towing due to untwisting and stretching caused by the tension in the warp from the weight and extra water resistance, plus the additional strain of heaving back from great depths. The warp used is 32mm diameter and weighs 385 kg. per 100 m. Once wound onto the winch under this amount of tension, the warp attempts to regain it's original diameter, pushing not only against adjacent layers of warp, causing damage to the outer wire strands but also squeezing it's interior fibre core and also placing massive pressure on the central drum core and the drum flanges.

4. Processing trials of argentines 1989 - 90

Once a substantial resource of argentines had been located in 1988, it was felt that the further development of the fishery should not only be directed at diversifying the fishing effort of larger vessels away from pressure stocks, but also provide a source of high quality material for human consumption. In 1989 a total of 465 tonnes was successfully landed to Killybegs Seafoods Ltd for pilot processing trials in the form of fillets and fish mince at a price of IR£150 per tonne to the vessels compared to IR£45 to 50 per tonne of fishmeal. These trials were partly funded under the FAR fisheries research programme of the EU. From a perspective of commercial viability, the two vessel owners required a minimum catch rate in excess of 200 tonnes per day at fishmeal prices, or 70 tonnes per day for human consumption. The latter would naturally minimize the impact of commercial fishing on the resource, and was considered the best alternative.

Filleting rates of between 41 and 43% were achieved during the trials with the only significant problem being the requirement for removing scales prior to machine filleting. The machine filleting rate achieved was an average rate of 125 fish/minute. Removing the scales proved less of a problem with fish that had been held on board in refrigerated sea water (RSW) than those which had been boxed in ice. The quality of the mince produced was judged on the whiteness of the flesh, and this was entirely dependant on the absence of the black belly membrane, skin or guts. Whilst pre-skinning would have been desirable, (no suitable skinning machines were available for these trials), mince yields of 29% of the whole fish were achieved, whilst still maintaining an excellent quality white mince. Trials at sea on the Veronica, gave yields of 38% for fillets and 28.5% by the mince stage and, whilst teething problems were evident, it was considered possible to produce 20 tonne of mince per day on an approximate basis of 1 tonne of mince from 3.5 tonne of whole fish. The market value of good quality mince was estimated at £1,100 per tonne.

5. Deep-water trawling 1992 - 93

During the 1988 to 1990 trials and despite the very large mesh sizes in the wings of pelagic trawls, interesting bycatches of black scabbard fish (*Aphanopus carbo*), grenadier (*Coryphaenoides rupestris*) and Portuguese shark (*Centroscymnus coelolepis*) were noted. The unexpected failure of the argentine fishery in 1990 and the associated high gear costs had dimmed the enthusiasm of many fishermen for exploratory fishing. However one determined Irish skipper-owner, Mr Jim Murrin of the 40 m. Mary-M (1800 hp.), assisted by the European Commission's Exploratory Fishing Voyage Scheme, commenced fishing for these species with demersal trawls in 1992, and terminating in October 1993. A report on this work has recently been sent to DG XIVof the Commission.

This exploratory fishing programme got off to a very poor start , as the author of this document can readily attest, having been onboard at the time. Damaged trawls and lost codends from coral and uncharted wrecks, crossed otter boards, and large catches of smoothheads (*Alepocephalus bairdii*) in the early voyages and for which there is no commercial value, nearly brought the entire programme to a halt. On one memorable occasion, when the vessel actually achieved the first ever viable daily catch rate of IR£4,500, west of the Porcupine Bank, but the success was swiftly negated by the complete destruction of a brand new £7,500 trawl the following day. A number of lessons can be drawn from exploratory fishing voyages such as these, summarized as follows.

Little or nothing is really known of the seasonal productivity of many potential deepwater fishing grounds along the edge of the eastern Atlantic slope. There are no suitable up to date fishing charts accurately indicating ground contours, substrate type or more importantly wrecks and areas of coral. This lack of information will influence gear damage and in some cases vessel safety and ultimately the commercial viability of deep-water fishing. There are even areas where the edge of the continental shelf slopes away so quickly that it is impossible to trawl along a depth contour. It may well be that trawling is not the most appropriate means of harvesting this fishery, but the threat to the environment posed by lost gillnets in deep-water, which is one obvious alternative, would need caution. The biological data, unavailable two years ago, is only now beginning to emerge and this urgently needs to be used at European Union and internationally for the management of deep-water stocks.

On the positive side, there is little doubt however, that based on the experience of the Mary-M, a viable trawl fishery exists for grenadiers, which formed 73% of the vessels landings. As the Mary-M trials continued and more experience was gained, profitability greatly improved, and particularly on the final voyages marketing confidence increased. Irish gear manufacturers have also benefited from the experience by the slow process of design and the adapting of existing trawls to suit the emergent deep water fishery and to minimize gear damage costs. The flexibility and inventiveness of the equipment supply in Europe is amply demonstrated by the thorough research and development which has now been put into the design of new winches by the French winch manufacturer Bopp S.A. suited to this specific fishery, which will overcome some of the problems described earlier in this paper.

During the course of these trials BIM deployed Scanmar spread, height, depth, temperature and codend sensors on the trawls, and as a result was able to compare the use of the equipment with normal commercial fishing in conventional depths. Useful information was assembled on gear settling times once the gear was shot and warp drag forces. An area which definitely needs much more dedicated research than this short trial would allow, is to study water temperature in relation to depth and species distribution. The Fisheries Research Centre have assembled a body of biological knowledge during this exploratory voyage, further augmented by a separate EU funded STRIDE research programme which has added much to the knowledge. It is hoped that in 1994, Irish State agencies may support deep-water longlining trials on Ireland's first ever commercial automated 30 m. longliner, the Sea Sparkle (Skipper William Harrigan), commissioned recently in a joint venture between BIM, the Norwegian company Mustad A/S and the vessel owner.

Hopefully this short paper has indicated the evolutionary progression in deep-water fishing that the Irish fishing industry and state agencies have being following since 1988. Much has been learned and much still remains to be discovered. However, it is true to say that the current state of knowledge, imperfect and incomplete though it may be, would not be as far advanced without the assistance of the European Commission's DG XIV, and this must be clearly acknowledged at this Workshop. It is regrettable that the Exploratory Fishing Voyage scheme has recently been dropped, but hopefully the Commission, in it's wisdom, will see that this type of work is not a wasted experience and a replacement research programme will emerge, targeting not just the biological parameters, but also the many technological issues requiring more research and development which have been highlighted at this Workshop.

Finally the author would like place on record the considerable role played by the European Commission's DG XIV Exploratory Fishing Voyage Scheme in the emergence of the deep-water fisheries of Ireland. The organisers of this Workshop must also be congratulated for hosting this very worthwhile conference and, gathering of experienced technical and biological researchers from all over the world. Perhaps this will help at last to unlock at least some of the dark secrets of the deep-water species fisheries and more importantly, point the way to a sustainable future for the deep ocean resources.

Figure 2. Trawler/purse seiner Atlantic Challenge, 65 m., 3260 hp.

Figure 3. 300 tonne codend of argentines (Argentina silus) alongside of the Western Endeavour

Figure 4. Pumping onboard a catch of argentines - Western Endeavour

FISHERIES POLICY AND THE SURVIVAL OF FISHING COMMUNITIES IN EASTERN CANADA

ROSEMARY E. OMMER

Institute of Social and Economic Research

Memorial University of Newfoundland

St. John's, Newfoundland A1B 5X7 Canada

ABSTRACT.

Since the introduction of deep-sea fishing technology in the 1970s, fisheries policy in Canada has increasingly favoured large corporate enterprises. The deep-sea fleets have been seen as highly efficient operators in contrast to the smaller, inshore operators, in which simple labour intensive methods are employed, and the catch rates are relatively low. The deep-sea fleets have replaced labour with capital intensity, and have been centralized as much as possible in a few ports. With the fall-out from the current moratoria on most fishing operations, however, such an approach is being re-examined since it is now considered likely that an over abundance of poorly-managed but technologically efficient fishing effort may be the principal cause of the collapse of the groundfish stocks in the north west Atlantic. Concerns about a fleet in which there has been excessive capital investment are matched by worries about the sustainability of the many small communities of coastal eastern Canada whose inshore fisheries have suffered in the general collapse of the stocks. There is a growing realisation that fisheries management and science cannot be separated from the social context in which they operate. Fisheries are ultimately about employment and the overall economic well-being of regions; not just industry, although this, of course, is related. A meaningful debate is necessary that brings together all the various stakeholders in this remarkably diffuse resource sector. The sustainability of stocks is related to the viability of provincial economies, fishing and processing firms and fishing communities. Failure to recognise this interdependence has led to a great deal of division and bitterness between those most concerned in the fishery. As deep water fishing on the continental slope becomes a regular part of the industry, and as new fisheries are found and developed, it is essential that the

A. G. Hopper (ed.), *Deep-Water Fisheries of the North Atlantic Oceanic Slope*, 307–322.
© 1995 *Kluwer Academic Publishers.*

mistakes of the past not be repeated. Regardless of how the fishery will be prosecuted and managed in the future, debates on the sustainability of fish stocks must ultimately include concern about the sustainability of fishing communities.

1. Introduction

This paper starts from the basic premise that the current outlook on the environment is inspired by paradigms which arise from our cultures, societies and economies. In the past this has been more readily comprehensible for social scientists than for natural scientists, since the social scientists are confronted by social constructs as a regular part of their work. But the world is changing, and those natural scientists who study the environment, the ecologists, are increasingly aware that social attitudes impinge on their work and that "one of the reasons for the problems with current methods of environmental management is the issue of scientific uncertainty . . . the radically different expectations and modes of operation that science and policy/ management have developed to deal with it" (Costanza, 1993; cf. Hutchings and Myers, 1994). Natural and social science have to come to grips jointly with the reality of the environment as embedded in society, economy and culture, for we face horrendous problems of environmental degradation at a time when public awareness is increasing. At the same time public appreciation of the complexities involved leaves much to be desired, and public expectations are at best unrealistic. Moreover, policy makers require information in order to formulate regulations, and thus demand certainties from the scientist where none exists. The end result is "frustration and poor communication . . . mixed messages in the media" and the danger that environmental issues become subject to manipulation "by political and economic interest groups" (Costanza, 1993; see also Steele et al., 1992). In terms of this Workshop, the social issues surrounding fisheries management are currently making a very difficult problem even worse.

The primary focus of fisheries resource management is taken, these days, to be the future sustainability of our fish stocks (Hutchings and Myers, 1994). This is no simple matter and current debate ranges all the way from protests that sustainability in economic terms is just too expensive, to the insistence that sustainability is an illusion (Ludwig et al., 1993; Rosenberg et al., 1993). We are reminded that we not only have to ". . . address fundamental economic biases against sustainability" but also remember that "biological, social and economic considerations [are] an integral part of the development of policies for renewable resource use" (Rosenberg et al., 1993). This is a very tall order, but it is essential, because fisheries management deals with the interface between ecology and society and not just the economy - the place where the needs and problems of people impinge upon the natural world in which we live and of which we are a part (Cronon, 1990). In fact to speak of the sustainability of fish stocks is to speak at one and the same time of the sustainability of human communities.

The specific matter that lies before this Workshop is how the exploitation of deep water oceanic slope fisheries are to be managed. The author believes that our attitudes towards exploitation of a new resource must take into account what has happened in the past, to both the fish and the fishing communities around our shores. In other words, in order to understand the situation in which we now find ourselves, and thus to be able to plan in an informed manner, we need to know where we have come from. That entails a history lesson of the eastern Canadian fisheries.

2. Evolution of Today's Fishing Industry in Eastern Canada

Fishing policy in the 20th century has increasingly supported an up to date capital-intensive year round industry rather than the labour-intensive seasonal inshore fishery that used to be the norm. In Canada, as elsewhere, we now have year-round search and kill technology capable of taking huge amounts of fish, and of pursuing them even into remote places to which they used to be able to retreat in cyclical downturns until the biomass recovered. The spawning seasons are no longer sacred (Harris, 1993). This is an aspect of our harvesting capacity that is crucially different from earlier technologies - there is, now, no place left to hide for disturbed stocks, even on the deep waters of the continental slope. From the point of view of both the resource and the communities it sustained, the use of such efficient killing technology may not have been the wisest choice to make, either in ecological or economic terms. A look at historical data on previous community practice in Atlantic Canada demonstrates this point. Prior to the advent of modern industrial methods of production in the fishery with their drive to achieve efficiency and economies of scale, there existed a balance between ecology and economy which sustained not just fish stocks but also the fishing communities around our shores. In Newfoundland the traditional inshore fishery was prosecuted until the middle of the twentieth century. The centrepiece of the economy for two centuries and more was the summer inshore cod fishery, which was supplemented by resources exploited at other seasons and in a variety of locations. The basic logic of this kind of resource exploitation was that its inherent flexibility rendered economic the settlement of places which would otherwise have been too "marginal" to support a year-round existence. Seasonal resources were seen as a strength, and not the problem they are considered to be today.

In fact, such communities practised a sophisticated approach of using a whole range of different local resources - some near, some far, some landward, some seaward. Indeed, a wide variety of resources were utilized to supplement the central summer fishery all round the shores of what is now Atlantic Canada, and elements of that old economy still persist in many places, although using rather different mixes of skills and resources today. This approach was also typical of coastal communities all round the North Atlantic rim - in Norway, Iceland, the Highlands of Scotland and Ireland.

Flexibility was the way in which such communities managed the ever present resource uncertainty. In Newfoundland at the turn of the century, climatic conditions are thought to have resulted in a decline of the cod stocks and, even at that time the response to decreased stocks was increased fishing effort (Head et al., 1993). A few years later, the rapid expansion in the use of the otter trawl in the offshore fishery also placed the stocks under extreme pressure (Hutchings and Myers, 1994), and this provided the first documentary record of inshore fishermen complaining about what we would now call environmental degradation (Ommer and Hiller, 1990). Such an upset of the resource base would have been disastrous for communities such as Grey River, on the south coast of Newfoundland (figure.8). With its mountainous hinterland and lack of flat terrain suitable for agriculture, this small coastal community is typical of many in the area. During the depression of the 1930s, one visiting government official said of Grey River that the local people knew how to live there, "but all the experts would starve to death". They adopted a flexible approach to their livelihood - they fished for cod, herring and salmon, and caught game in the highland wilderness area behind the mountains (Handcock, pers. comm.). This was a long-term strategy, genuinely sustainable of both the resource base and the people. Significantly, in terms of the modern debates on uncertainty in the literature on sustainability, this traditional culture built flexibility into its *modus vivendi* (Costanza, 1993). It is also worth noting that, unlike the usual perception of western cultures, such communities did not so much treat marine resources as a "commons", but managed them at the community level (Hardin, 1968; Meyer and Helfman, 1993). This intuitive understanding of the limitations of nature to provide food is repeated in many cultures around the North Atlantic and indeed throughout the world.

Ecosystems operate at the local scale. The environment is made up of assemblages of the component units which are interdependent, and when taken together, they are sustainable. When an assemblage is disturbed, adjustments have to take place to maintain the balance of the system. Traditional outport communities (the term applied to rural fishing villages in Newfoundland) operated this way - people (the "top predator" in the system) could, and did, switch their prey, and they also knew that any given system had limits to its level of exploitation. Figures 1 to 3 demonstrate the point for three settlements on the south coast of Newfoundland: Fortune, Gaultois and Harbour Breton. Figure 1 shows the number of marine resources exploited by the three communities from 1884 to 1921, the years covering the 1890s climatic stress and the introduction of the otter trawl. Figure 2 shows the populations of the communities over these years. They were all relatively stable, and Harbour Breton even showed some growth, a result of prosecuting a bait fishery (herring) for the growing Nova Scotian and American offshore banks fishery at that time. If, however, we look below the surface of the total population statistics and examine the labour force cohorts (Figure 3) we see that stability overall was maintained partly by the temporary outmigration of young people in difficult years, and return migration when things improved. The point is that these settlements responded to resource fluctuation and economic uncertainty by employing a strategy which interwove multiple resource exploitation with migration so that the population of the community never went beyond the carrying capacity of the local resource base.

These communities have been selected as examples because they had, and still have, seriously restricted landward resource bases - something which is now making their

NUMBER OF SPECIES EXPLOITED

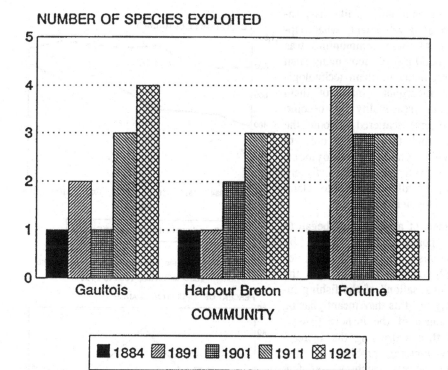

Figure 1. The number of marine resources exploited in Fortune, Gaultois and Harbour Breton ,1884 to 1921

continued existence less and less tenable. What has happened?: Figures 1-3 are typical of outport life in its "undisturbed" condition. After Confederation with Canada in 1949, the modern fishing industry began to expand in Newfoundland, with increasing industrialization of the offshore and relative neglect of the inshore fishery. With the introduction of deep water fishing technology in the 1970s and the establishment of the Canadian 200-mile limit at the end of that decade, Canadian fisheries policy increasingly supported large corporate fishing enterprises, marked by industrial structures, high technology, and high productivity (measured as output per worker). These enterprises were seen as much more efficient than the smaller less intensive inshore operations. In effect, resource management was being increasingly subjected to the requirements of free-market commercial industrial operations, a policy option that is coming to be recognised as less than successful in terms of resource sustainability (Meyer and Helfman, 1993; Ludwig et al., 1993). By the late 1970s the offshore fishery had become a highly capital-intensive industry and the inshore fishery was being increasingly identified as problematic - out of date, inefficient, seasonal, relatively unskilled, and the employer of the last resort for a province burdened with chronic unemployment. There was some truth to that too, for the inshore sector had been neglected in terms of economic development, and the inshore communities likewise. Instead of development, what happened to these communities was the building of too many fish plants, many of them technologically inadequate, many of them seasonal (seasonality had become a problem), scattered all round the coast and there really only to provide enough employment, along with the inshore fishery itself, to allow people to claim unemployment insurance.

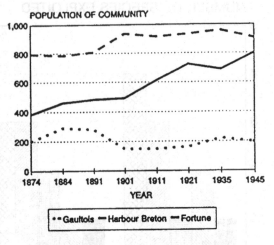

Figure 2. Population change: Fortune ,Gaultois and Harbour Breton, 1884 to 1921

As fears of a stock crisis began to rise, policy makers, backed by economists, began to speak of finally coming to grips with the "rationalisation" of the fishing industry. By this they meant phasing out much of the inshore fishery, and thus addressing the increasingly pressing problem of "too many vessels catching too few fish". That is how the problem was perceived . . . and the language is important here. It was inshore fishermen who were seen as the problem, not too much fishing *effort*, which could have been either fishermen or fishing vessels. That way of thinking, coupled with inadequate biological data and inadequate comprehension by policy makers of the

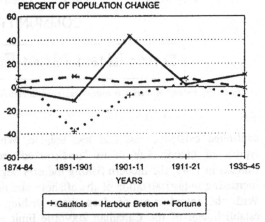

Figure 3. Labour force change : Fortune, Gaultois and Harbour Breton, 1884 to 1921

difficulties faced by scientists of Canada's Department of Fisheries and Oceans (DFO), all of it fuelled by anxiety over spiralling unemployment, led to the ongoing encouragement of over-capitalization in both the processing and harvesting sectors of the industry. The immediate consequence of that, of course, was that investors needed to have a return on their investment, which translated directly into a need to increase landings. Table 1 shows the landings for all Atlantic Canada for the period 1978 to 1993, including the fact that after 1979 catches were never again able to reach allocations. Table 2 shows the increase in the number of processing plants (Government of Canada, 1993), and Figure 4 charts fish catch against those data. Taken together, these data demonstrate clearly that fishing effort in eastern Canada (not just Newfoundland) was responding primarily to socio-economic requirements for growth, with catches spiralling upwards in the wake of increased technological, but not ecological efficiency. What in fact was happening was that the industrial system was building in decreased flexibility in the name of increased industrial efficiency. Policy makers bought the "General Motors" logic which says that what is good for the firm is good for the region and the nation ....the environment was excluded from this logic. The environment was, simply,..... out there to be used and at no time was it seriously considered that we were interfering in an ecosystem whose dynamics were poorly understood. Most importantly, we never thought in terms of being the top predator in that ecosystem, with a technology which we were smart

TABLE 1. Canadian Atlantic groundfish allocations and catches, for the years 1978-93. Entries are thousands of tonnes.

YEAR	ALL GROUNDFISH ALLOCATION	CATCH	ALL COD ALLOCATION	CATCH	NORTHERN COD ALLOCATION	CATCH
1978	472	535	204	271	100	102
1979	562	634	270	359	130	131
1980	705	615	353	400	155	147
1981	790	741	400	422	185	133
1982	924	775	490	508	215	211
1983	997	728	561	505	240	214
1984	1,005	700	553	466	246	208
1985	1,003	738	576	477	250	193
1986	973	748	530	475	250	207
1987	969	723	512	458	247	209
1988	985	688	523	461	266	245
1989	942	652	478	422	235	215
1990	812	604	408	384	197	188
1991	812	572	399	311	188	133
1992	808	418	333	182	120	21
1993	512	N/A	121	N/A	Moratorium	

From Government of Canada (1993), Appendix C, DFO Resource Allocation, p. 124.

TABLE 2. The number of fish processing plants registered with DFO for Canada's Atlantic provinces in the years 1981-1991.

YEAR	TOTAL	Nfld	PEI	NS	NB	Quebec
1981	700	225	42	213	101	119
1982	685	215	42	225	114	89
1983	670	205	43	237	127	58
1984	724	212	48	256	140	68
1985	788	213	55	278	163	79
1986	840	228	60	327	135	90
1987	890	244	56	307	177	106
1988	991	252	65	343	207	124
1989	975	256	76	345	187	111
1990	1018	268	75	347	190	138
1991	1063	281	75	348	194	165

Nfld = Newfoundland, PEI = Prince Edward Island, NS = Nova Scotia, NB = New Brunswick.

Source; Government of Canada (1993), Appendix C, DFO Inspection Data, p. 146.

314

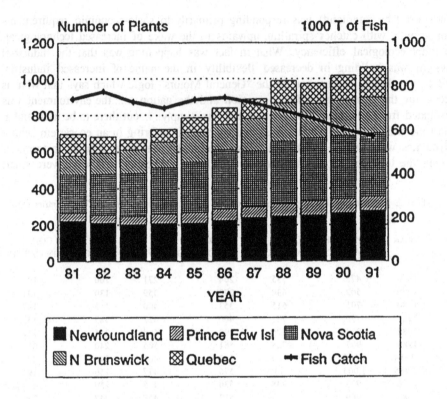

Figure 4. The number of processing plants and the fish catch by year, Atlantic Canada 1981 to 1991

enough to create but not intelligent enough to manage (Steele et al., 1992; Harris, 1993).

3. Communities and Their Ecology

Outport communities have now become, in the modern industrial fishing industry, single industry towns whose single resource has failed. Yet, in the past, these same communities could boast that, although life was never easy, "you'd never starve here". Indeed, up until the 1980s Newfoundlanders traditionally went back to the outports, because there always existed there, the capacity to survive, using the so-called "food fishery" and the informal pluralist economy that has already been briefly touched upon: an economy which had been crafted in such a way as to deal with chronic uncertainty, which worked with seasonality, and which was designed to prevent the vulnerability we associate with single industry mining towns. Now, in March 1994, the food fishery is prohibited for the first time in Newfoundland's history. The crisis that stares us in the face is one of massive impoverishment. It is not just a question of "commercial" or even "biological"extinction of a marine fish stock. It is a question of an ecosystem being put under such stress that it - including the human economy which is part of it - is in process of collapsing. In Newfoundland, it is a matter of the destruction of the basic underpinnings of the culture of a people. This is not merely the collapse of 50% of the jobs in the primary manufacturing sector, 25% of those jobs in the goods-producing sector, or the wiping out of 16% of the jobs left for the total workforce in a province where the official unemployment rate already is around 20% (Government of Canada, 1993). People are now afraid that they may be looking at the end of rural Newfoundland.

It is now necessary to step back and consider the larger structural processes behind these events. The strategy that the outports used, and which social scientists call "occupational pluralism" is, in theoretical terms, the opposite strategy to industrial specialization. It has several great strengths, not least of which is that it is ecologically sensitive (Thornton, 1980; Newfoundland Royal Commission on Employment and Unemployment, 1986; Ommer and Hiller, 1990). It has to be because such a strategy demands a sophisticated knowledge of local ecology and the ability to exploit local resources in such a way as to integrate them rationally into a seasonal round of activities. The examples cited earlier in section 2 are, of course, pre-industrial - but this is not a strategy that needs to be confined to pre-industrial times. What the industrial revolution did, ecologically speaking, was to move us further and further back in the production process away from its natural resource base. Economists say that land ceased to be the principal factor of production, a role that was taken over by capital. Think of that ecologically for a moment, because it captures very nicely the loss of awareness of our roots in nature, and the shift in our thinking from these roots to economic growth - more wealth, more money, more investment . . . and more technology to make even more of that possible. We became, in the process, increasingly remote from the ecosystem of which we are a part; food webs ceased to be a tangible reality for urban dwellers; ecology became a scientific field of research rather than a social science, and it ceased to be something which was a daily awareness for ordinary people. They no longer thought in terms of the surrounding environment, for nature had to all intents and purposes been conquered. Or so it seemed, until recently. The current crisis in the groundfish stocks of the northwest Atlantic has occurred at least in part because ecology, economy and capitalist ideology are in conflict. This is because the nation states who prosecute those stocks are enduring prolonged economic stress as competition in an increasingly global economy threatens to undermine industrial production. In our fishing industries, the response of increasing productivity by using more efficient harvesting technology, has been responsible to some as yet unidentified degree for serious depletion of the stocks (Hutchings and Myers, 1994). The logic of the response is embedded in our belief system which tells us that economic *growth* is essential to national well-being.

The fallacy of that kind of thinking, which has been basic to capitalism since the industrial revolution, is now being demonstrated in our resource sectors. The idea that growth has to be maintained in order to remain competitive must be rethought. At

316

present we do not have enough time to pause in the headlong rush to sell more and more of our natural resources to consider what might be happening to them as a result. We are beginning to be forced to face the unpleasant, and ideologically unpopular, idea that "not all economic modes are ecologically sustainable" (Worster, 1988). It is time to concede that we have been so caught up in the pursuit of economic growth that we are destroying that upon which we depend for that self-same growth . . . and, of course, the fishery is only one resource sector where this is happening. It is time to think about economic development, rather than simple growth, and that means sustainable management of our resource base. In the fisheries, as in other resource sectors, industrially-generated environmental degradation (and sometimes downright destruction) remains a mostly unaccounted cost to industry. In other words environmental damage is not something which, if firms inflict it, they then have to repair it automatically at their own expense; nor do they normally unilaterally build in environmentally-safe technology. It has not been part of industrial capitalist business culture to accept such costs as routinely associated with the cost of doing business, and it is still usually a matter for litigation when disasters do occur (McEvoy, 1986).

It is to call into question this kind of thinking that some ecologists are putting forward recommendations to distrust claims of sustainability and confront uncertainty

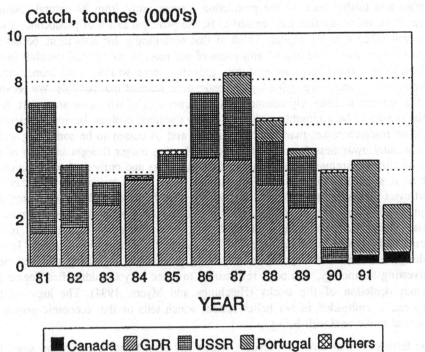

Figure 5. Who caught roundnose grenadier? Catches by country, 1981 to 1992

by shifting the burden of proof from the public to the parties that stand to gain from resource use through devices like "environmental assurance bonds" posted to cover potential damage (Costanza, 1993). These issues are not exclusive to Canada and all states around the North Atlantic are having to re-think their attitudes to restoring the balance in favour of the long term survival of marine resources. As Figure 5 shows, the deep water fisheries resources of the north-west Atlantic already provide high-protein food, not only for Canadians, but for other nations too. A great Canadian economic historian once said of the inshore and offshore fisheries that they are "inherently divisive", and he was absolutely right (Innis, 1954). They have set, and continue today to set, nation against nation, region against region, industry against industry, industry against labour, community against community and even academic discipline against academic discipline.

But this ecological crisis must concern us all; we need to find ways to become partners rather than remain as separate players with an inadequate appreciation of each other's concerns. Fisheries are ultimately about people: their employment, their sustenance, the economic well-being of regions, but also the way of life. The meaningful debate that needs to be held is one that brings together the interests and the knowledge of all the various key players in the marine resource sector. The sustainability of stocks is related to the sustainability of communities, industries and regional economies. Failure to recognize this interdependence in a concrete way, paying it only lip service, is what has led to the current crisis in eastern Canada, and to a great deal of division and bitterness between all concerned, including other countries.

4. Deep Water Fisheries and the Future

With the opening of deep water fishing on the slope as a regular part of the fisheries of the North Atlantic rim, there is a danger that these fisheries will be seen as a way out of the current economic and stock crisis in Canada (cf. Duthie and Marsden, this volume). Given the excess capacity in the industry and the political pressure that can be created thereby, there will be considerable temptation to turn to exploitation of these fisheries with undue haste before enough is known to ensure their safety. We could make the same mistakes all over again if we fall prey to the inherent divisiveness of the fisheries once more. Often, sadly, the divisiveness is the result of goodwill combined with desperation. The goodwill comes from recognising that people in single industry communities have to be helped to survive somehow - another fish plant licence, even, if there is nothing else to hand. The desperation comes from there being nothing else to hand, and there being no clear consensus on how to move from one failing resource into either another, or alternatively into a more diversified economy. Some of the solutions to Newfoundland's community crises will have to come through landward economic diversification, but others are likely to come from crafting a new fishery out of the old, and through development of the continental slope fisheries. How that might be done has to concern us all. In this respect, Costanza's (1993) directive about natural scientists makes sense: we should rely, he advises, on "ecologists and other scientists to recognise the edges

318

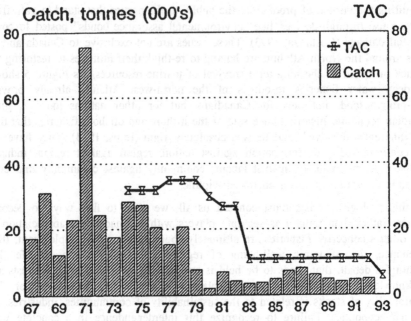

Figure 6. Catch and TACs, roundnose grenadier,1967 to 1993

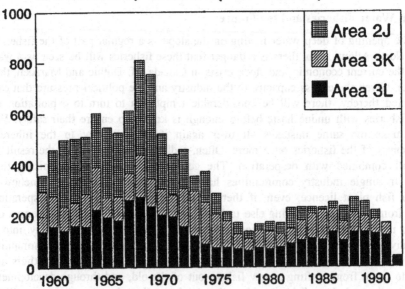

Figure 7. The commercial collapse of the northern cod stock, 1960 to 1992

beyond which we have inadequate knowledge and the worst cases of environmental stress", but we should not "rely on them to solve the problems themselves". "Research", he adds, "should be 'policy-linked' and 'edge-focused'".

This does not mean merely that fisheries scientists will be obliged to suggest quota levels in the same old way as before. Of course there are going to be issues of technology, issues of quotas, fleet and boat size and gear type. However, the overall approach to the management of the deep water fisheries will need to take into account what has happened in the past - for good and for ill - and incorporate the lessons from the past appropriately into guidelines for the future. At the very least, natural and social scientists will need to work together to make it very clear to policy makers that another act of overfishing driven by desperation will produce only another stock crisis. In other words, it is vital that scientists make clear what the present and future problems of exploiting the deep water stocks are, and especially the dangers of overfishing and the use of non-selective techniques.

Figures 6 and 7, are well known to many fisheries scientists, and they illustrate what can happen when biological research and management practice do not work together. Figure 6 is a capsule history of the roundnose grenadier fishery in Canadian waters (Atkinson, this volume), and is a clear example of the lack of coordination between policy, society and science. The TAC bears little or no resemblance to actual catches, and was clearly not the determining factor in what should have happened: it was reactive, not proactive. Society and its policy makers have to understand the long-term implication of data such as this, or the result shown in Figure 7 can be expected to occur in the new deep water fishery, (cf. Clark, this volume; McCormick, this volume). We all have joint responsibility for the disaster that is currently being played out on the beautiful Island of Newfoundland, and we have a joint responsibility to work towards ensuring the lessons of the past are learnt by not mismanaging the deep water stocks in the future. There are "new forms of resource and environmental management where uncertainty and surprises become an integral part of an anticipated set of adaptive responses" (Walters, 1986; Holling, 1993). These are complex, operate at a variety of scales, and recognise that biological "systems require policies and actions that not only satisfy social objectives but . . . also achieve continually modified understanding of the evolving conditions and provide flexibility . . . Science, policy and management then become inextricably linked" (Holling, 1993). If not, then the divisiveness continues, the blame goes round and round . . . and meanwhile species such as the roundnose grenadier and the Greenland halibut (Bowering and Brodie, this volume), and places such as Gaultois and Fortune fade away and die.

There are some searching questions to be asked in taking a new approach to fisheries management for which we do not yet have clear answers.

1) What precisely does international management of fisheries mean? Does it mean managing the resource for maximum profit while it lasts, or does it mean maximising the yield of food fish, or does it mean managing for the well being of the community

320

as a whole and especially for those depend on the sea?

2) In the same context what does best use of the resources mean? Is it for corporate profit, regional wealth, or both?

3) Should technology be allowed to produce more expensive and efficient equipment which then has to be paid for by more fish, which in turn destroys the ecosystem in which it is designed to be used? Could more appropriate technology be devised and the scale of operations in deep-water be regulated to ensure survival of the resources?

The author is grateful for being allowed to speak at this Workshop, but has only shown a fraction of the thinking which is now being heard in the social sciences. The historical data, presented by the author are inadequate for the new analytical treatment imposed on them here, but they show how, in the future, historians and natural scientists can work together and enter into a new kind of research to tackle these problems. There is much to learn and the task may seem intimidating but it is exciting. Human beings possess the intellectual skills to create scientific technologies powerful enough to destroy environments. We need now to develop jointly the wisdom to devise ways to work with the environments on which we rely and of which we are all a part.

Figure 8. Grey River Harbour, Newfoundland

" Only when the last tree has died, and the last river has
been poisoned, and the last fish caught, will we realise that
we cannot eat money"

Native American wisdom saying

Acknowledgements

This work was supported in part by a grant from Canada's Eco-Research Tri-Council, and the cost travel to take part in the Workshop was aided by a grant from the Workshop's Organising Committee. The author is grateful to both sources, and also to Dr. Richard Haedrich of Memorial University for advice on this paper and, indeed, on many aspects of oceanography and fisheries science.

References

Atkinson, D.B. (this volume) 'The biology and fishery of roundnose grenadier (*Coryphaenoides rupestris* Gunnerus, 1765) in the northwest Atlantic', *in*: A.G. Hopper (ed.), Deep Water Fisheries of the North Atlantic Oceanic Slope, Kluwer Academic Publishers, Dordrecht.

Bowering, W.R. and Brodie, W.B. (this volume) 'Greenland halibut (*Reinhardtius hippoglossoides*): A review of the dynamics of its distribution and fisheries off Eastern Canada and Greenland, *in*: A.G. Hopper (ed.), Deep Water Fisheries of the North Atlantic Oceanic Slope, Kluwer Academic Publishers, Dordrecht.

Clark, M. (this volume) 'Experience with management of orange roughy (*Hoplostethus atlanticus*) in New Zealand waters, and the effects of commercial fishing on stocks over the period 1980-1993', *in*: A.G. Hopper (ed.), Deep Water Fisheries of the North Atlantic Oceanic Slope, Kluwer Academic Publishers, Dordrecht.

Costanza, R. (1993) 'Developing ecological research that is relevant for achieving sustainability', Ecological Applications, 3(4), 579-581.

Cronon, W. (1990) 'Modes of profit and production: placing nature in history', Journal of American History, 76(4), 1123.

Duthie, A. and Marsden, A. (this volume) 'Canadian Experience: Deep Water Fishing', *in*: A.G. Hopper (ed.), Deep Water Fisheries of the North Atlantic Oceanic Slope, Kluwer Academic Publishers, Dordrecht.

Gordon, J.D.M., Merrett, N.R. and Haedrich, R.L. (this volume) 'Environmental and biological aspects of slope-dwelling fishes'of the North Atlantic, *in*: A.G. Hopper (ed.), Deep Water Fisheries of the North Atlantic Oceanic Slope, Kluwer Academic Publishers, Dordrecht.

Government of Canada (1993) 'Charting a New Course: Towards the Fishery of the Future', Task Force on Incomes and Adjustment in the Atlantic Fishery, Ottawa.

Gunnarsson, G. (this volume) 'Developments in deep-water trawling', *in*: A.G. Hopper (ed.), Deep Water Fisheries of the North Atlantic Oceanic Slope, Kluwer Academic Publishers, Dordrecht.

Handcock, Gordon, personal communication, Memorial University of Newfoundland.

Hardin, G. (1968) 'The tragedy of the commons', Science, 162, 1243-1248.

Harris, L. (1993) 'Seeking equilibrium: an historical glance at aspects of the Newfoundland fisheries', *in*: K. Storey (ed.), The Newfoundland Groundfish Fisheries: Defining the Reality, Conference Proceedings,

322

Institute of Social and Economic Research, Memorial University, St. John's, pp. 1-8.

Head, C.G., Ommer, R.E. and Thornton, P.A. (1993) 'Canadian Fisheries, 1850-1900', Plate 37 *in*: R.L. Gentilcore (ed.), Historical Atlas of Canada, volume II, University of Toronto Press, Toronto.

Heal, G. (1994) 'Environmental Economics', paper presented to the Social Sciences and the Environment Conference, February, Ottawa.

Holling, C.S. (1993) 'Investing in research for sustainability', Ecological Adaptations, 3(4), 552-555.

Hutchings, J.A. and Myers, R.A. (1994) 'What can be learned from the collapse of a renewable resource? Atlantic cod, *Gadus morhua*, of Newfoundland and Labrador', Canadian Journal of Fisheries and Aquatic Science, in press.

Innis, H. (1954) The Cod Fisheries; the History of an International Economy, Toronto (rev.ed.).

Ludwig, D., Hilborn, R. and Walters, C. (1993) 'Uncertainty, resource exploitation and conservation: lessons from history', Science, 260, 17, 36.

McCormick, R. (this volume) 'The Irish experience of deepwater fishing in the NE Atlantic', *in*: A.G. Hopper (ed.), Deep Water Fisheries of the North Atlantic Oceanic Slope, Kluwer Academic Publishers, Dordrecht.

Meyer, J.L. and Helfman, G.S. (1993) 'The ecological basis of sustainability', Ecological Applications, 3(4), 569-571.

McEvoy, A.F. (1986) The Fisherman's Problem: Ecology and Law in the California Fisheries, 1850-1980, New York.

Newfoundland Royal Commission on Employment and Unemployment (1986), Building on Our Strengths, St. John's.

Ommer, R.E. and Hiller, J.K. (1990) 'Historical Background Report on the Canada France Maritime Boundary Arbitration', Government of Canada (Departments of Justice and External Affairs).

Rosenberg, A.A., Fogarty, M.J., Sissenwine, M.P., Beddington, J.R. and Shepherd, J.G. (1993) 'Achieving sustainable use of renewable resources', Science, 262, 828-829.

Steele, D.H., Andersen, R., and Green, J.M. (1992) 'The managed commercial annihilation of Northern Cod', Newfoundland Studies, 8, no. 1, 34-68.

Thornton, P.A. (1980) 'Dynamic Equilibrium: Population, Ecology and Economy in the Strait of Belle Isle", Ph.D. thesis, University of Aberdeen.

Walters, C.J. (1986) Adaptive Management of Renewable Resources, McGraw-Hill, New York.

Worster, D. (1988) The Ends of the Earth: Perspectives on Modern Environmental History, Cambridge University Press, Cambridge.

LINE FISHING FOR BLACK SCABBARDFISH (*Aphanopus carbo* Lowe, 1839) AND OTHER DEEP WATER SPECIES IN THE EASTERN MID ATLANTIC TO THE NORTH OF MADEIRA

ROGÉLIA MARTINS and CARLOS FERREIRA
Instituto Português de Investigação Maritima (IPIMAR),
Avenida de Brasilia,
1400 Lisboa,
Portugal.

ABSTRACT

Between 1980 and 1986 nine cruises were carried out on board the Research Vessel NORUEGA to study the black scabbardfish fishery in the eastern central Atlantic area around the island of Madeira to search for new fishing grounds and to introduce new fishing technologies.

The results of these studies showed that drifting vertical longlines made from monofilament are more easily worked and have greater durability, as well as having a higher catch efficiency than those traditionally used by Madeira fishermen.

The seamounts with a highest abundance of black scabbardfish were Seine, Lion and Susan.

It is believed that the improved results in black scabbardfish landings in Madeira since 1982 have been due to the fact that the fishermen have replaced the traditional hemp drifting vertical longline by the monofilament drifting vertical longline and some vessels with better equipment have started to fish on other fishing grounds with higher yields.

A. G. Hopper (ed.), Deep-Water Fisheries of the North Atlantic Oceanic Slope, 323–335.
© 1995 Kluwer Academic Publishers.

1.Introduction

The black scabbardfish (*Aphanopus carbo* Lowe, 1839) fishery has been an important fishery for Madeira Island for more than a century. This species has a long established place in the diet of the local population.

The black scabbard fish are found in the waters adjacent to Madeira between 700 and 1300m depth (inside the "Nucleus of Mediterranean Water"), and they seem to concentrate mainly near the bottom of an accentuated slope (Pissara et al., 1983; Martins et al., 1990).

In 1979 the Portuguese Institute for Marine Research (IPIMAR, formerly INIP) was requested by the Regional Government of Madeira to carry out the first studies of the black scabbardfish fishery, in order to evaluate the possibilities of introducing technological improvements in fishing gear and methods, and to search for new fishing grounds so as to increase the annual catch of this important species. These studies were carried out on board the Research Vessel NORUEGA between 1979 and 1986.The fishing operations were carried out using drifting vertical longlines. Some of the results have been published by INIP (Anon, 1980; 1982; 1984a; 1984b; and 1984c.)

2. Fishing Operations

2.1. AREAS SURVEYED

The areas surveyed (Fig. 1) during the nine cruises were located around the islands of Madeira, Desertas, Porto Santo and Selvagens and also the seamounts Ampere, Gorringe, Seine, Lion, Unicorn, Josephine and Susan. On some of these seamounts (Josephine and Susan) experiments were carried out only during one cruise. In all cases the hauls were made in depths of 1300 to 1400 m

2.2. FISHING GEAR - LONGLINE

A longline is normally made up of several lines connected to each other.

During the experiments two types of drifting vertical longlines were used. The traditional longline (Fig. 2) is known in Madeira Islands as "Aparelho da espada" and is made entirely of hemp fibre (sisal) and divided in two parts: the upper part or buoy rope, and the lower part or main line with hooks (Leite, 1988).

The other longline used (Fig. 3), was designed and made using new manufactured synthetic materials (monofilament) and based on the longline traditionally used by the fishermen of Câmara de Lobos.

2.3.BAIT

Different types of bait were used during the experiments, but most were flying squids (Table 1), and the most suitable for black scabbardfish were the orangeback flying squid (*Ommastrephes pteropus* (Steenstrup, 1855)) and the neon flying squid (*Ommastrephes bartami*, Lesueur, 1821).

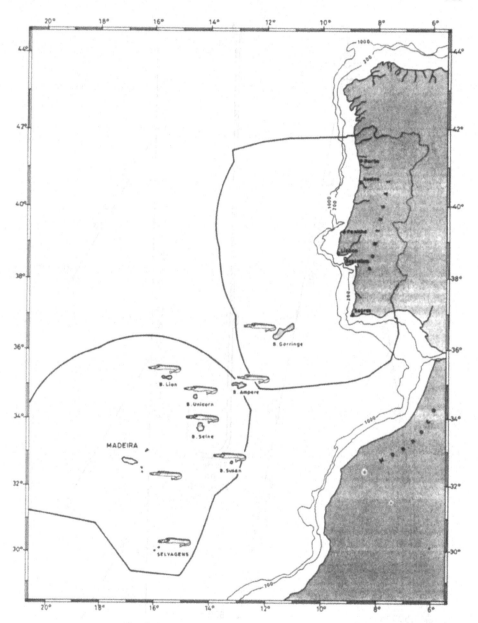

Figure 1. Areas surveyed by Research Vessel NORUEGA - 1980 to 1986

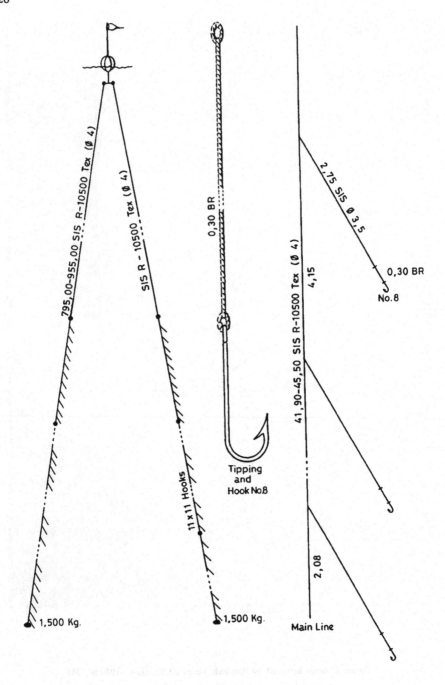

Figure 2. Traditional drifting vertical longline (Anon 1982 ,INIP Report)

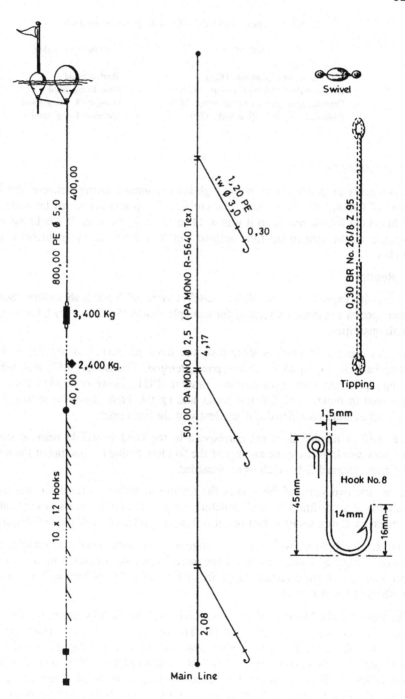

Figure 3. Monofilament drifting vertical longline (Anon 1984c ,INIP Report)

TABLE 1. - List of main bait used in drifting vertical longlines

SPECIES	COMMON NAME
Illex illecebrosus (Lesueur, 1821)	Shortfin squid
Ommastrephes bartami (Lesueur, 1821)	Neon flying squid
Ommastrephes pteropus (Steenstrup, 1855)	Orangeback flying squid
Todarodes sagittatus (Lamarck, 1799)	European flying squid

2.4. IMMERSION TIME

Traditionally longlines are set out at nightfall and hauled during darkness after 8 to 9 hours of fishing. The immersion time during the experiments varied between 7 and 15 hours but for the most part it was of 11 to 13 hours duration. It was found that an increase in the immersion time seemed not to correspond to an increase in the catches.

3 . Results

During the experiments the highest catches were of black scabbardfish. Some of other species appeared frequently, for example sharks (*Squalidae spp*), but mostly in small quantities.

For the capture of black scabbardfish the most effective bait seemed to be the orangeback flying squid (*Ommastrephes pteropus*, Steenstrup, 1855) and the neon flying squid (*Ommastrephes bartami*, Lesseur, 1821). However to obtain the yields measured in numbers of fish/100 hooks and kg/100 hooks from the surveyed areas the catches were considered independently of the baits used.

The yield in number does not correspond to the yield in weight because the total numbers caught include the number of the fish lost during the hauling of the longline and also damaged fish which were discarded.

In the first two years of the studies the traditional drifting vertical longline and the monofilament drifting vertical longline were set in four areas using only the orangeback flying squid as bait because this was preferred by Madeira fishermen.

The results obtained with the two longlines - monofilament and traditional- are shown in Figures. 4 and 5 are show the more favourable benefits from the use of the monofilament drifting vertical longline, with better yields in both kg/100 hooks and numbers of fish/100 hooks.

The yields of the black scabbardfish (weight and number) by areas and cruises are shown in Figures. 6,7,8,9,10 and 11. The best yields obtained from the Seine Seamount, were mainly in November 1980 and April 1985 (Figures 7 and 10) and from the Lion Seamount during May 1983 and April 1985 (Figures 6 and 9). The best yields of all areas surveyed were obtained from the Susan Seamount, in April 1985 (Figures 6 and 9). It is considered that these banks are important black scabbardfish fishing grounds.

The lowest yields were obtained in the Madeira area, perhaps due the proximity of the fishing grounds to the island, and where the fishing effort is probably much higher than the other areas surveyed.

The black scabbardfish landings in Madeira, in the period from 1960 to 1980 were about 900 to 1200 tons (Leite, 1988), and these increased very much in the following years (Figure 12). This can be explained partly by the increased use of monofilament drifting vertical longlines by the most of the Madeira fishermen, which occurred during 1982 (Leite, 1988), and also by the search for new fishing grounds (seamounts with good yields) by those fishermen having boats with better equipment for the longer voyages.

TABLE 2. - List of species caught by drifting vertical longlines

FAMILY	SPECIES	COMMON NAME
HEXANCHIDAE	*Hexanchus griseus* (Bonaterre, 1788)	Bluntnose sixgill shark
LAMNIDAE	*Lamna nasus* (Bonaterre, 1788)	Porbeagle
SQUALIDAE	*Centrophorus granulosus* (Scheneider, 1801)	Granulose shark
	Centrophorus squamosus (Bonnaterre, 1788)	Leafscale gulper shark
	Centroscymnus coelolepis Boc. & Cap., 1864	Siki/Portuguese shark
	Dalathias licha (Bonnaterre, 1788)	Darkie Charlie
	Deania calceus Lowe, 1839	Birdbeak dogfish
	Etmopterus spinax (Linnaeus, 1758)	Black spiny shark
MORIDAE	*Mora moro* (Risso, 1810)	Morid cod
POLYMIXIIDAE	*Polymixia nobilis* Lowe, 1836	Black beardfish
BERYCIDAE	*Beryx decadactylus* Cuvier, 1829	Red bream
APOGONIDAE	*Epigonus telescopus* (Risso, 1810)	Big eye
CHIASMODONTIDAE	*Chiasmodon niger* Johnson, 1863	Black swallower
GEMPYLIDAE	*Ruvettus pretiosus* Cocco, 1829	Oilfish
TRICHIURIDAE	*Aphanopus carbo* Lowe, 1839	Black scabbardfish
	Benthodesmus elongatus (Steindachner, 1891)	Frostfish

330

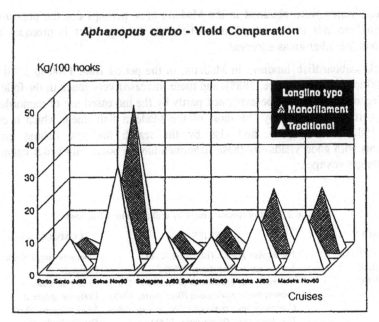

Figure 4. Comparison of yields - kg. / 100 hooks - monofilament v traditional

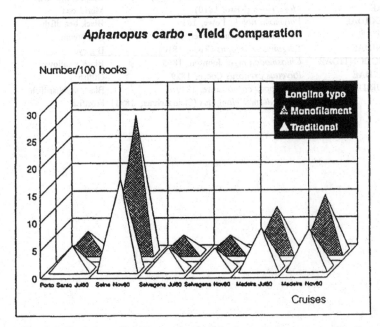

Figure 5. Comparison of yields - number / 100 hooks - monofilament v traditional

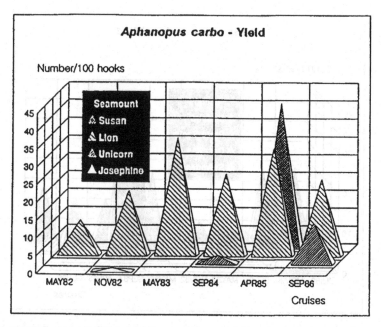

Figure 6. Comparison of yields - number/ 100 hooks - Susan, Lion, Unicorn and Josephine Seamounts

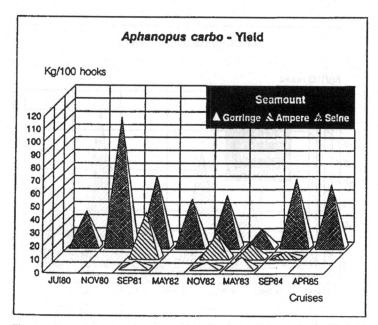

Figure 7. Comparison of yields - kg. / 100 hooks - Gorringe, Ampere and Seine Seamounts

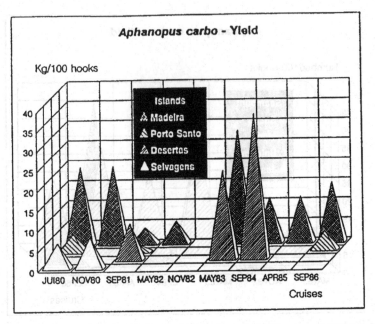

Figure 8. Comparison of yields - kg. / 100 hooks - Madeira, Porto Santo, Desertas and Selvagens Islands

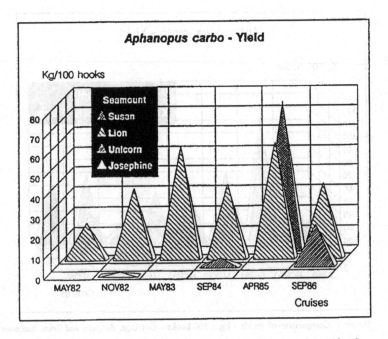

Figure 9. Comparison of yields - kg. / 100 hooks - Susan, Lion,Unicorn and Josephine Seamounts

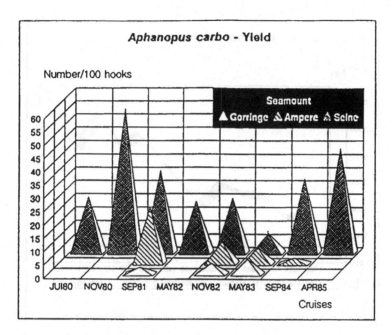

Figure 10. Comparison of yields - number / 100 hooks - Gorringe, Ampere and Seine Seamounts

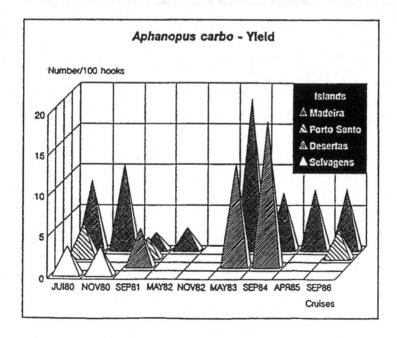

Figure 11. Comparison of yields - number / 100 hooks - Madeira, Porto Santo, Desertas and Selvagens Islands

334

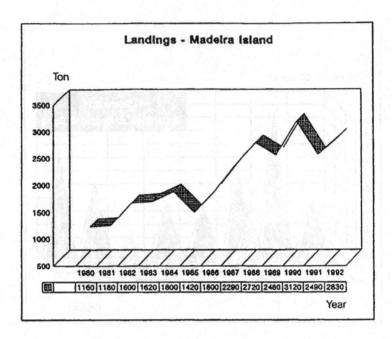

Figure 12. Landings of black scabbard fish at Madeira Island during the period 1980 and 1992

References

Anon. (1980) Programa de Apoio às Pescas na Madeira - I Cruzeiro de Reconhecimento de Pesca e Oceanografia 020080779, *Relat. Tec. Adm. INIP*, Lisboa 3, 141p.

Anon. (1982) Programa de Apoio às Pescas na Madeira - II Cruzeiro de Reconhecimento de Pesca e Oceanografia 020170680, Cruzeiro de Reconhecimento de Pesca de Pesca e Oceanografia 020241180, *Relat. INIP*, Lisboa 11, 220p.

Anon (1984a) Programa de Apoio às Pescas na Madeira - III Cruzeiro de Reconhecimento de Pesca e Oceanografia 020330981, *Relat. INIP*, Lisboa 22, 132p.

Anon. (1984b) Programa de Apoio às Pescas na Madeira - IV Cruzeiro de Reconhecimento de Pesca e Oceanografia 020390582, *Relat. INIP*, Lisboa 25, 125p.

Anon. (1984c) Programa de Apoio às Pescas na Madeira - V Cruzeiro de Reconhecimento de Pesca e Oceanografia 020451182, *Relat. INIP*, Lisboa 31, 124p.

Leite, A. M. (1988) The deep-sea fishery of the black scabbardfish (*Aphanopus carbo* Lowe, 1839) in Madeira Island Waters, *Proc. World Symposium on Fishing Gear and Fishing Vessel Design*, Marine Inst. St. John's, Newfoundland, Canada, 240-243.

Martins, M. R. *et al.* (1989) Portuguese fishery of black scabbardfish (*Aphanopus carbo* Lowe, 1983) off Sesimbra waters, ICES C.M. 1989/G:38, 14p.

Martins, M. R. *et al.* (1990) Pescarias de Profundidade, Peixe Espada Preto. *10ª Semana das Pescas dos Açores, Relatório 1990*, Faial, 199-206.

Pissarra, J. L. *et al.* (1983) Caracterização Oceanográfica na Região da Madeira: Determinação das massas de água no "Núcleo de Água Mediterrânica", *Bol. Inst. Nac. Invest. Pescas*, Lisboa 10, 65-80.

FRENCH EXPLOITATION OF THE DEEP- WATER FISHERIES OF THE NORTH ATLANTIC

A. CHARUAU[1], H. DUPOUY[1] and P. LORANCE[2]

[1]*IFREMER*
8, rue François Toullec
56100 Lorient

[2]*IFREMER*
150, quai Gambetta
62321 Boulogne sur mer

ABSTRACT

French commercial fishermen have been engaged in a deep-water fishery for blue ling (Molva dypterygia) since 1973, and in the early days the roundnose grenadier (Coryphaenoides rupestris) was considered a by-catch species. As a consequence of the decline in catches of the more traditional gadoid species on the continental shelf, increased effort from 1989 onwards, has been directed at the slope fisheries in the north east Atlantic . Today there are commercial catches of blue ling, roundnose grenadier, orange roughy (Hoplostethus atlanticus), black scabbard fish (Aphanopus carbo) and Portuguese sharks (mainly Centroscymnus coelolepis). The fishing vessels engaged in this fishery range from 24 m. to 60 m. in length and 500 kw. to 2000 kw. in power.

Much work needs to be done on the biology of the slope species, but so far in France only blue ling, grenadier and orange roughy have been studied in any detail. Time series of growth, ageing and yield per area have been started but the data is still too incomplete to make a proper analysis. There is a general concern that trawling on the slope for these species will be a high ecological risk to the benthic communities and the coral formations. It is also a fishery directed at virgin stock, and these could quickly become over-exploited.

A. G. Hopper (ed.), Deep-Water Fisheries of the North Atlantic Oceanic Slope, 337–356.
© *1995 Kluwer Academic Publishers.*

1. Introduction

1.1. PRESENT EXPLOITATION

The fisheries of the continental slope have been exploited by the French fleets since 1973, mainly for blue ling *(Molva dypterygia)*. More recently the fisheries were extended for roundnose grenadier *(Coryphaenoides rupestris)* in 1989, and for orange roughy *(Hoplostethus atlanticus)* in 1991. In the past, the two first named species occured regularly in the catches of the trawlers fishing on the edge of the continental shelf, and blue ling became a target species, whilst the grenadier was discarded. The fillets of grenadier were first introduced on the market in 1990 and were quickly appreciated by the consumers. These new fisheries were developed after the decrease of the yield of the gadoid species - mainly cod *(Gadus morhua)* and saithe *(Pollachius virens)* - on the continental shelf, in ICES divisions VIa and Vb. The blue ling is a seasonal fishery from March to June but the effort devoted to that species is now much reduced, and the landings are decreasing from year to year (Figure. 1). At the present time, the five main species caught during deep-water voyages are blue ling, roundnose grenadier, black scabbardfish *(Aphanopus carbo)*, orange roughy and Portuguese sharks (mainly *Centroscymnus coelolepis*).

TABLE 1. Landings by French vessels of the main deep-water species

Year	Orange Roughy	Grenadier	Black Scabbardfish	Black Sharks	Blue Ling
1988	0	0	0	21	9870
1989	8	2727	311	57	9246
1990	19	7501	1524	562	7928
1991	4952	10164	2912	1449	5912
1992	4548	12219	5428	3573	4923
1993*	(2100)	(12000)			

* Preliminary figures only

1.2 THE FLEETS

The vessels exploiting the deep-water species are of various designs and range from 24 m. to 60 m. in length and with powers from 500 kw to 2000 kw .

During the last three years, four separate French fleets have been adapted for fishing for part of the time in deepwater. It is seldom a constant activity. These fleets are:

a) A fleet of 6 trawlers of 50 m. to 55 m. and 1700 kw to 2000 kw, operating from Boulogne sur Mer. In the past these vessels used to work in the North Sea fishing for gadoid species, and from time to time made voyages to the west of Scotland. These vessels are mostly under 20 years of age. They are well equipped, and they can attain fishing depths of 1200 m to 1700 m They were the first to start with a systematic search for orange roughy.

b) A fleet of 10 trawlers of 50 m to 60 m, with powers from 1300 kw to 1500 kw from Lorient, specialising in fishing for gadoids, mainly saithe, blue ling and cod. They are older than the Boulogne fleet and are equipped to fish on the continental shelf, in medium and shallow waters. They have difficulties in attaining very deep water, but from time to time, they make incursions on to the slope to catch grenadier up to 1000 m. deep, but they cannot easily reach the concentrations of orange roughy in depths below 1000 m.

c) A small fleet from Concarneau consisting of 2 trawlers of 30 m. and 35 m. with engine powers of 700 kw. These are modern and well equipped vessels and have for a some time, engaged in trawling on the edge of the shelf for monkfish (*Lophius* spp.), hake and megrim.

d) A fleet of 4 small trawlers from Le Guilvinec of 24 to 26 m. with engine powers of 450 kw. These vessels are the result of very sophisticated technical innovations including special winches for deep water, and are probably the best example of a design of small vessel suitable for variations in fishing methods and target species. They are able to fish up down 1900 m. depth, and achieve good profitability.

2. The biology of the deep-water species

The biology of most of the species caught in deep water is not yet well understood and there is a need for important new research and development. The conditions of the environment influence the behaviour and the physiology of the fish species. The fishing grounds in many instances are completely untouched, and the trawl introduces a considerable disturbance to the benthic communities, and also to the substratum made up of fossilized materials as corals.

In France, only three species have been the subject of recent research studies. The blue ling has been studied for 6 years, the grenadier for 4 years and, the orange roughy only very recently. A common feature of these species is the great difficulty in reading the otoliths to obtain age estimates of the fish. For example black scabbardfishes are landed without heads and it is not possible to initiate, at present, studies on the ageing of landed specimens. As for the small sharks, which are landed as a mixture of four separate species (*Centroscymnus coelolepsis, Centroscymnus crepidater, Centrophorus squamosus and Scymnodon ringens*), it is not possible to make classical studies, and the work is limited only to observations on the variations in the landings of species and weight.

2.1 BLUE LING (*Molva dypterygia*)

At the beginning of the exploitation of the blue ling fishery in the 1970s, the fishing grounds were concentrated in ICES Divisions Vb and VIa. After 1979, Division VIb was also fished, and in 1985, the total international production of blue ling was the same from each of the three divisions. After a small increase in production in 1986, there has been a constant decrease up to the present time (Figure 1 and Table 3).

2.1.1 *Growth of blue ling and age compositions.*
The recruitment at the fishery of the youngest fish of this species starts about 6 or 7

years-old at lengths of about 75 cm. Up to 11 years, the growth is relatively rapid, but there are differences for males and females. For a 12 year old fish, the length of males is about 96 cm and for females 108 cm. The oldest females measured were 18 years-old, and were 130 cm. in length whilst the males, were 106 cm. at the same age (Moguedet, 1988). The continuing diagnosis of the ages of blue ling is a difficult task and an international agreement on the methods is needed. Improvements to the techniques used in France are now being studied, by using an acid etching of the calcite layers aimed at bringing out the different rings corresponding to seasonal growth. The first diagnosis of this work is given in Figure 2.

2.1.2 *Yield by area.*
During each year for the last 10 years since 1984, and in each area there has been a steady decrease in catches and yields of blue ling. To study the changes in yields, the voyages were classified according to the rate of the catches for a given species and the level of 15% was chosen to discriminate the effective target species of the voyage. The results are given in figure 3. For ICES area Vb, there is a continuing decrease, and it is noted that there is a very good consistency between the index from Faroese groundfish survey (Figure 4. (Anon. 1993)) and the French CPUE (Figure 3). The huge yield in 1984 followed by a complete change in the pattern of the fishery from 1985 to 1992 is more an index of a migratory behaviour of the blue ling than a sign of an over-exploitation.

2.1.3 *Assessment.*
A pseudo-cohort analysis was made assuming the hypothesis of a constant catchability corrected by the annual fishing effort and with recruitment at equilibrium. The numbers of fishes were the mean of the numbers over the years from 1988 to 1992. The results presented are very optimistic with very low F value of 0.25 and a Y/R curve showing no F_{max} (figure 5).

2.2. THE ROUNDNOSE GRENADIER (*Coryphaenoides rupestris*)

The roundnose grenadier has been present in the catches of deep water trawlers for many years, and trials were made some years ago to introduce the fillets of roundnose grenadiers well as those of *Chimerea monstrosa.* on the French market The trials were not successful and, when the blue ling fishery developed in 1973, all the grenadiers were discarded at sea. There is no remaining data on these discard practices, but the present stock is no longer to be considered as virgin. There were no preliminary studies of the grenadier, and the sampling method adopted was that usually associated with the gadoid species, and for the first landings at Lorient the lengths were measured and otoliths taken on monthly samples, (Dupouy et Kergoat, 1992) .Throughout the whole of 1993, an evaluation of the small fishes discarded from catches of these deep-water trawlers was carried out.

2.2.1 *Growth of the roundnose grenadier.*
It is not possible to use the total length as standard length because the tails of landed specimens are cut off by the fishermen before processing and storage in ice. The official and commonly used length is the pre-anal length, between the anus and the

tip of the snout (Jensen, 1976). Nonetheless the pre-anal length, is not always easy to measure on commercial samples due to the evisceration of the fish, and as a matter of routine a second length is systematically recorded, representing the length, between the dorsal fin and the tip of the snout. Relationships between total length with respect to the pre-anal and dorsal lengths were calculated.

The first investigations on growth in relation to length were made by IFREMER at Lorient, without reference of sex, and showed that the extreme ages in the commercial catches are in the range 2 to 27 years:

-for 2 year-olds the pre-anal length is 2.5 cm. for a total length of 9 cm.
-for 10 year-olds respectively 13 and 57 cm.
-for 15 year-olds respectively 15 and 66 cm.
-for 20 year-olds respectively 18 and 79 cm
-for 25 year-olds respectively 22 and 94 cm

Systematic ageing of roundnose grenadiers is now included in the programme of commercial sampling of these species landed from ICES sub-areas VI and VII, (Figure 6). The data base is still modest and covers only the period 1990 to 1993. An ogive curve of maturity of grenadiers was also calculated using data collected at sea. The following text table gives this data.

AGE	1	2	3	4	5	6	7	8	9	10	11	12	13	14	15
%MATURE	0	0	0	0	7	26	48	60	67	70	72	73	75	84	100

2.2.2 *Yield by area.*

The fisheries statistics are not yet known for 1993. For the period 1989 to 1992, the results are not sufficiently clear for a logical interpretation (Figure 7). The grenadier is seldom found as a by-catch of the blue ling fishery in the upper water stratums, or of the orange roughy fishery, in the lower stratums, . When the data is chosen according to the area, or the aggregation of the trips, the results can conclude either a decrease or an increase in the yield. When the trips are chosen according to a percentage of the main species caught, with a limit to 15%, there is a tendency to a slight increase or to stability. In sub-area VIa, and VIb, and using cumulative landings for all four fleets of vessels the apparent trend in landings is increasing. But for the same fleets, the trend is decreasing when the yield is calculated on the most productive ICES rectangles (Figures 8a & 8b).

The market imposes a self-limitation of the catches, according to demand, and that limitation can appear either as discards of the smallest sizes of fish or by a limitation of fishing effort on productive areas.

2.2.3 *Assessment.*

The time series are too short to produce a complete analysis of the population of roundnose grenadier. As for blue ling, an analysis of pseudo-cohort was carried out.

342

The diagnosis again is very optimistic, with fishing mortality (Figure 9a) by age being very low, and a yield per recruit curve (Y/R) increasing under the hypothesis of the increase of effort (Figure 9b). These preliminary assessments show that the equilibrium landings would be approximately 8 000 tonnes for the ICES subarea VI.

2.3 ORANGE ROUGHY (*Hoplostethus atlanticus*)

Orange roughy is a new species in the landings from the deep-water trawlers. It first appeared in the catches of these vessels as recently as 1989. More than simply a new species, it became a real hope for a fleet for which the traditionally exploited stocks were much reduced because of the over-fishing of gadoids. The first landings were in 1991 with 5,000 tonnes appearing in the combined fishery statistics of Boulogne and Lorient. In 1992, the landings reduced to 4,500 tonnes and will probably be nearer to 2,900 tonnes in 1993.

The first research work carried out on orange roughy concerned mainly the growth and the sexual maturity (Lorance et al., 1993).

2.3.1 *Growth of orange roughy.*
The following preliminary studies have been made for describing the growth phenomena in orange roughy.

a). Allometric growth of the otolith - Relationships were calculated between the total length of the fish and the weight and the length of the otolith (Figure 10a & b). In both cases, it is obvious that for a length greater than 50 cm, the growth of the otolith does not follow the law of the growth of the fish. In reality, it is very difficult, and probably impossible to read the otoliths of a fish of a length greater than 40 cm.

b). Reading otoliths. -Trials were made of reading the otoliths of the youngest specimens in the catches. The results are given in Table 2. For fish of 30 cm and larger, the dispersion is so wide that it is absolutely impossible to determine the age of a fish, with the range of the diagnosis being + or - 5 years.

c). Ageing by radiometry. - The nuclei of the otoliths do not furnish enough material to make a significant measure for radiometry measurement, and the real age calculated is at best the mean age of the otolith. Since the calcite layers have not all the same volume, the interpretation of the result will describe the growth of the otolith and not the age of the fish.

3. Conclusions

The real problem for the fishing fleets exploiting the deep water stocks is to obtain steady yields. It is obvious that they are changing their target species according to the abundance of a stock (figure 12), but at the present the only deep-water stock which could give steady catches is the roundnose grenadier for which the market is not able to absorb any increases in landings. The success of orange roughy is promising but the production is not yet at the level of the demand. Moreover in 1993, and at the beginning of the 1994 fishing year, the vessels able to fish in deep water have returned towards the traditional species, mainly saithe.

Because the deep-water fishery takes place on virgin grounds, where the benthic fauna and corals are numerous, the ecological risk and their gradual destruction cannot be ignored. On the other hand, for each of these stocks, it is an accumulated biomass, with very old slow growing fish, and the exploitation, if heavy, could lead quickly to a situation of overfishing with a very slow recovery process of the stocks. This is specially true in the case of black sharks which have a low fecundity.

There is much work to be done on age determination of these deep-water species, and this is best achieved by international collaboration between scientific experts in this field. An international aging workshop to develop a consensus of methodology would be an important first step.

There is a great lack of knowledge about the stock structure of the three species discussed in this paper, and much could be learned from international collaboration and sharing of present and past scientific data. It is vital that we do not over exploit these resources and this can easily happen if too much effort is applied too soon.

TABLE 2. Age keys of young *Hoplosethus atlanticus*

length (cm)	5	6	7	8	9	10	11	12	13	14	15	16	17	18	19	20	21	22	23	24	25	26	27	28	29	30	31	32	33	total
18			1																											1
19				1																										1
20			1				1																							2
21	1			1	2	2																								6
22			3	1	3	2			2																					11
23					2	4	1	3	3	4	2			1																20
24				1	3	2	2		2	1	2					1														14
25					1	1	1	5	1	1						1														11
26					1	1		1	1				1																	5
27								3	1			2																		6
28							1	1																						2
29							1	1			1		1							1										5
30								2		2	2		1			1				1	2									11
31								1				1		1				1		1	1									6
32						1			2			1			1	1				1		1								8
33												1			2															3
34																														0
35															1				1											2
36											1							1	2											4
37							1														1									2
38									1						1				1	2	1						1			7
39													1			1														2
40													1																	1
41																1	1	1										1		4
42																														0
43																					1	1								2
																														136

TABLE 3. International landings of blue ling from ICES divisions Vb, VIa, & VIb 1973 to 1992 -tonnes (see also Figure 1)

	Vb	VIa	VIb	Total
1973	5400	9550	0	14950
1974	3300	12650	0	15950
1975	6500	6000	0	12500
1976	6000	8000	0	14000
1977	8000	9500	0	17500
1978	5400	7100	0	12500
1979	4800	4351	835	9986
1980	10020	2907	9361	22288
1981	5027	3685	4483	13195
1982	6457	3624	831	10912
1983	5725	5375	333	11433
1984	8094	3886	3457	15437
1985	6054	5808	7343	19205
1986	7781	7688	4099	19568
1987	6640	9497	491	16628
1988	9601	6383	2601	18585
1989	5189	7161	1607	13597
1990	3549	3829	1185	8563
1991	2391	3452	2498	8341
1992	4674	3512	1586	9772
Mean	6030	6197	2035	14263

345

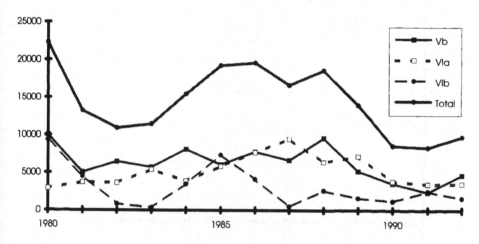

figure 1

international landings of blue ling in tonnes

Figure 1. International landings of blue ling in tonnes from ICES divisions Vb, VIa and VIb
from 1973 to 1992.

346

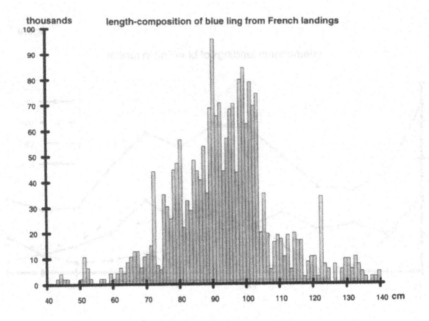

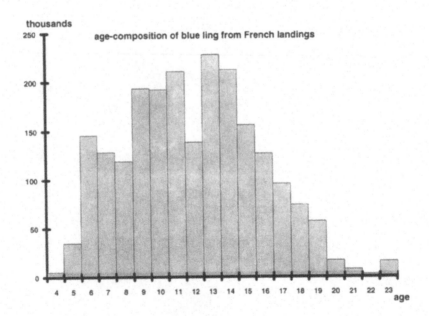

Figure 2b. Blue ling age composition from French landings from ICES divisions
Vb, VIa-b and VIIb-k

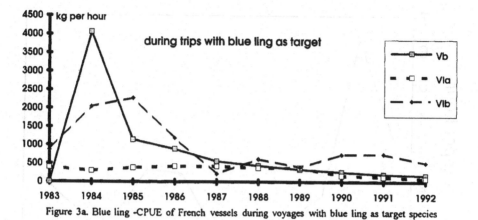

Figure 3a. Blue ling -CPUE of French vessels during voyages with blue ling as target species

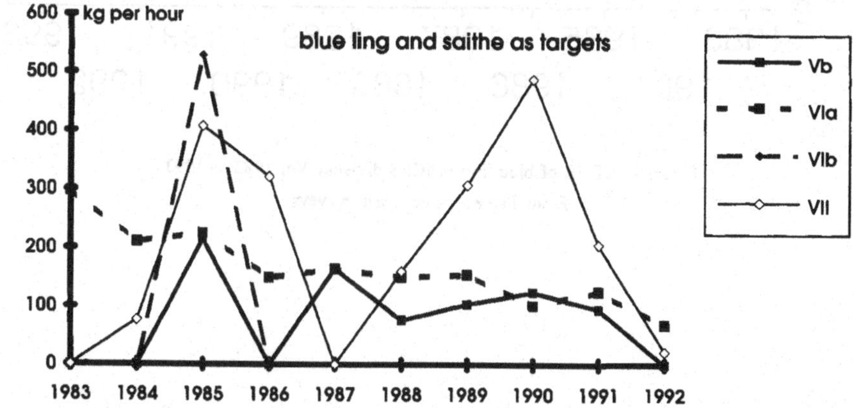

Figure 3b. Blue ling -CPUE of French vessels during voyages with blue ling and saithe as target species

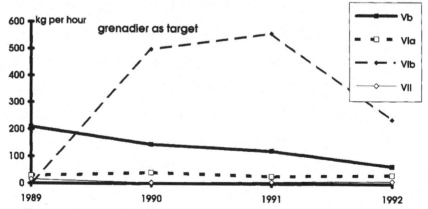

Figure 3c. Blue ling -CPUE of French vessels during voyages with grenadier as target species

348

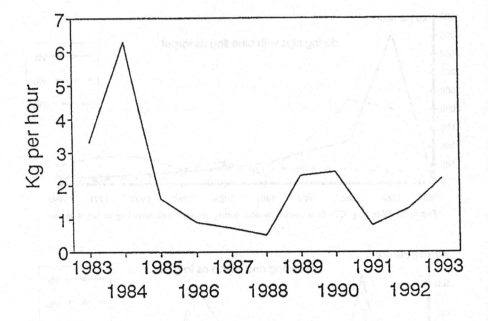

Figure 4. CPUE of blue ling in ICES division Vb, 1983 to 1993
From Faroese groundfish surveys

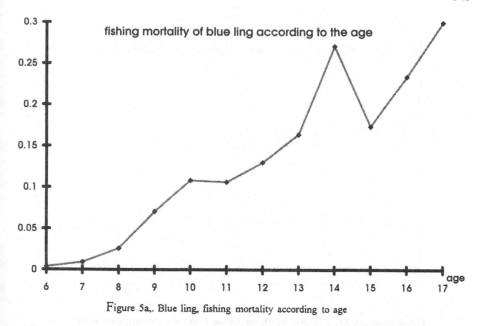

Figure 5a,. Blue ling, fishing mortality according to age

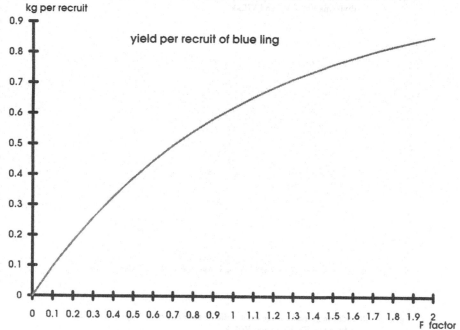

Figure 5b,. Blue ling, yield per recruit

350

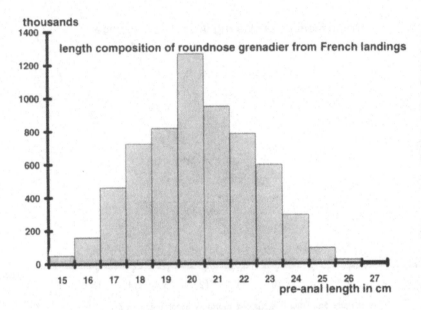

Figure 6a. Roundnose grenadier, length composition from French landings from ICES divisions Vb, VIa-b and VIIb-k

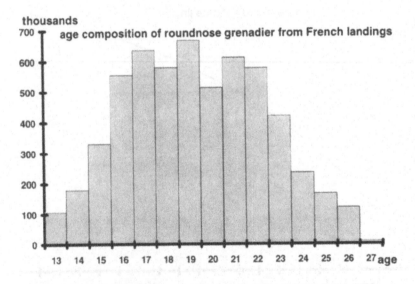

Figure 6b. Roundnose grenadier, age composition from French landings from ICES divisions Vb, VIa-b and VIIb-k

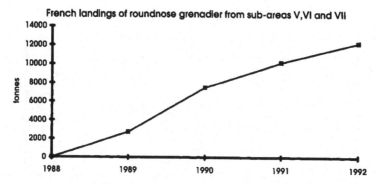

Figure 7a. Roundnose grenadier- total landings by French vessels from ICES divisions V, VI and VII 1988 to 1992

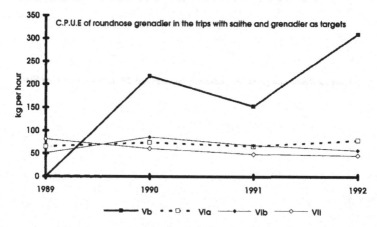

Figure 7b. Roundnose grenadier- CPUE from voyages with saithe and grenadier as target species

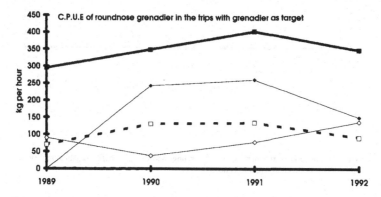

Figure 7c. Roundnose grenadier- CPUE from voyages with grenadier as target species

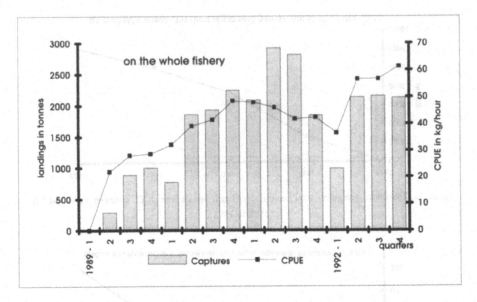

Figure 8a. Roundnose grenadier - quarterly landings and CPUE in ICES divisions VIa-b

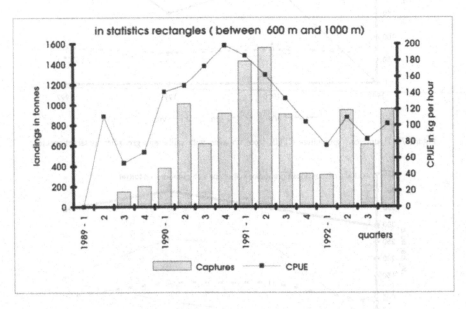

Figure 8b. Roundnose grenadier - quarterly landings and CPUE in ICES statistical rectangles
between 600 m and 1000 m

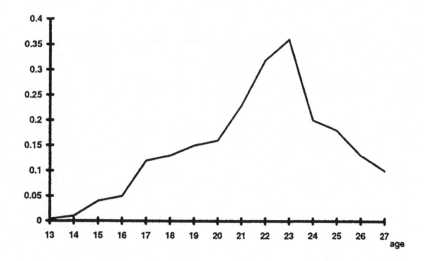

Figure 9a. Roundnose grenadier - fishing mortality in ICES division VIa

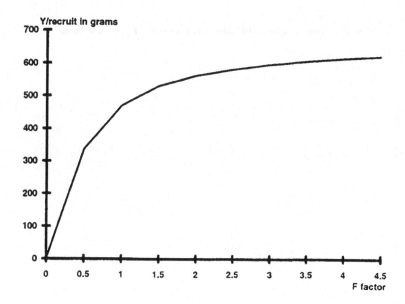

Figure 9 b. Roundnose grenadier - long term yield

354

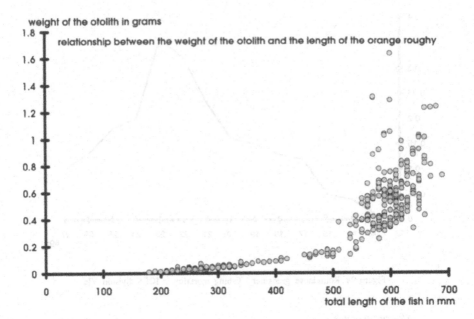

Figure 10a. Orange roughy - relationship between weight of otolith and length

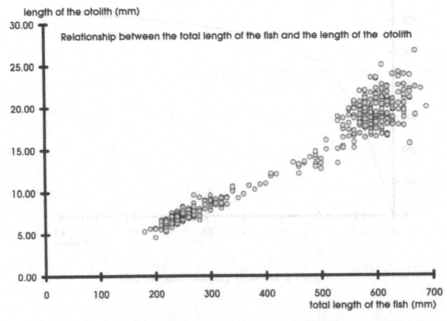

Figure 10b. Orange roughy - relationship between total length of fish and length of otolith

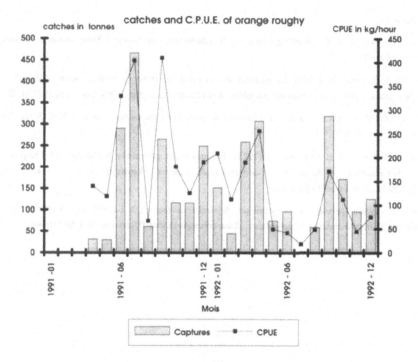

Figure 11. Orange roughy - catches and CPUE

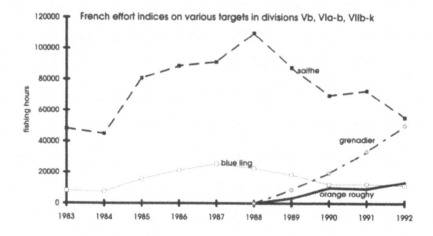

Figure 12. Effort indices on various target species in ICES divisions Vb, VIa-b and VIIb-k

Bibliography

Anon, (1993).- Report of the Working Group on the Assessment of Northern Shelf Demersal Stocks. -C.M.1993/Assess:20.

Dupouy, H. et Kergoat, B, (1992). La pêcherie de grenadier de roche (*Coryphaenoides rupestris*) de l'ouest de l'Ecosse : production, mortalité par pêche et rendement par recrue. ICES, Doc. C.M.1992/G:40.

Jensen, J.M., (1976) Length measurements of roundnose grenadier (*Macrourus rupestris*). ICNAF res. doc 76/VI/93, Ser. No. 3913.

Lorance, P., Dupouy, H. et Charuau, A., (1993). Estimation des paramétres biologiques de l'empereur (*Hoplostethus atlanticus*) et du grenadier (*Coryphaenoides rupestris*). Report in execution of a contract between IFREMER and the EU (in press).

Moguedet, P. (1988) Approche de la dynamique de stocks accessoires; le cas des lingues (Molva spp.) exploitées par la flottille industrielle lorientaise. Thèse présentée devant l'Université de Lille (France)

RUSSIAN (USSR) FISHERIES RESEARCH IN DEEP WATERS (BELOW 500 m.) IN THE NORTH ATLANTIC

F. M. TROYANOVSKY & S. F. LISOVSKY

Polar Research Institute of Marine Fisheries and Oceanography, (PINRO),
6 Knipovich Street,
183763,Murmansk,
Russia

ABSTRACT

Russian investigations of the deep-water species of the Atlantic commenced in 1963 and have continued to 1993. The first investigations were carried out by fishing vessels and research vessels on Greenland halibut (Reinhardtius hippoglossoides) in the Barents Sea, and then on witch flounder (Glyptocephalus cynoglossus), in the western Atlantic. In 1967-68 a commercial fishery for roundnose grenadier (Coryphaenoides rupestris) and Greenland halibut started on the slope in the north west Atlantic. At this time there was also a fishery at Iceland for Greenland halibut. Since 1973 Russian vessels have been fishing on the ridges and underwater rises in the open areas of the North Atlantic for a variety of deep-water species

On these cruises the environmental conditions of temperature and salinity were recorded together with the biological data and bathometric and hydrographic information down to 2000 m. depth. Biological sampling was done with both mid-water trawls, bottom trawls, longlines and traps. Fish behaviour patterns and distribution have been studied using a manned submersible vehicle.

Some of the results of these surveys are given together with estimates of the commercial catches between 1967 and 1991. In the peak year of 1971, 82,000 tonnes of roundnose grenadier were caught. The paper also makes some estimates of the biomass and suggests possible commercial catch levels of the main species of marketable value, which can be caught in the north east Atlantic outside of the economic zones of the coastal states

A. G. Hopper (ed.), Deep-Water Fisheries of the North Atlantic Oceanic Slope, 357–365.
© *1995 Kluwer Academic Publishers.*

358

1. Introduction

Development of fish resources in deep waters of the North Atlantic became necessary due to the increased intensity of fishing and the reduction in stocks on the continental shelf around the Atlantic basin. Scientific research and surveys along with industrial development were carried out for the first time on the deep-water stocks which had no precedents in World fisheries. The work encountered new difficulties in research and it differed from similar research on the traditional fishing grounds in the waters of the continental shelf. It was necessary to revise the existing opinions of scarcity of fish stocks in these depths, and to introduce into the Russian fisheries those species of fish which had not been exploited previously, or had been poorly utilized. The investigations were carried out both in deep-water and in the epipelagic and mesopelagic zones. Only the results from those investigations carried out below 500 m. are considered in the paper.

2. Methods

Russian fisheries investigations of deep-water fish were carried out on existing research and commercial vessels. To effect this the re-equipping of the technical and scientific equipment on the vessels, and special training of specialists were necessary, and a set of scientific investigations was worked out, directed towards the development of entirely new methodologies for studying the aggregations, the forecasting of migration and spawning patterns, and the fisheries management of deep-water fish.

The environmental parameters of temperature, salinity and biological productivity, the biology, distribution and conditions for formation of fish aggregations, together with the relief of underwater mounts, were studied during these investigations. The manned submersible vehicle SEVER-2 , operated on underwater mounts in the North Atlantic down to 2000 m. depth, and largely contributed to studying the conditions for fish aggregation formation. Biological sampling was done down to the 2000 m. depth contour with mid-water and bottom trawls, traps and longlines. Trawling was by the traditional two warp system and an experimental one-wire scheme. Several hauls were done below 4000 m. depth. Scientific investigations and commercial fishing were mainly performed by vessels with a engine power of 1500 to 3000 kW. The work carried out by the vessels of the Ministry of Fisheries of the USSR (Fisheries Committee of the Russian Federation) is given in Table 1.

3. Results

3.1. GENERAL SCIENTIFIC DATA

The most important factors, determining the biological productivity and the characteristics of distribution and behaviour of deep-water fish, are the main and secondary polar fronts, the eddies and their transformations and the quasi-stationary gyres (Loctionov, 1983; Popova, 1977). Most frequently these species were observed near the underwater rises, interacting with the water masses of quasi-stationary currents.

TABLE 1. Deepwater fisheries investigations conducted by Russia (USSR) in the North Atlantic.

SURVEY AREA	TARGET SPECIES	YEARS	No. OF CRUISES	DURATION
north west Atlantic	Greenland halibut roundnose grenadier witch flounder	1963-93	118	303 months
north east Atlantic excluding the Barents Sea	roundnose grenadier *Sebastes mentella* alfonsino orange roughy tusk etc.	1972-93	101	334 months

Underwater seamounts may be separate summits or part of a broken-up undersea massif, rising, by as much as 1000 m. above the general level of the northern Mid-Atlantic Ridge. Smaller undersea rises are also interesting for the deep-water fishery, (Lavrov,1979; Shreider et al., 1986).

Large-scale investigations of the biological resources, carried out by research workers from both the academic and applied institutes of the USSR, described not only the species composition of deep-water ichthyofauna, but also identified the most promising species for commercial fishing, and considered the regularities of biological productivity formation. They also developed detailed fishing charts and plotboards of the best surveyed underwater seamounts, determined the food and technical value of the individual species, and the best means of processing and utilization.

Due to difficult sea-bed conditions and the peculiarities of bathometric distribution of deep-water fish, there were difficulties in adapting the traditional methods for studying ichthyofauna, and some unique expeditions using the manned submersible vehicle SEVER-2 were conducted by the Polar Research Institute of Marine Fisheries and Oceanography (PINRO). During these investigations some additional knowledge of the ichthyofauna composition was obtained, and the behaviour and distribution of the main species were studied. The methods for biomass estimation of these deep-water stocks were developed (Zaferman et al., 1991; Shibanov et al., 1987, 1988; Shibanov and Yarovoy, 1988; Shibanov and Kalugin, 1989).

In mid 1970s, prior to introduction of the 200-mile economic zones by all Atlantic coastal states, reconnaissance fish finding surveys were carried out over almost the whole of continental slope of Europe from the Faroe-Icelandic ridge to North Africa. Species composition of ichthyofauna was elucidated, and the most promising areas and species for a commercial fishery were defined. An experimental fishery was carried out on the most abundant deepwater species:- roundnose grenadier (*Coryphaenoides rupestris*), blue ling (*Molva dypterygia*), ling *(Molva molva)*, alfonsino (*Beryx splendens*) Baird's smoothead (*Alepocephalus bairdii*), orange roughy (*Hoplostethus atlanticus*), cardinal fish (*Epigonus telescopus*) deepwater sharks (*Squalus centroscyllium, Etmopterus scyliorhinus, Apristurus, Pseudotriakis* etc), . Table 3 includes results from this fishery.

From the results from the investigations undertaken, the ichthyofauna from the north east Atlantic continental slope, and the underwater mounts of the Mid-Atlantic Ridge includes about 180 to 200 fish species from 46 families, pertaining to the demersal and bottom-pelagic ichthyofauna (Kukuev et al., 1980; Gushchin et al., 1981; Kukuev, 1991; Parin, 1987, 1988; Parin and Golovan',1976; Shcherbachev et al., 1985). The most abundant and promising species for fisheries development are the representatives from the families of *Macrouridae, Berycidae, Alepocephalidae, Moridae, Gadidae, Trachichthyidae, Squalidae, Scorpaenidae, Lophiidae* The life cycle, distribution, behaviour, stock status of roundnose grenadier *(Coryphaenoides rupestris)*, ocean redfish *(Sebastes mentella)*, beryx alfonsino *(Beryx splendens)* and some other species have been studied to the full

3.2 ROUNDNOSE GRENADIER (*Coryphaenoides rupestris*)[1]

Analysis of the biological data on roundnose grenadier suggests the hypothesis that it is a single population having a complicated structure consisting of a system of dependent and independent sub-populations inhabiting the North Atlantic (Figure 1). The population area is on the mid-Atlantic Ridge, and the sub-populations inhabiting this area play the main role in the population reproductive capacity.[1]

The populations inhabiting the north west Atlantic and those along the continental slope of the European states are dependent populations, and they maintain their abundance due to the transportation by currents of eggs and larvae from the spawning area on the Mid-Atlantic Ridge. For this reason the distribution areas of the roundnose grenadier are areas of species eviction.

Since the drift of fish at a very early stage of development to the north west Atlantic occurs under unfavourable conditions physiological variations in their organisms retard further development of the gonital glands. For this reason the sub-population in the north west Atlantic could be referred to as pseudo-populations and the areas of settling and the concentration of specimens to an area of sterile eviction

To judge by the population structure a commercial fishery for roundnose grenadier should be developed more intensively in these species eviction (spawning) areas. Overfishing should not be allowed on the Mid-Atlantic Ridge since it will diminish the reproductive capability of the population and adversely effect the species abundance over the whole North Atlantic.

---

1 Editor's note: Two papers within this volume give conflicting views on the stock structure of *Coryphaenoides rupestris*. Troyanovsky and Lisovsky; cf. this section, based mainly on Russian research and hypotheses from 1963 to the 1980s argue for a single stock throughout the North Atlantic. Atkinson; cf. the scientific section of this volume, using more recent data and a re-assessment of earlier research work argues that this hypothesis is no longer tenable. This highights the continuing debate on this species .

No data on a spawning period has been determined for the life cycle of roundnose grenadier from the north west Atlantic. Well pronounced seasonal migrations are typical of the roundnose grenadier in this area. In winter the fish move to deep-waters, and in spring they move to the upper part of the slope where fattening takes place. These migrations are of a trophic nature and take place by following the zooplankton feed, and shifting through the season. Searching for aggregations of roundnose grenadier and the commercial fishing activities are carried out during the fish fattening period on the upper part of the continental slope.

Roundnose grenadier seasonal migrations are poorly pronounced on the Mid-Atlantic Ridge and the rises in the north east Atlantic. Searching for and fishing the aggregations are carried out on a round-the-year basis. However, the best results are in the summer and autumn when spawning takes place and density of fish aggregations are at the the highest.

The methods for searching and fishing the roundnose grenadier were developed by PINRO scientists and fishermen from an understanding of the specific peculiarities of the sea-bed topography in each of the areas where the aggregations concentrate. They are also based on an understanding of the daily vertical migrations, the distribution and behaviour of fish and how these vary depending on the influence of hydrographic and hydrobiological processes (Troyanovsky, (1972); Troyanovsky and Kosobutsky, 1974).

3.3. GREENLAND HALIBUT (*Reinhardtius hippoglossoides*)

The area of distribution for Greenland halibut in the North Atlantic extends from extreme north eastern areas of the Barents Sea (Franz Josef Land Island) to the USA continental shelf, and is comparable to roundnose grenadier area in scale. However, by contrast with the roundnose grenadier, three independent populations of Greenland halibut inhabit this area;- the Norwegian-Barents Sea, the Icelandic and the Greenland-Canadian populations (Pechenik and Troyanovsky, 1970; Nizovtsev and Troyanovsky, 1970) (Figure. 2).

The structure of the Norwegian-Barents Sea and Icelandic populations is a simple one. Within the regions occupied by these populations only one spawning ground and clearly pronounced shelf areas of juvenile feeding, and the fattening of adult specimens on the continental slope are known.

The Greenland-Canadian population on the other hand is a complicated one with a stepped model of the population structure. The main spawning grounds are in the southern part of the Greenland-Canadian rapid (threshold). Additionally, halibut, which permanently inhabit the Davis Strait, and specimens from subpopulations, existing along the Labrador and Newfoundland slope, migrate to this area to spawn.

According to results from the PINRO investigations there are no less than four such subpopulations. The relations between them takes place in the prespawning period in December and January each year. At this time specimens having attained maturation migrate from the area of their previous localization northward where they spawn. During several years of such step-by-step migrations they reach the main spawning ground. The period of spawning on the main and subpopulation spawning grounds coincides, taking place in February to March. The depths of the spawning grounds are below 1200 m. Young halibut from the subpopulation spawning grounds recruit to the total stream of young fish on Canadian continental shelf, and are transported by currents from the main spawning ground of the population in the Greenland-Canadian threshold.

The peculiarities of the structure of the Greenland-Canadian population of Greenland halibut result in a constantly high abundance of new recruits to the fishery, and this counteracts the possible adverse effects of natural factors.

The surveys and the fishing season for Greenland halibut are carried out in the fattening areas of prespawning and postspawning migrations with due allowance for the peculiarities of the life cycle. Fishing on aggregations of Greenland halibut on spawning grounds occur only in the Barents Sea; in other areas they occur in deep water, and become inaccessible for trawling.

Greenland halibut are not detectable by acoustic fish-finding devices. Therefore, the understanding of the fishery for this species has depended on knowledge of the distribution and behaviour at the different stages of the life cycle in relation to the feeding and environmental conditions (Troyanovsky and Chumakov, 1992).

The location of aggregations and development of trawl and long-line fisheries for Greenland halibut outside the 200 mile EEZ, to the west of Iceland, on the Hatton-Rockall Plateau, and in the rapids in the Norwegian and Greenland Seas is promising.

3.4. OTHER SPECIES

The spawning grounds of witch flounder (*Glyptocephalus cynoglossus*), found in Divisions 2J and 3K, belong to the population inhabiting the shelf of these areas. Knowledge of the peculiarities of the witch flounder life cycle have allowed a successful fishery on spawning fish aggregations to take place there for a long time (Pechenik, 1970).

When investigating deep waters to the west of the British Isles, near the Faeroe Islands and Iceland quantities of Baird's smoothead (*Alepocephalus bairdii*), black scabbard fish (*Aphanopus carbo*), cardinal fish (*Epigonus telescopius*), orange roughy (*Hoplostethus atlanticus*) and deepwater sharks occurred (the catches of the latter were from 2 to 15 tonne per haul). In 1977-1978 a fishery for blue ling (*Molva dypterygia*) and ling (*Molva molva*) was conducted on the Outer-Baileys Bank. A casual fishery for alfonsino, orange roughy and black scabbard fish is carried out around the southern seamounts of the Mid-Atlantic Ridge and Corner Rise (NAFO Subarea 6G). An experimental long-line fishery for tusk (*Brosme brosme*) was carried out on the seamounts of the Mid-Atlantic Ridge as far as south as 48°N.

According to preliminary estimates by Russian scientists the possible catch of deep-water fish from the north east Atlantic open areas amounts 125,000 to 250,000 tonnes (Table 2).

TABLE 2. Estimate of the possible catch of deep-water fish by Russian (USSR) vessels in the north east Atlantic open areas.

SPECIES	BIOMASS	POSSIBLE CATCH
	tonnes x 1000	tonnes x 1000
Roundnose grenadier	400-500	25-50
Alfonsino	200-400	10-50
Orange roughy		10-30
Black scabbard fish		10-20
Cardinal fish		10-20
Tusk		10-20
Deepwater sharks		50-60

4. Summary of Russian catches of deep-water fish in the North Atlantic

The Russian catches of the main commercial deep-water fishes between 1967 and 1991 are summarised in Table 3 and sub-divided into geographical regions

TABLE 3. Catch of deepwater fish from the North Atlantic by Russian (USSR) vessels, 1967 TO 1991

AREA	ROUNDNOSE GRENADIER		GREENLAND HALIBUT		ALFONSINO	MOLVA
	NWA[1]	MAR[2]	NWA[1]	NEA[3]	N.Atlantic[4]	NEA[2]
1967	15.9		4.3	5.7		
1968	32.7		10.2	3.4		
1969	15.0		10.2	19.8		
1970	29.4		8.0	35.6		
1971	82.5		10.9	54.3		
1972	31.8		19.8	16.2		
1973	19.6	0.2	12.8	8.6		
1974	35.8	17.1	19.2	17.0		
1975	27.9	29.7	29.7	20.4		
1976	28.2	4.5	17.7	16.6	3.4	
1977	17.0	8.7	8.7	15.0	0.7	12.6
1978	17.8	10.0	5.6	14.7	0.2	0.8
1979	7.3	4.1	2.9	10.3	1.4	0.3
1980	1.1	11.9	1.8	7.7	0.3	0.2
1981	5.7	4.1	7.0	9.3	0.6	
1982	2.7	0.7	5.0	12.4	0.3	
1983	1.0	7.8	4.7	15.2	0.2	
1984	0.2	4.4	0.5	15.2	0.2	
1985	1.0	5.7	0.3	10.2		
1986	2.8	9.0	0.8	12.2	0.3	
1987	2.7	10.5	6.7	9.7	2.0	
1988	2.0	10.0	1.1	9.4		
1989	2.2	8.0	1.1	8.8		
1990	0.6	2.3	8.9	4.3		
1991	0.3	2.9	11.4	2.5		

Notes: [1] =North west Atlantic [2] = Mid-Atlantic Ridge [3] = North east Atlantic [4] = Atlantic Ocean as a whole

364

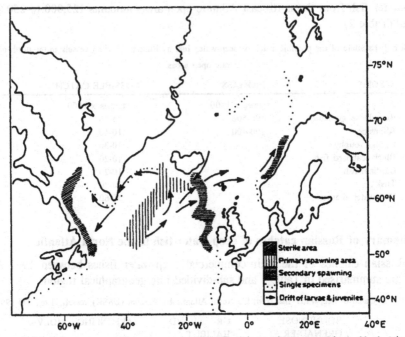

Figure 1. Population structure of roundnose grenadier (*Coryphaenoides rupestris*) in the North Atlantic

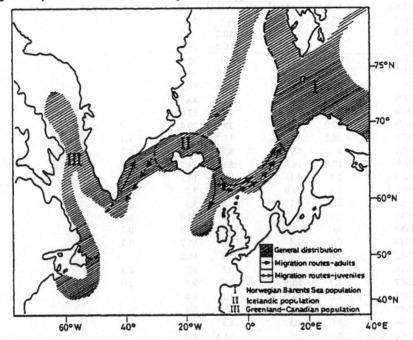

Figure 2. Distribution of Greenland halibut(*Reinhardtius hippoglossoides*) in the North Atlantic

References

Gushchin,A.V., Kukuev, E.I. 1980. On composition of ichthyophauna of the northern Mid-Atlantic Ridge. In: Fish of the open ocean.- Sbornik nauchnykh trudov.- Nauka Press, Moscow, pp. 36-40 (in Russian).

Kukuev,E.I.1991. Ichthyofauna of underwater elevations of boreal and subtropical zones of the North Atlantic. In: Biological resources of talassobathyal zone of the World Ocean.-Sbornik nauchnykh trudov.-VNIRO, Moscow, pp.15-39 (in Russian).

Kukuev,E.I.,Gushchin,A.V.,Gomolitsky,V.D.,Miloradov,G.K.1980. Methodical materials for indicating fish from the North Atlantic open areas. AtlantNIRO, Kaliningrad, 145 p. (in Russian).

Lavrov,V.M.1979. Geology of the Mid-Atlantic Ridge. Nauka Press, Moscow, 143 p. (in Russian).

Loctionov,Yu.A.1983. Working scheme of location of the main productive zones by oceanographic data from the Mid-Atlantic Ridge area. In: Oceanographic basis of formation of biological productivity in the North Atlantic.-Kaliningrad, 3: pp28-37 (in Russian).

Nizovtsev,G.P., Troyanovsky, F.M. 1970.Biological characteristic and distribution of Greenland halibut *Reinhardtiushippoglossoides* Walbaum. Commercial concentrations in Iceland area.- Materials of fisheries investigations of the Northern Basin.-16(2): pp.50-59 (in Russian).

Parin,N.V.1988. Fish of the open ocean. -Nauka Press, Moscow, 271p.(in Russian).

Parin,N.V., Golovan',G.A. 1976. Pelagic deepwater fish of the families typical of the open ocean .- Proceedings of the Institute of Oceanography of the USSR Academy of Sciences. Nauka Press, Moscow, pp.104: 237-276 (in Russian).

Parin,N.V. 1987. System of oceanic ichthyocenosis and its commercial potential. In 'Biological resources of the open ocean'.- Proceedings of the Institute of Oceanography of the USSR Academy of Sciences, Moscow, pp.138-163 (in Russian).

Popova, M.N.1977. On water circulation in the area of the Mid-Atlantic Ridge. In: Voprosy promyslovoy okeanologii Mirovogo Okeana. (Theses of Reports of the IVth All- Union Conference on commercial oceanology).-Murmansk, pp.30-33 (in Russian).

Pechenik, L.N.,Troyanovsky, F.M. 1970. Some biological characteristics of Greenland halibut on the continental slope of Iceland.- Rybnoe khozyaistvo, 2: 4-6 (in Russian).

Pechenik,L.N.,Troyanovsky, F.M.1970. Witch flounder. -Rybnoye khozyaistvo, 1:p.13-14 (in Russian).

Shiibanov,V.N., Kalugin,A.N., Yarovoy,A.S. 1989. On methods for estimating roundnose grenadier. -In: Biological resources of meso-and bathypelagic of the North Atlantic open area.-Sbornik nauchnykh trudov PINRO.-Murmansk,pp. 156-165 (in Russian).

Shibanov,V.N.,Yarovoy, A.S.1988. Determination of reflectivity in roundnose grenadier and macroplankton using "SEVER-2" submersible vehicle.- Rybnoye khozyaistvo, 3:pp.74-76 (in Russian).

Shibanov,V.N., Kalugin,A.N., Bondarev,V.V.1988. Estimation of biomass of bottom ichthyocenosis on the underwater rocks in the Azores area using "SEVER-2" submersible vehicle.- Instrumental methods for estimating the stocks of commercial objects.-Sbornik nauchnykh trudov PINRO,Murmansk, p.65-74 (in Russian).

References

NORWEGIAN EXPERIENCE OF DEEP-WATER FISHING WITH LONGLINES

HANS EDVARD OLSEN,
Division of Experimental Fishing,
Directorate of Fisheries,
PO Box 185,
N-5002 Bergen,
Norway

ABSTRACT

Experimental fishing with long lines for deep-water fish was carried out in four areas of the North Atlantic by staff of the Directorate of Fisheries, Bergen. These were off North Norway in 1990, west of the British Isles and along the Reykjanes Ridge in 1991 and finally on Hatton Bank in 1992.

Along the edge of Tromsoeyflaket off North Norway the total catch for 48 hours fishing was 7,900 kg of roughhead grenadier(Macrourus berglax) and 2,000 kg of other fish. The catch was taken on 26,400 hooks in depths ranging from 350 to 500 fathoms (640 to 915 m.)

However, the sale of the catches of roughhead grenadier to the fish processors was difficult, and the price was too low to give a adequate profit, and the fishermen were not interested in starting this fishery on a commercial basis.

Along the Reykjanes Ridge the experiments were carried out between depths from 280 to 520 fathoms (512 to 950 m.). The catches consisted of tusk, blue ling, halibut, red fish and sharks. The bottom conditions were rough and as a consequence the vessel suffered loss of fishing gear.

West of the British Isles and along the Hatton Bank the fishery was carried out between depths from 250 to 600 fathoms (458 to 1100 m.). Besides the types of fish usually caught with long lines west of the British Isles, commercial quantities of mora and sharks were caught. It was, however once again, difficult to obtain a profitable fishery mainly owing to low prices and marketing problems. Plans for experimental fishing with long lines in 1993 were not carried out owing to these difficulties.

A. G. Hopper (ed.), Deep-Water Fisheries of the North Atlantic Oceanic Slope, 367–373.
© *1995 Kluwer Academic Publishers.*

1. Experimental fishing for roughhead grenadier off north Norway in 1990

Reducing catch levels for cod led to the proposal of experimental fishing with long lines for roughhead grenadier *(Macrourus berglax)* along the continental slope off north Norway. The purpose was to find out whether fishing for roughhead grenadier with by-catches of tusk *(Brosme brosme)* and Greenland halibut *(Reinhardtius hippoglossoides)* might result in a profitable fishery. It was also important to keep within the by-catch limits for cod of 10%.

The experiments were carried out in two periods. The first trip was made in February-March 1990 and the second trip in June 1990. To carry out the experiments the "Værland", an autoline vessel of 32 m. length and an engine of 425 h.p. was chartered.

Figure. 1 shows the areas where the fishing experiments were carried out. South west of the Lofoten Islands the catches were up to 227 kg live weight of roughhead grenadier per 1000 hooks. At the edge of the Tromsoeyflaket the best catches were 300 kg per 1000 hooks. The catching depths were between 350 to 500 fathoms (640 to 925 m.).

It should be mentioned that a catch of 300 kg of roughhead grenadier per 1000 hooks within 350-500 fathoms depth, is an acceptable catch. This may also be considered as the maximum catch possible as the fish have an average weight just above 1 kg. It represents an average catch of one fish on every third hook.

Over a 48 hours period of fishing along the edge of the Tromsoeyflaket the total catch was 7,900 kg of roughhead grenadier and 2,000 kg of other fish. This catch was taken on 26,400 hooks. The by-catch of cod was 8%.

However, selling the catch of roughhead grenadier was difficult, and the price was only NOK. 3 per kg. of gutted and headed fish, and NOK. 14 for the fillet. These prices for roughhead grenadier caught by long lines, were not sufficient to result in a profitable fishery. For this reason there was little interest in starting commercial fishery for this species in 1993.

A profitable fishery for roughhead grenadier would also require labour-saving technical improvements for processing the catch at sea. Because the fillet yield of the fish is low, the price roughly estimated, ought to be between NOK. 10 and NOK. 15 per kg. gutted and headed fish and NOK. 30 per kg. for the fillet. However, the profitability also depends on the by-catches of other species.

2. Experimental fishing for ling and other species west of the British Isles and along the Reykjanes Ridge in 1991

The purpose of this voyage was to carry out experimental fishing with long lines on less known fishing grounds and mainly in greater depths for ling *(Molva molva)*, blue ling *(Molva dipterygia)* and tusk. It was of special interest to make use of by-catches of unknown species to obtain a more profitable fishery for the long line fleet. The voyage was carried out in May and June 1991. When carrying out the experiments, it was necessary to consider the regulations in force regarding by-catches limits and species in the different areas.

The "Fjellmøy" a modern autoline vessel of 37 m. length with an engine of 880 h.p. was chartered for the trip. She had a total crew of 12. The long lines were constructed of 7 and 7.6 mm main lines and branch lines with a length of 23 in(58 cm.). The hooks were of the Mustad type Easy-bait No. 12/0. and the bait used was squid, herring and mackerel.

Figure. 2 shows the areas investigated on this voyage. West of the British Isles and along the Hatton Bank long lines were set out between depths from 250 to 600 fathoms (457 to 1100 m.) . Besides the types of fish usually caught by long lines west of the British Isles, commercial quantities of the following species were also caught:

Greenland halibut, grenadiers, mora *(Mora moro)*,

Greater forkbeard *(Phycis blennoides)*,

Portuguese dogfish *(Centroscymnus coelolepis)*,

Leafscale gulper shark *(Lepidorhinus squamosus)*,

Greater lanternshark *(Etmopterus pusillus)* and

Birdbeak dogfish *(Deania calceus)*.

Of the areas investigated west of the British Isles, Hatton Bank was considered as the most promising area within a depth range between 450 and 520 fathoms (823 to 950 m.). In 24 hours experimental fishing along the northern edge, the catch was 6,340 kg mainly mora, tusk and blue ling. The catch was taken on 18,200 hooks. In addition to this a considerable quantity of sharks were caught. It should be mentioned that along the south western slope there were some losses of fishing gear apparently a result of shark bite.

Along the Reykjanes Ridge the experimental fishing commenced 240 nautical miles south west of Rekyjanes between depths from 280 to 520 fathoms (512 to 950 m.). The catches consisted of tusk, blue ling, halibut, red fish and sharks. The survey continued south westward along the Ridge. However, the bottom conditions were very rough and the vessel suffered loss of fishing gear. The weather conditions were excellent during the trip.

Samples were collected of the different unknown species for processing experiments and marketing.

3. Experimental fishing on Hatton Bank in 1992

This voyage was a continuation of the experiments started in 1991. The task was to bring back catches of mora, grenadier and greater forkbeard to test the market for these species. Samples of sharks and shark liver also were required for marketing tests.

For this voyage a charter was arranged with the owners of the autoline vessel "Hordagutt". The "Hordagutt" was 37 m. long and had an engine of 690 h.p. She was equipped with the same fishing gear as the previously mentioned vessels. The trip was carried out in May-June 1992.

On the Hatton Bank a full commercial fishery was carried out lasting 11 days, and took place along the northern edge of the bank mainly between 400 and 500 fathoms (732 to 915 m.) depth (Figure. 3).

In this period 197,000 hooks were set. The catch which was retained on board amounted to a total of 70,000 kg live weight, and comprising 11 different species. Of this total about 30,000 kg was mora. In addition large quantities of sharks were returned to the sea.

With the prices fixed for the trip, the value of the total delivered catch was about NOK. 350,000. The results from the test marketing indicate that gutted and headed mora are valued at NOK. 12, per kg. It should also be mentioned that in 1992 a sum of 2 million NOK. was granted by the Government to finance the processing and marketing trials.

Although in quantity the catch was acceptable, it was difficult to demonstrate a profitable fishery. There are several difficulties connected to fishing with long lines in large depths. There is the reduced capacity of hooks, but there is also the influence of weather conditions that causes losses of fish.

The onboard processing of sharks also require technical improvements for skinning and gutting. It was, however, difficult to sell the sharks because the maximum permitted levels of mercury in the fish were exceeded.

4. Planning of experimental fishing with longlines in 1993

For 1993 plans were made for a new voyage to Hatton Bank to continue the fishing experiments, but in the circumstances there was very little interest by the fishermen for this fishery. The reason was the low prices and marketing difficulties for deep-water species. Questions too were raised regarding the EU quotas and these were of importance in the decision making process. For all these reasons the planned trip to Hatton Bank in 1993 was cancelled.

The results of the efforts of developing the new fishery was evaluated by a Steering Committee in the spring of 1993. The Steering Committee was of the opinion that a fishery for the new species might not give sufficient profit. Nevertheless many of these species appear as by-catches in Norwegian waters, and for this reason there should be an understanding of how to make best use of these by-catches.

For 1993 the Directorate of Fisheries also had made plans for another voyage for experimental fishing for roughhead grenadier along the continental slopes off Norway. The justification that there had been an indication of higher prices for this species and access to new technical equipment for processing at sea that gave encouragement for a profitable fishery . However, the necessary confirmation to purchase the equipment was not approved and the prices for the roughhead grenadier was not sustained, therefore this plan was also cancelled.

References

Olsen H.E. 1990. Forsøksfiske etter isgalt med liner. Fiskeridirektoratet, Biblioteket. Rapportar. Nr. 1 og 2 1990: 31-37.

Myklebust N. & Olsen H.E. 1991. Forsøksfisk med liner etter lange m.v. Fiskeridirektoratet, Biblioteket. Rapportar. Nr. 1 og 2 1991: 67-87.

Sele S.A. & Olsen H.E. 1992. Forsøksfiske etter nye fiskefelt. Fiskeridirektoratet, Biblioteket. Rapportar. Nr. 1 og 2 1992: 39-50.

372

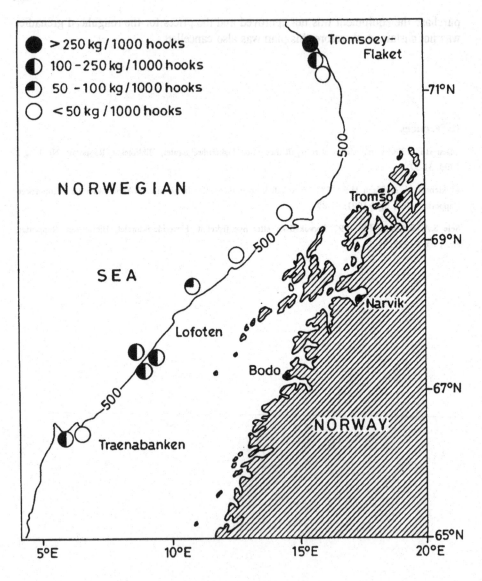

Figure 1. M/S VÆRLAND 1990 experimental fishing off north Norway for roughhead grenadier during March and April

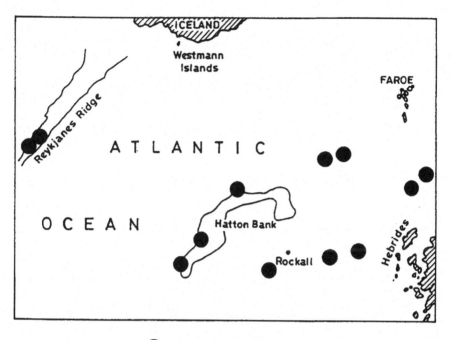

Figure 2. M/S FJELLMØY 1991 experimental fishing grounds in the North Atlantic, during May and June

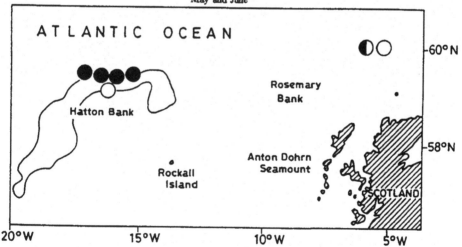

Figure 3. M/S HORDAGUTT 1992 experimental fishing in the North Atlantic for Mora, during May and June

SECTION 3   Technical papers

EXPERIMENTAL UTILISATION AND MARKETING OF BY-CATCHES AND DEEP WATER SPECIES IN ICELAND

THORSTEINSSON, H.P. & VALDIMARSSON, G.
Icelandic Fisheries Laboratories,
Reykjavík,
Iceland.

ABSTRACT

The importance of fisheries in the economy of Iceland is stressed. The annual catch is about 1.5 million tonnes of which about 1.0 million are reduced to fish meal, but the total export value is about 1.1 billion US$ annually. Better utilisation of the catch and the exploitation of unconventional species are important research tasks. The By-Catch Bank was established in 1990 for this purpose. It guarantees minimum prices for under utilised species which are then frozen and then test marketed in collaboration with Icelandic processors. This has been successful in achieving utilisation of these new species and has built up good relations with fishermen, producers and exporters. The importance of good handling of these fish is emphasised.

A. G. Hopper (ed.), Deep-Water Fisheries of the North Atlantic Oceanic Slope, 377–384.
© 1995 *Kluwer Academic Publishers.*

1. Introduction

The rich fishing grounds around Iceland provide an annual catch of about 1.5 million tonnes of which about 1.0 million tonnes are pelagic fish which are mostly reduced to fishmeal. The total annual export value of the fish products is about 1.1 billion US $, accounting for about 3.5% of the total world trade in fishery products. The most important fish species in Icelandic waters are the gadoids, i.e. cod, haddock, and saithe. Ocean perch, Greenland halibut, herring, capelin and pink shrimp are also very significant. The Icelandic fishing fleet and the processing plants are highly mechanised as demonstrated by the fact that only 13% of the country's work force is engaged in fishing and fish processing.

Protecting the fish stocks from over-fishing has for the last few decades been a major issue in the fishery policy of Iceland. In 1984 fixed transferable quotas were introduced for all the major species. Assigning every fishing vessel with a given annual catch quota has significantly increased awareness among fishermen and processors towards better utilisation of the catch and exploitation of unconventional species from deeper waters outside the 200 mile Exclusive Economic Zone (EEZ). Bringing small quantities of various by-catch species to the processing plants has not been considered economical, especially when using ice for preservation. With introduction of the freezer trawlers in 1982 this situation has changed. Freezing on board has made it possible to freeze smaller quantities of unconventional species for later thawing and processing. The introduction of auction markets for fresh fish (since 1986) has also significantly encouraged the emergence of small specialised processors concentrating on particular species. This paper describes a special four year programme, organised by the Ministry of Fisheries in conjunction with industry, aimed at encouraging exploitation of unconventional fish species in Iceland, including the deep water species.

2. New Species in the Commercial Fisheries of Iceland

Complex factors determine the utilisation of new fish species and the most important factors are the available catch techniques, together with processing and marketing. There also has to be a connection to local culinary culture. Ready access to large stocks of well known species obviously discourages exploitation of new ones. There are many examples where foreign vessels have traditionally caught particular species, ignored by the Iceland fishing fleet, but this is changing. An example is monkfish which is now considered a good catch, but a few decades ago it had no value. Today it is one of the most popular species offered at the local restaurants. Table 1 gives in chronological order a list of new entries to the Icelandic catch fauna. Behind each entry there is a story of entrepreneurship, successes as well as failures.

TABLE 1. Examples of New Entries to the Icelandic Catch

YEAR OF ENTRY	SPECIES[1]
1930	Red fish
1958	Norway lobster
1969	Greenland halibut
1964	Capelin
1983	Deep-sea shrimp
1986	Witch/Pole dab
1989	Deep-sea rosefish
1990	Starry ray
1991	Greater silver smelt
1992	Orange roughy

A recent example of under-utilised species in Iceland waters today are the various dogfish. From Table 2 it could for example be deduced that the Piked dogfish *(Squalus acanthias)* stock in Icelandic waters is very limited, whereas in many other northern waters they are significant. Research by the Marine Research Institute in Reykjavík shows, however, that dogfish are common in Icelandic waters, but until recently it was rarely landed by Icelandic fishermen.

TABLE 2. Catches of Piked Dogfish in Some Countries

COUNTRY	1987	1988	1989	1990	1991
Denmark	1394	1494	1086	1364	1246
France	13523	9892	9892	9892	3449
Ireland	8706	5612	3062	3000	2500
Norway	3614	4139	5329	8105	9627
UK	6754	5894	4369	4384	2749
Scotland	8044	7864	7463	7781	8536
Canada	3767	5314	2370	3174	3167
USA	2738	2968	4477	14838	13334
Iceland	5	4	17	15	53

TONNES

[1] See Appendix 1

380

3. The By-Catch Bank

The By-Catch Bank was established in 1989 as part of a four year programme aimed at better utilisation of catches. Its objectives were to encourage utilisation of under exploited species, including smaller amounts of by-catches. The main strategy of the bank was to guarantee the fishermen a minimum price for the whole frozen fish and to maintain frozen stocks for redistribution. Contacts were established with marketing firms for test marketing. Pricing of the fish was kept at a level high enough for the fishermen to find it worthwhile to bring the fish ashore, but low enough to prevent competition with the normal fish buyers. A programme was launched for keeping the industry informed about these activities through a newsletter, reporting on the species and amounts in the bank, as well as how the markets were responding. Also, some special events were organised, such as "Strange Fish Week" where 20 restaurants throughout the country offered dishes made from selected species. Fishermen and industry were specially invited to try these delicacies. Recipes were also published in special brochures. It is fair to say that these activities caught much attention by the national media as chefs and guests were interviewed, and culinary columnists made their comments etc. This proved to be a very effective way to convey information about possibilities in harvesting these species and their consumption in other countries.

To follow up on these efforts a special information folder was prepared and distributed, particularly to the freezer trawler fleet. It contained essential information on about 20 species, with information for easy identification of the fish, packaging and labelling.

TABLE 3. Species Received by the By-Catch Bank 1989-1993

SPECIES[2]	TONNES
Starry ray	138
Greater silver smelt	95
Piked dogfish	68
Grenadiers	50
Cod <50cm	13
American plaice	10
Megrim	4
Norway haddock	4
Black dogfish	2
Portuguese shark	1
Black scabbard fish	1
Other species[3]	2
Total	388

[2] See Appendix 1

[3] North Atlantic codling, Birdbeak dogfish, Leafscale gulper shark, Roughnose grenadier, Bairdii, smooth head, Spine eel, Agassiz.

4. Results of the 4 Year Programme 1989-1993

Table 3 shows the quantities and the species received by the Bank from 1989 to 1993. In all, some 10 species were received in quantities in excess of one tonne. A marketing programme was made for 8 species. Positive results were obtained for 5 of these:- Starry ray, Great Silver smelt, Grenadiers, Piked dogfish and Portuguese shark. Today it is considered that these species can be caught and traded through normal channels except for the last mentioned, which is termed up to now as "promising".

Table 4 shows catch figures for 11 unconventional fish species from Icelandic waters in the years 1989 and in 1992. From the table it can be seen that there has been a very significant increase in the utilisation of the traditionally underexploited species over a period of 4 years. It should be emphasised, however, that it is always difficult to evaluate objectively the outcome of such a programme or to what extent it inspired and encouraged new ventures in this field. Firstly, the quota system has put general pressure on the fishermen to find new species which are not limited by the quota, and certainly the By-Catch Bank could not claim to be directly involved in the success of some of the species listed in the table.

TABLE 4. Unconventional Species in 1989 and 1992

SPECIES[4]	1989 TONNES	1989 VALUE[5]	1992 TONNES	1992 VALUE[6]
Grenadiers (roughhead & roundnose)	2	3	210	70
Starry ray	99	12	317	58
Piked dogfish	16	6	181	162
Greater silver smelt	8	4	657	177
Deep-sea rosefish	1374	764	13845	9290
Lemon sole	802	1383	915	1495
Witch/Pole dab	2269	1941	2564	3325
Megrim	344	1531	246	125
American plaice	565	206	1468	1034
Rabbit-fish/rat-fish	0	0	106	22
Orange roughy	0	0	382	800
Other species	0	0	14	3
Total	5479		20905	

---

[4] See Appendix 1

[5] US$ (x1000)

[6] US$ (x1000)

The marketing aspect of these species was quite interesting. It showed how complex and specialised the fish market really is. Each species, and the different sizes and cuts have their own market niches. Also, the general supply-demand situation on the European fish markets has a great effect on the possibility of their marketing. When traditional species are lacking the market is clearly more receptive towards experimenting with "new" species. Some 10 different companies took part in the marketing effort in addition to the two large export organisations for frozen fish. The species being offered are in most cases of relatively low value, and it takes a lot of work to get information and contacts. Also, because very little is known about the availability of these species the exporters may be betting on a low value, low volume item. As they receive no financial assistance in these activities there was a considerable risk involved for them. In many cases technical difficulties emerged. Buyers wanted detailed information about the chemical composition of these species, particularly the presence of heavy metals. Some of these species such as sharks contain above acceptable levels of mercury. Keeping quality during frozen storage also came into consideration. Species with a fat layer under the skin, such as grenadiers tended to become rancid during relatively short frozen storage. Thus, the grenadiers had to be deep skinned after filleting. It can be concluded that even though the programme has cost about 250,000 US$ net, it has paid off.

5. The Future

There is no doubt that the programme was very important for the industry in terms of giving general information about the new species, their possible uses, etc. One of the key problems regarding these species is our lack of knowledge about their individual TAC's, and processing techniques.

To summarise, these are the main observations regarding the programme:

1. A demonstration programme of this type successfully leads to utilisation of "new" species.

2. Of particular importance are:

    - Good relations with both fishermen, producers and exporters.
    - Stimulation of information flow.
    - Demonstrating that the "strange" species are good food.

3. Much emphasis must be put on proper handling of the fish at all stages.

4. Flow of information to all parties is very important.

References

Jónsson, Gunnar (1992). Íslenskir fiskar, 2nd Ed. Reykavík: Fjölva útgáfa.

Fisheries Association of Iceland (1990&1992). Útvegur (Fishery statistics). Reykjavík: Fishery Association of Iceland.

Pálsson, Thorgeir (1993). Fiskvinnslan 1/93 Fiskiðn.

Ríkharðsson, Jón H. (1992). Vinnsluskip-fullnýting sjávarfangs Icelandic Fisheries Laboratories: 31. Rit.

FAO Yearbook 1991. Fishery statistics, catch and landings Vol. 72.

APPENDIX 1
Fish Names in English, Latin and Icelandic

ENGLISH	LATIN	ICELANDIC
Aggassiz, smooth head	*Alepocephalus agassizii*	Berhaus
American plaice	*Hippoglossoides platessoides limandoides*	Skrápflúra
Birdbeak dogfish	*Deania calceus*	Flatnefur
Black dogfish	*Centroscyllium fabricii*	Svartháfur
Black scabbard fish	*Aphanophus carbo*	Stinglax
Capelin	*Mallotus villosus*	Loðna
Deep-sea rosefish	*Sebastes mentella*	Úthafskarfi
Deep-sea shrimp	*Pandalus borealis*	Úthafsrækja
Greater silver smelt	*Argentina silus*	Gulllax
Greenland halibut	*Reinhardtius hippoglossoides*	Grálúða
Leafscale gulper shark	*Centrophorus squamosus*	Rauðháfur
Lemon sole	*Microstomus kitt*	Þykkvalúra
Megrim	*Lepidorhombus whiffiagonis*	Stórkjafta
Monkfish	*Lophius sp.*	
North-Atlantic codling	*Lepidion eques*	Bláriddari
Norway haddock	*Sebastes viviparus*	Litli karfi
Norway lobster	*Nephrops norvegicus*	Humar
Orange roughy	*Hoplostethus atlanticus*	Búrfiskur
Piked dogfish	*Squalus acanthias*	Háfur
Portuguese shark	*Centroscymnus coelolepis*	Gljáháfur
Rabbit-fish	*Chimaera monstrosa*	Geirnyt
Red fish	*Sebastes marinus*	karfi
Rough head grenadier	*Macrourus berglax*	Snarphali
Roughnose grenadier	*Trachyrhynchus murrayi*	Langhalabróðir
Roundnose grenadier	*Coryphaenoides rupestris*	Slétthali
Smooth-head	*Alepocephalus bairdii*	Gjölnir
Spine eel	*Notacanthas chemnitzii*	Broddabakur
Starry ray	*Raja radiata*	Tindaskata
Witch/Pole dab	*Glyptocephalus cynoglossus*	Langlúra

DEEP-WATER TRAWLING TECHNIQUES USED BY ICELANDIC FISHERMEN

GUÐMUNDER GUNNARSSON
Hampiðjan HF,
Reykjavik,
Iceland.

ABSTRACT

Fisheries are the mainstay of Iceland's economy, accounting for about 80% of total national export revenue and worth about US $ 1.1 billion. This paper deals with the development of deep-water trawling technology in recent years. From the Icelandic fleet's total landings some 64% are produced from bottom and mid-water trawling and both methods are now being used to exploit the new species found on the oceanic slope.

In the paper is a description of a new model of high capacity and high powered trawler which Icelandic operators have been introducing over the past two years, together with a description of its bottom trawl and the fishing techniques currently used by the fishermen. A description of the more commonly used Granton type trawl is given to illustrate how the size and weight of the gear has changed recently.

There follows a survey of midwater trawling which has been on the increase both within and outside of the Icelandic EEZ over the last five years. Major advances have been made in the design of these trawls and in particular with the giant 'Gloria' trawl which has been designed in close collaboration with the fishermen and has been particularly successful for ocean redfish (Sebastes mentella). It is recorded that in 1993 sixteen Icelandic freezer trawlers caught 20,000 tonnes of this species.

A. G. Hopper (ed.), Deep-Water Fisheries of the North Atlantic Oceanic Slope, 385–395.
© *1995 Kluwer Academic Publishers.*

1. Iceland's trawler fleet

There are now a total of 108 stern trawlers in the Icelandic fishing fleet with the oldest dating from 1972. Recently the fleet has been modernising with a new generation of freezer trawlers, which are now typically of 60 to 70 m. in length, up to 4000 hp and in excess of 2000 GRT. These vessels are capable of fishing in depths greater than 1000 m.

One such vessel is the freezer trawler Vigri (RE 71) which was delivered to the Ögurvik trawler company on 25th. September 1992 by the Flekkefjord Slipp & Maskinfabrik A/S Shipyard of Norway, (Figure 1.).

The technical specification of the Vigri is as follows:-

Length overall	67.00 m.
Breadth	13.00 m.
GRT	2201
Main engine	Wartsila 4076 hp.
Propeller	3.80 m. dia.
Freezer hold capacity	1070 m³ equal to 800 tonnes of frozen fish
Fuel capacity	450 m³
Winch systems:	
2 trawl winches	40 tonne pull each
4 bridle winches	15 tonne pull each
2 gilson winches	18 tonne pull each
2 codend gilson winches	10 tonne pull each
2 midwater trawl winches	35 tonne pull each

All the deck machinery was provided by Brattvåg A/S of Norway.

Figure 1. Freezer trawler Vigri

In addition the Vigri has a very comprehensive package of fish detection and navigation equipment including 2 Furuno radars, 2 Atlas depth sounders, a Furuno headline sensor, a Simrad headline sonar and scanner 'trawl eye' with sensors on the trawl. For precision navigation the vessel has its navigation aids linked to a Macsea integrated navigation system.

The processing deck is technologically advanced and has a high level of automation. It contains nine processing machines for gutting and filleting, and the freezing capacity is 80 tonnes in each 24 hour period. The fish hold capacity is 800 tonnes of frozen product.

The crew of 26 work round-the-clock in 6 hour shifts. Each voyage lasts between 4 and 6 weeks. During the first year of operations in 1993 the total landed catch was 4488 tonnes caught over 316 fishing days and valued at US$ 6.6 million.

2. Bottom trawl design and operation

2.1 THE GRANTON TRAWL AND THE NEW TRAWL USED BY THE VIGRI

The conventional two panel Granton trawl has been a favourite with trawlermen for many years, because of its suitability for fishing on rough ground, its straightforward design and it is easy to fish with under tough conditions such as strong currents, steep slopes and a rocky seabed. It is also very easy to manoeuvre on a hilly seabed where the trawl has to be swept down between steep slopes or gullies to reach the fish. The conventional Granton trawl (Figure 2.) measures 60 m. in circumference with a 30 m. headline, 18.3 m. fishing line and a 60 m. footrope with rockhopper discs (Figure 3.).

More recently the new and larger vessels such as the Vigri have required a two panel trawl based on the Granton but up to 60% greater in size, and with an increased weight of ground gear to take the trawl into deep-water. The new type of trawl now in use is 90 to 100 m. in circumference with a 70 m. headline, 40 m. fishing line and a footrope of up to 120 m. with rockhopper discs.

The total weight of the gear used by the Vigri is as follows:-

2 panel bottom trawl	1,200 kg.
Rockhopper gear	5,000 kg.
Bridles	800 kg.
Otter boards	4,200 kg each
Trawl warps	2 X 3200 m. of 32 mm. dia, 23,400 kg.
Total weight of gear	38,800 kg.

2.2 NETTING

Polyethylene netting is used for the trawls with a specific gravity of 0.96, making it buoyant in sea water. This causes the netting to float upwards if it comes fast on a seabed obstruction, and it is therefore less likely to be damaged than the heavy netting with a specific gravity of about 1.05 which was used formerly. The twine thickness is from 4 to 6 mm. and the inside mesh size is 135 to 155 mm. The total

388

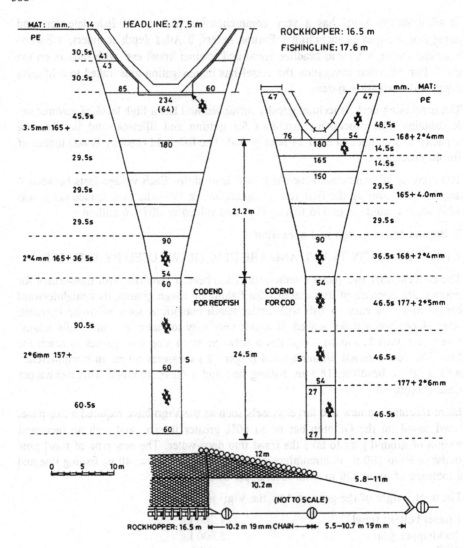

Figure 2. Conventional Granton trawl

weight of netting in a bottom trawl ranges from 500 to 800 kg.

2.3 FOOTROPE AND BOTTOM GEAR

To protect the netting when the trawl is towed along the seabed, the lower part of the trawl is fitted with bobbins or alternatively rockhopper discs which are made from sections cut from heavy duty tractor tyres. Rockhoppers are generally preferred to steel bobbins by most trawlermen since they are considered to give a smoother motion across the seabed and greatly reduce the amount of rock and other seabed debris entering the trawl. The rockhopper is more quiet during towing and this is especially important when fishing for pollack and saithe as these fish are easily scared from the path of a trawl by noise. Using rockhoppers also extends the life of the netting as there is less crush and abrasion damage to the netting during hauling and shooting of the gear, and the crews consider it to be less dangerous than the steel bobbins.

A 40 m. rockhopper section weighs around 4800 kg. although the gear on the Vigri is slightly heavier than this. This is equivalent to 1500 kg. when under the water. The heavy weight holds the trawl hard down on the seabed which is important when fishing in strong currents and gullies with steep sides. The trawl is generally aimed at species which do not move more than 3 to 5 m. off the bottom.

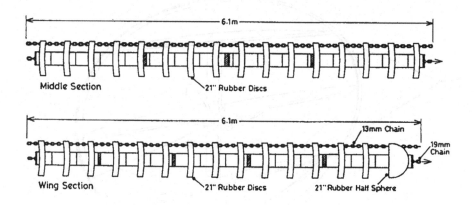

Figure 3. Rockhopper ground gear

2.4 OTTERBOARDS

The otter boards spread open the trawl mouth. Oval boards are preferred such as those made by Polyice or Ögurvik and the sizes are 8 to 11m^2 and weighing between 2400 and 4200 kg. each, depending on the size of the trawl and the power of the vessel. Figure 4 shows typical boards in use.

The otter boards need to be sturdily and robustly constructed to withstand the hydrodynamic and other forces acting on them as they are towed along the seabed. In one way they act as a pathfinder for the trawl because if the otter board is prevented from getting past or over an obstruction then the netting is given some degree of protection from extensive damage. Because of the considerable frictional force acting on the boards they are fitted with manganese steel shoes.

The trawl warps are secured to the otter boards through a bracket fixed to the towing face of the board, whilst on the back of the boards are backstrops secured to the sweeplines which link the trawl to the boards. The combination of spreading forces on the board and the tension applied through the sweep wire open the net during towing.

The trawl must be fitted with the correct boards for its size otherwise the fishing performance will suffer. If they are too small then they will be unable to provide the necessary force to open the trawl fully.

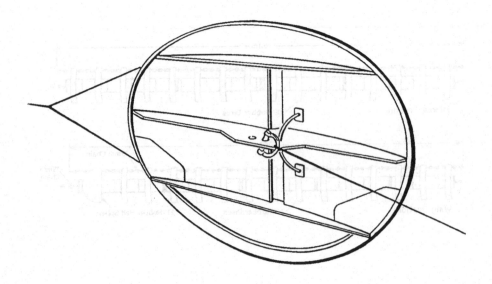

Figure 4. Hinriksson Polyice single slot cambered oval otter board

3. Towing procedures

During fishing an average of 4 to 6 hauls are made in the space of 24 h. Before the introduction of electronic catch sensors and trawl monitors a depth to warp length ratio was 1:3 but this has been reduced to between 1:1.8 and 1:2.5 depending on the depth. The extra warp length in earlier times was necessary to be sure that the trawl was settled properly on the seabed. Sensors fitted to the otter boards, headline and codend allow the warp length to be adjusted to give the optimum headline height and net spread with only a minimum of warp. These sensors have significantly improved fishing efficiency and cut fishing gear costs by showing the fishing captain if the trawl has come fast on the bottom, or has been torn. They also show whether or not there are fish entering the cod end corresponding with the information displayed on the echo sounder.

New advanced navigation systems such as GPS and video chart plotters have also given greater precision to the towing process, and made it easier to locate lost trawls and avoid seabed obstructions. All these systems are brought together into a single information display known as the integrated navigation system (INS) which has become widely used by Icelandic fishermen

During towing the trawl covers quite a large area. For example towing at 3.5 knots for 4 h. with an otterboard spread of 100 m. will give a swept area of 2.5 km². Towing depths are continually increasing and recently fishermen have been regularly fishing for grenadier and orange roughy down to 1500 m. and occasionally 1700 m. Only the largest and most powerful vessels can fish at these depths such as the Vigri with its twin 40 tonne winches and 3200 m. of 32 mm. dia. steel warp with a breaking load of 54 tonnes

4. The Gloria midwater trawl

4.1.DESIGN

This is a trawl designed specifically for fishing oceanic redfish (*Sebastes mentella*) which are found throughout the water column from about 200 to 900 m. mainly down the Reykjanes Ridge and the Irminger Sea.

The first trials with the Gloria midwater trawl were made in April 1989 by two Icelandic freezer trawlers, the Sjóli and the Haldur Kristjánsson both of which had requested a midwater trawl with a larger opening than was commonly used at that time. The trawl had to be capable of being towed at relatively high speeds of 4 to 5 knots for the widely dispersed oceanic redfish. At an early stage Hampidjan opted to set the front meshes at 32 m. mesh size in order to reduce the towing resistance of the trawl and to keep the quantities of netting used to reasonable limits.

Today the Gloria trawls are manufactured in sizes ranging from 544 m. to 3072 m. measured around the circumference of the stretched netting at the widest part of the opening of the trawl mouth. The front meshes of these giant trawls are up to 128m., although the most common size in use is the 64 m. mesh size (Figure 5)

Various models of the trawl have been tested in the Flume Tanks in Denmark and Canada to adjust and modify the shape of the netting, and to adjust the weights and bridle lengths to give optimum performance and net geometry.

The trawler captains and the net makers have worked closely together with Hampdjan's designers to develop the Gloria. It has been an expensive and slow process to optimize the trawl for easier handling both at sea and ashore. Great care has to be taken in the design and production of a trawl on this scale because the huge size multiplies the effects of any mistakes, and of course the cost of subsequent corrections can be very expensive. Various types of ropes have been tried and today braided ropes are preferred made from nylon and the super fibre Dyneema which because of its high breaking strength is used for the 128 m. mesh. The Gloria trawl has four equal panels and each panel is colour coded for ease of identification.

4.2. FOOTROPE WEIGHTS

Weights are not used on the lower wingends of the Gloria trawl as is common practice in many midwater trawls. The weight which is needed to open the trawl downwards is distributed evenly along the footrope by using different sizes of chains. These have proved to be much more easy to handle than wingend clump weights, and the shooting and hauling of the trawl is faster and there is less danger of accidents to the crew.

4.3. OTTER BOARDS

The most commonly used otter board is a modified Superkrub of 9 to 13 m^2 with an extra foil and weighing between 2800 and 3600 kg. These boards have been modified to have a bigger camber and a higher aspect ratio, and they are made in Iceland by Polyice or Hinriksson. Others are imported from Morgère in France or from Poland. The Morgère type is shown in Figure 6.

4.4. SWEEPLINES

The sweepline lengths have to be kept constant. On the earlier versions these were 225 m. with a 20 m. extension in the lower bridle to achieve a better spread of the trawl. With the introduction of larger vessels up to 4000 hp. the sweepline length has been increased to 300 m. or more. Backstrops or pennants are now being used with chains instead of wire which gives a much longer life.

4.5. FISHING SEASON

The fishing seasons for oceanic redfish are from March to July and from September into December depending latterly upon weather conditions. The best season is from May to June. The main fishing area is 350 miles south west from the Iceland EEZ and lying in the Irminger Sea at the southern end of the Reykjanes Ridge.

In the summer fishery fishing is carried out mostly during daylight with only small catches during darkness. Fishing generally starts between 5 and 6 am and continues until 6 to 8 pm.

The fish is widely dispersed mainly between 300 and 600 m. depth range and the footrope of the trawl must be at least 70 m. below the soundings indicated on the echo sounder if it is to catch fish. Shoals seem to dive due to disturbances of the otter boards, bridles and propeller noise as the trawler approaches them. It has been found that a trawling speed of 3 to 3.2 knots gives the best results, and not the higher speeds of 4 to 4.5 knots originally thought necessary. The towing time depends on the density of the shoal.

Most of the modern trawlers can process 40 tonnes of redfish per 24 h. yielding about 20 tonne of frozen product. The fishing effort by Icelandic vessels for this species has been increasing dramatically since 1989 (Thorsteinsson and Valdimarsson, 1994) and this trend will continue as long as there is a demand for this species and it remains profitable for the large freezer trawlers.

In 1993 sixteen Icelandic freezer trawlers caught 20,000 tonnes of oceanic redfish in 42 fishing trips between early April and the middle of July valued at US$ 13.7 million. The bulk of this was exported to Japan as headed and gutted products.

The quantity has been increasing every year and this is expected to continue within the capabilities of the resource. With the reductions in cod quotas for Icelandic fishermen within the EEZ these large trawlers now have no option but to seek alternative resources. The oceanic redfish will play an important part along with other under-utilised species such as blue ling, orange roughy and grenadier.

References

Magnússon J. and Magnússon J. (This volume), The distribution, relative abundance and biology of the deep-sea fishes of the I Icelandic Slope and Reykjanes Ridge; in A.G.Hopper (ed), Deep-water fisheries of the North Atlantic Oceanic Slope, (proceedings of the NATO Advanced Research Workshop, March 1994); Kluwer, Dordrecht, The Netherlands.

Reinert J. (This volume), Deep-water resources in Faroese waters to the south, southwest and west of the Faroes - a preliminary account; in A.G.Hopper (ed), Deep-water fisheries of the North Atlantic Oceanic Slope, (proceedings of the NATO Advanced Research Workshop, March 1994); Kluwer, Dordrecht, The Netherlands.

Thorsteinsson H.P. and Valdimarsson G. (This volume), Experimental utilisation and marketing of by-catch and deep-water species in Iceland; in A.G.Hopper (ed), Deep-water fisheries of the North Atlantic Oceanic Slope, (proceedings of the NATO Advanced Research Workshop, March 1994); Kluwer, Dordrecht, The Netherlands.

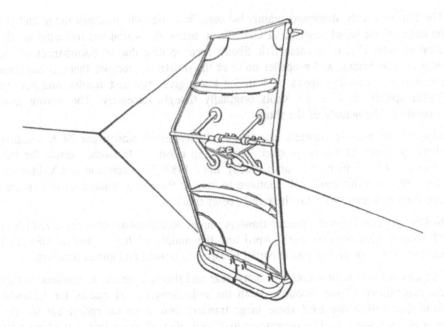

Figure 6. Morgère high aspect ratio cambered otterboard

Figure 7. A large catch of oceanic redfish (*Sebastes mentella*)

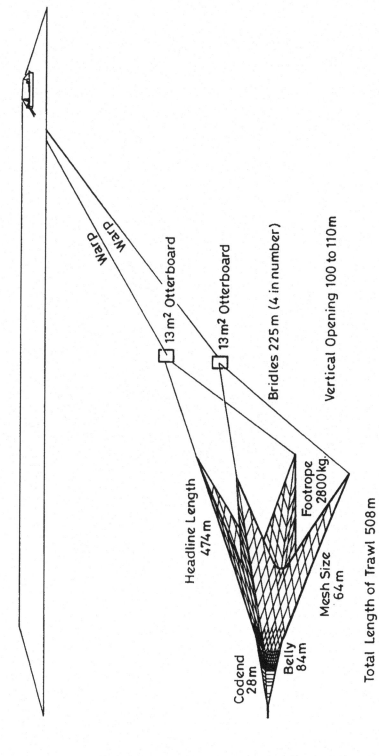

Warp

Warp

13 m² Otterboard

13 m² Otterboard

Bridles 225 m (4 in number)

Vertical Opening 100 to 110m

Headline Length
474 m

Footrope
2800 kg.

Mesh Size
64 m

Total Length of Trawl 508 m

Codend
28 m

Belly
84 m

Figure 5. Gloria midwater trawl

CANADIAN EXPERIENCE: DEEP WATER FISHING GILLNETTING IN THE NORTHWEST ATLANTIC OCEAN

ANDREW DUTHIE & ALAN MARSDEN

Department of Fisheries and Oceans,
200 Kent Street,
Ottawa, Ontario,
Canada, K1A 0E6.

ABSTRACT

In 1993, commercial inshore gillnet vessels continued exploratory fisheries for Greenland halibut in the north west Atlantic, off the coasts of Newfoundland and Labrador. Catch rates doubled those of fisheries conducted in 1992; five vessels fishing to maximum depths landed 518,790 kg of Greenland halibut and 4,300 kg of roughhead grenadier. In the 1992 season fishing effort, fishing depths and fishing gear for 22 inshore gillnet vessels in the length range 16.5 to 19.5 m. were monitored in the north west Atlantic. Greenland halibut (Reinhardtius hippoglossoides) catches were analysed to identify catch rates, the average fish size, by-catch quantities and the mesh size versus size of fish caught. Recommendations included the expansion of deep-water gillnet fishing, detailed stock assessment studies, analysis of fishing pressure impact, and marketing and utilisation of the significant by-catch of roughhead grenadier.

1. Introduction

Diminishing catch rates in the traditional inshore Greenland halibut areas and a growing presence of foreign fishing vessels working in deep water off the continental shelf and fishing for this species prompted the initial inshore deep-water exploratory fisheries. Since 1989 several exploratory deep-water surveys for Greenland halibut have revealed the potential for commercial inshore gillnet fisheries in the north west Atlantic. Subsequent developments of commercial fisheries in different areas have been ongoing since 1992.

This paper provides a description of:

a) The 1989, 1990 and 1991 deep water fisheries for Greenland halibut and

A. G. Hopper (ed.), Deep-Water Fisheries of the North Atlantic Oceanic Slope, 397–405.

398

roughhead grenadier by gillnet vessels.

b) The 1992 development of a deep-water Greenland halibut fishery by gillnetters.

c)1993 Northern Labrador harvesting and processing review for Greenland halibut.

2. The 1989,1990 and 1991 deep water fisheries for Greenland halibut and roughhead grenadier by gillnet vessels

Exploratory fisheries from 1989 to 1991 revealed the potential for a viable commercial fishery for Greenland halibut and the significant by-catch of roughhead grenadier.

The fisheries took place off the north east coast of Newfoundland between 49° 37' 40"N to 51° 00' 00"N latitude and 49° 30' 00"W to 50° 30' 00"W longitude, an area of about 6500 km². Depths ranged from 300 to 750 fathoms (549 to 1372 m.). Gillnet mesh sizes between 13.9 to 22.8 cm were used to determine the best catch rates.

Substantial quantities of Greenland halibut were landed in 1990 and 1991. In addition, a 15% to 30% by-catch of roughhead grenadier was landed. Gillnet catch rates varied between 45 kg and 68 kg per net for a four to eight day soak time. The

Figure 1 Typical Canadian gill net vessel- SUSAN & SISTERS

most productive depths ranged between 350 and 750 fathoms (640 to 1372 m.).

2.1 SUMMARY OF 1989 FISHERY

A 19.2 m. gillnet vessel fished four gillnet sets in depths ranging from 390 to 530 fathoms (713 to 970 m.). The sets consisted of two fleets of 25 nets, each 50 fathoms (90 m.) long and 25 meshes deep using 15.2 and 16.5 cm mesh. A total of 2,818 kg of Greenland halibut was caught, and, as a result of the smaller mesh size used, 1,762 kg of roughhead grenadier was also harvested. The average soak time was less than 24 hours and the average weight of Greenland halibut was 28 kg per net.

2.2 SUMMARY OF 1990 FISHERY

In 1990, in an expanded exploratory fishing program, four vessels ranging from 13.5 to 19.2 m., fished with gillnets at depths of 350 to 420 fathoms (640 to 768 m.). Out of a total of eight trips the average weight for each trip was 150,00 kg, comprising 75% Greenland halibut and 25% roughhead grenadier. The soak time was 3.5 days, and the average total weight of Greenland halibut per net was 42 kg. Mesh sizes were 15.2 to 20.3 cm. The larger mesh reduced the by-catch of roughhead grenadier.

In addition, a deep water Greenland halibut survey was undertaken by the Fisheries Development Division. Two 19.2 m. vessels fished 16 days using mesh sizes from 13.9 to 20.3 cm, in depths of 450 to 750 fathoms (823 to 1372 m.). The catch from the 2100 sq. nautical mile survey area totalled 60,706 kg of Greenland halibut and a significant by-catch of roughhead grenadier. The large mesh size tended to reduce undesirable species such as crabs and skate. The best overall mesh size was 19 cm where an average of 56 kg of large Greenland halibut per net were taken The average overall weight was 46 kg. the analysis showed that the average length and weight caught in the 19 and 20.3 cm mesh was higher than that for other mesh sizes. The soak time was four days.

2.3 SUMMARY OF 1991 FISHERY

In 1991, up to seven 19.2 m. vessels fished depths of 346 to 595 fathoms (633 to 1088 m.). All but one used the standard 25 mesh deep gillnet with mesh sizes ranging from 13.9 to 20.3 cm. Catch statistics indicated that the larger mesh sizes gave the best results. An average catch rate of 54 kg per net was achieved using 17.8 to 20.3 cm as compared to an average of 40.5 kg per net with 13.9 cm mesh. The average soak time of the nets during this survey was seven to eight days.

3. The 1992 development of a deep-water Greenland halibut fishery by inshore gillnetters

The 1992 deep water Greenland halibut project was initiated by the Underutilised Species Program of the Canada/Newfoundland Inshore Fisheries Development Agreement administered by the Canadian Department of Fisheries and Oceans.

The fishing depths and fishing gear for 22 inshore gillnet vessels (16.5 to 19.5 m.) were monitored in the north west Atlantic. In an effort to assess the performance of deep water gillnets, four experiments were identified and tested, which included variations in extended mesh and hanging ratios, the effects of net twisting and the use of traditional gear.

Catches, fishing effort, positions, water depths, set and hauling times and dates, number of nets, mesh size, hanging ratios and discards were recorded. These were then analysed to identify catch rates, average fish size, by-catch quantities, mesh size versus size of fish caught and other gear parameters versus the catch relationships.

Approximately 2.7 million kg of Greenland halibut were harvested off the north east coast of Newfoundland. The highest catch rates were from a depth range of 600 to 649 fathoms (1098 to 1187 m.).

3.1 PROJECT DATA

Project duration: May to October

Number of vessels: 22 vessels

Number of trips: 160

Nets hauled: 59,293 nets

Depth:400 to 800 fathoms (732 to 1464 m.) most of the catch and the highest catch rates were from a depth range of 600 to 649 fathoms (1098 to 1187 m.).

Catch: 2.7 million kg of Greenland halibut representing 80% of the catch; and roughhead grenadier 13.5% of the catch.

Average catch rate: 4.3 kg per net per day

Soak time: Average soak time 9 days. The analysis of information on Greenland halibut catches and discards by soak time indicates that the average Greenland halibut catch rate for gillnets is 10 kg per net per day, for a soak time of 1 to 2.9 days,but decreasing as the soak time increased.

Discards: 2% of the total catch was discarded. The discard rate increased sharply with soak times of 13 to 15 days, thus diminishing the quality and landed value.

Gear damage: Catch rates declined gradually as the condition of nets deteriorated, but not to the same extent as the soak time factor.

3.2 GEAR EXPERIMENTS

Working in deeper water than usual with stronger tides and with longer fleets of netting can cause gear problems. As a result, experiments were conducted with new and modified nets to determine if catch rates could be increased. Traditional gillnets used by Newfoundland fishermen have a hanging ratio of about 50% and the webbing is 25 meshes deep. 50,000 of the 59,000 nets used in the 1992 project were traditional nets.

3.3 NET TWISTING

Net twisting occurs when the head rope, webbing (mesh) and, or the footrope is wrapped around each other or twisted together. Some nets experienced as much as 50% twisting per fleet. No conclusive cause or a solution to net twisting was found. Gillnets used in deep water should have a footrope which is 4 to 5 fathoms longer than the headrope, not necessarily to reduce net twisting, but to ensure that the strain on the gear is born by the strongest part of the net which is the headrope.

3.4 HANGING RATIOS

The majority of the fishermen in this project preferred to use a 50% hanging ratio. Others experimented with a "gathered" hanging ratio of 33% and others with a "stretched one" of 66%. The overall catch rates were the same for each (41.4 kg per net). Observations on the gathered and stretched nets were:-

i) 33% hung nets:

It was difficult and time-consuming to remove Greenland halibut; and there were increased catches of smaller fish and other unwanted by-catch species;

The gillnet webs of 100 fathoms (183 m.) long resulted in a net that was only 33 fathoms (60 m.) long when hung at 33%. By comparison at a 50% hanging ratio the net is 50 fathoms (91 m.), thus a smaller ocean bottom area is covered;

Gear damage less frequent when compared to 50% or 66% hung nets.

ii) 66% hung nets:

There is extra strain and this resulted in increased net damage;

Catch rates improved, but these decreased after the first two hauls due to net damage;

By-catches were smaller;

It was easier to remove Greenland halibut.

3.5 EXTENDED MESH NETS

Most of the nets had 25 meshes in the depth but about 15% had been extended to 35 meshes. The observations were:-

When these nets were used with 33% and 50% hung ratios, it was difficult and time-consuming to remove fish, especially for the 33% ratio;

An increase of unwanted by-catch occurred (skate, blue hake) with 33% and 50%, and more so with 33%;

There was an increase in the cost of webs;

The extended mesh nets with 50% had the highest catch rate (6.13 kg per net per day).

3.6 BIOLOGICAL SAMPLING OF CATCH

To broaden the biological database and to assist with gear selection, processing and marketing strategies, a percentage of the catch was sampled. Results showed that:

Greenland halibut sampled from the 20.3 cm mesh nets were 1.5 times longer (average 11.02 cm) and weighed 3.5 times more (average 3.6 kg) than those caught in 13.9 and 15.2 cm mesh (see Figure 2);

Male/female ratios were identified and the female percentage ranged from 60% in 15.2 cm mesh nets to 93% in 20.3 cm mesh nets (see Figure 3).

4. 1993 Northern Labrador harvesting and processing review for Greenland halibut

In 1993, seven commercial inshore gillnet vessels conducted exploratory fisheries for Greenland halibut in the north west Atlantic, off the coast of Newfoundland and Labrador. Fishing activity took place 145 to 193 km. from shore in depths up to 800 fathoms (1464 m.). The project lasted from the end of July to October 21st 1993.

Two of the seven vessels were 33 m. middle-distance vessels which landed approximately 405,000 kg of Greenland halibut, of which 34% were over 4.5 kg and 52% were 2.25 to 4.5 kg. Five of these vessels landed 518,790 kg of Greenland halibut and 4,300 kg of roughhead grenadier.

Overall the seven vessels made 36 trips, had on board 364 fleets and 16,409 nets.

The catch rate of 9.5 kg per net per day in these fisheries doubled the 1992 average catch rate of 4.3 kg.

The average soak time decreased from the average in 1992 of 9 days to 5.5 days.

The catch rates were good in relatively shallow water with 8.6 kg per net per day in 300 to 399 fathoms (549 to 730m.).

5. Other Deep Water Fisheries in Canada

Although new exploratory deep-water fisheries with otter trawls are underway in the Pacific and with gillnets in the Atlantic, there are two successful deep water fisheries already well-established in Canada.

5.1 PACIFIC OCEAN SABLEFISH (BLACK COD) FISHERY

Harvested using traps (see Figure 4), sablefish off the British Columbia coast are fished for at depths of 300 to 400 fathoms (549 to 732 m.). The majority of fishing activity takes place in the deep canyons off the continental shelf on the west coast of Vancouver Island. Other fishing occurs along the coastline in deep inlets. This is a very valuable fishery and approximately twenty-four vessels harvest 5,000 tonnes of fish which is worth approximately Can$25 million.

5.2 BAFFIN ISLAND GREENLAND HALIBUT FISHERY

Longlining in the Arctic combines traditional native American practice with modern gear technology (Figure 5) for a successful deep water under-ice fishery harvesting close to 387,000 kg of Greenland halibut annually. The fishery operates at depths of 300 to 380 fathoms (549 to 695 m.) in Cumberland Sound and is carried out during the winter months when the ice is from 1.5 to 3 m. thick. It is commonly called "ice-platform fishing". This fishery brings in substantial revenue to the local economy and annually employs about 200 people.

A 180 m. mainline (or headline) with a "kite" is lowered through an ice hole. The "kite" is a 18 to 20 gauge piece of sheet metal 76 cm to 61 cm. Because of its shape the kite travels through the water, ultimately sinking to the bottom and stretching the line 180 m. across the ocean floor. Short, 63.5 cm lines - gangions - are attached to the line at 1.8 m. cm intervals. An appropriate size hook for the species is attached to each gangion. Generally there are 100 hooks fastened to the gangions on the mainline.

It is the practice to leave the line for 1½ to 2 hours and it is then hauled in with a mechanical or manual hauler. A mechanical hauler can retrieve 720 m. of line in 10 to 15 minutes compared to the two hour operation of a hand operated gurdy.

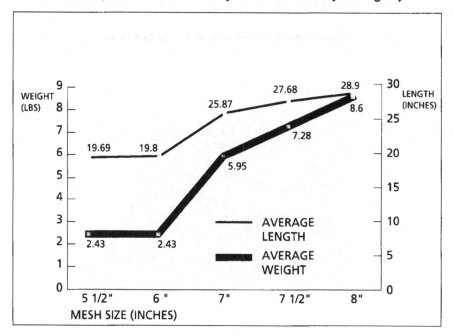

Figure 2 Greenland halibut- 1992 trials - average weight and length by mesh size

404

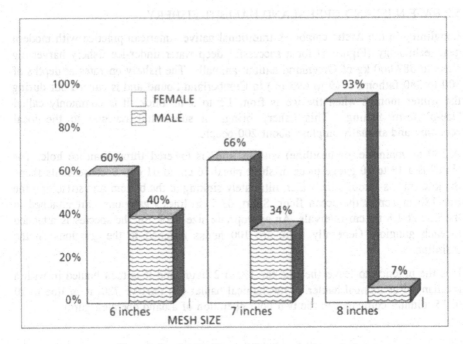

Figure 3 Greenland halibut male/female ratios by mesh size

Figure 4 Sablefish (black cod) trap

405

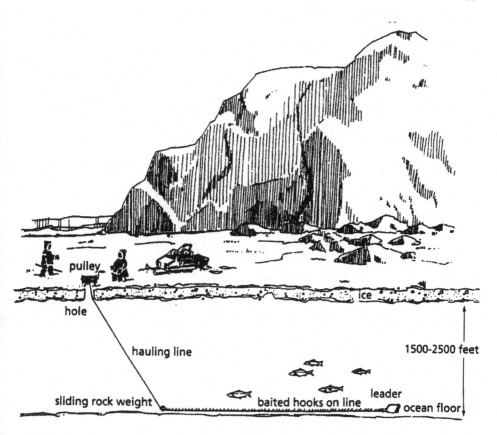

Figure 5 Gear technology for Arctic under-ice fishery

SECTION 4   Summary of workshop sessions

WORKSHOP SESSIONS - SUMMARY OF CONCLUSIONS AND RECOMMENDATIONS

ALAN G.HOPPER
Secretary to the Workshop

ABSTRACT

This paper summarises the collective thinking of the delegates to this Workshop on the future of these important resources and the work which needs to be done in research and technical development if they are to be exploited in a sustainable manner in the future. The findings are presented under four headings;- international management; processing, marketing and end-use; technical development and finally the opportunities for joint research and development.

There were no illusions as to the difficulties of managing these stocks as there is still a shortage of adequate data to provide a rational basis for management decisions. Moreover most stocks lie outside of the jurisdiction of national and community EEZs. Nevertheless the general concern of all the delegates was that these resources must be treated with extreme caution, and two examples were given of deep-water stocks fished already to below a level from which they may never recover.

It may well be some time before intensive fishing of these stocks takes place as there is only a limited market for very few of the species, and more work is needed to make the markets more aware of them and to establish their food value.

In the field of gear research the preferred method is still trawling but in the longer term alternative methods may prove more cost effective. The high cost of trawling in deep water is another barrier to a rapid development of this fishery.

The almost prohibitive costs of research in deep water, especially for individual country programmes was recognized and it was agreed that there was much to gain by collaborative work between institutions and small groups of countries. Sixteen separate research tasks were identified by the delegates for this type of collaborative work.

A. G. Hopper (ed.), Deep-Water Fisheries of the North Atlantic Oceanic Slope, 409–416.
© *1995 Kluwer Academic Publishers.*

1. Introduction

The final two days of the Workshop were spent in discussions of the papers which had been presented, and the implications for the future of these fragile and little understood resources.

During these two days there were four workshop sessions which were fully attended by all delegates. Each was under the direction of a chairman and a rapporteur who in the fifth and final session read out prepared statements on the conclusions and recommendations of the individual workshop sessions. These were then endorsed by the delegates as a whole. These findings are summarised in the following sections.

2. Workshop I. International management of deep-water resources

Chairman. Malcolm Clark, New Zealand.
Rapporteur. Philip Macmullen, United Kingdom.

The application of existing management strategies such as TAC which have been based on target mortality levels in multi-species shelf fisheries has, in general, had only limited success in northern hemisphere fisheries. The greater vulnerability of the deep-water species to overfishing necessitates effective management from the outset. It is simply not adequate to extend current practices and strategies to these fish. Instead the development of new fisheries must include investigations of alternative management techniques, or at least the improvement of existing techniques. Workshop I recommended the following:

2.1. Because of the characteristics of low productivity - high vulnerability - slow recovery pertaining to these species there is an urgent need to set in place research and management structures to protect them from irresponsible commercial exploitation.

2.2. It was agreed that as an interim management structure:

a) Individual nations should develop strategies and excercise management over those deep water resources within their own 200 mile EEZ

b) ICES and NAFO scientific committees should continue to give scientific advice on management for the deep-water stocks of the North Atlantic.

c) NAFO and NEAFC are existing management bodies able to undertake co-ordinated management of these stocks in the North West Atlantic.

2.3. The management objectives of any new or interim strategies need to be clearly stated. Are they to provide long term sustainability of the resource, or a short term economic yield or are they for some other purpose?

2.4. Consideration needs to be given to the most appropriate form of management for the species and what is currently known about them, and the associated assemblage. The choice may be to fix the TAC by area, effort control, restrictions on gear type and gear construction and seasonal closures.

2.5. There is a special need for integrated fisheries management, where the effects of fishing on one stock has serious repercussions on inter-related or dependent species in the same assemblage.

2.6. The views of other interest groups (scientific, industry, community and environmental) should be included in the management advice process especially where the effects of management of the target species can have wide ranging impacts on the ecosystem as a whole.`

2.7. Some of the high seas fisheries in international waters will be particularly at risk and there are no current management regimes covering these stocks. The UN should be given every support to progress ongoing UNCLOS resolutions and to prepare a new code of conduct for high seas fisheries.

2.8. There should be enforcement of existing international agreements to ensure that they are fully effective.

2.9. Adaptive management strategies are an appropriate method of gaining information rapidly. This strategy permits the fishing of a stock at a certain level during which time the changes in the stock structure are monitored. These changes are compared with expectations based on scientific predictions and the results are then used to adapt the catch levels and the management measures as appropriate. Such strategies need careful planning and close support of the fishermen so that the effects of fishing can be measured, the discarded and retained species identified and the circumstances surrounding the fishing operations noted such as depth, position and gear type. Good data from these sources can be used to estimate the size and status of the stocks. Even so because of the slow growth of most of these species the impact of fishing may not be detectable until many years after it has commenced

2.10. Scientists agreed that the ecosystems in which these species live are very different from the shelf ecosystem and that extensive and ill considered exploitation of these deep-water stocks will put at risk the future of this important resource. A cautious approach to development must be adopted.

3. Workshop II. Processing, marketing and end-use.

Chairman. Alan G. Hopper, United Kingdom
Rapporteur. Grimur Valdimarsson, Iceland.

3.1. The discussion on this subject was based mainly on the experience of Icelandic, New Zealand and Norwegian delegates who had some first hand knowledge of the species in the market place. The findings are summarised as follows:

3.2. The deep-water species are not yet in demand by the main fish markets of the world. There would seem to be an adequate supply of traditional species or acceptable substitutes even though an increasing number of these are now imported from parts of the world outside the North Atlantic.

3.3. The experience of the Canadian Atlantic cod fishery where there has been a catastrophic collapse of the resource in recent years should be a salutary lesson to all

412

concerned. Further failures of this type will increase pressures on new species and resources including those from deep-water.

3.4. On the upper slope of the Atlantic Basin there are 300 to 400 species variably distributed throughout the whole region. To date less than 10 have been considered to be marketable.

3.5. There is little to be gained from catching the deep-water species and then expecting the international market to accept them as alternatives to the more popular fish unless there is a planned and sustained promotional campaign. In the event of severe shortages of the traditional fish species customers would more likely turn to alternative protein foods such as red meat, chicken and processed meats rather than take a risk with an untried fish.

3.6. The success in New Zealand with orange roughy since the early 1980s. probably can be used as a model for a promotion campaign. Here a specific marketing plan from catching through processing to the consumer was devised in detail. The target market was the US market for quality white fish fillets and the ingredients of success were: a) a known resource, b) established processing techniques, c) a well researched consumer market with a high volume, and d) a commitment to succeed. Once the market had accepted the basic product then there was the opportunity to introduce new added value products.

3.7. The new market opportunities in which deep-water fish may find a niche are Russia where there is a large unsatisfied market and SE Asia where the market is growing in both volume and value.

3.8. Many of the names used to describe the deep-water fish are unattractive, eg. rat-tail fish,catfish, dogfish, rabbit fish and black scabbard fish etc. These names are not the language of a promotional campaign, nor are the Latin scientific names. Care however has to be exercised in the market place in choosing new names as it will cause confusion. For example in France the orange roughy is called the emperor fish.

3.9. It was suggested that collaborative research and market development should be undertaken on one North Atlantic species with good market potential and which is known to exist in volume and which has reasonable all year round availability. This would involve research into quality control and product development in close collaboration with processors and international traders. Marketing of deep-water species will take a long time before results can be expected.

3.10. It was also suggested that as well as food value, many of the deep-water species may have value in the field of medicine and pharmaceuticals.

4. Workshop III. Technical development - gear and vessels.

Chairman. John E. Tumilty United Kingdom
Rapporteur. Richard McCormick, Ireland.

4.1. Trawling is likely to be the principal method of fishing in deep water for the foreseeable future. At the present time vessels engaged in deep-water trawling range from 25 m. in length, 900 hp. to 67 m. in length, 4000 hp. The conclusions of the Russian delegates based on many years experience are that a vessel of 55 to 60 m. in length and 4000 hp. is necessary and will require 4000 m. of warp length on each winch. The ocean environment imposes considerable demands on the design of vessels and their equipment

It was generally agreed that no one specific method of fishing was wholly satisfactory in deep-water for all seabed conditions and all species. Pelagic trawling (mid-water trawling) was suitable for *Beryx* spp.. Longlining had proved to be highly selective, but some of the key species such as orange roughy and roundnose grenadier could not be caught with baited hooks. Gill netting was a low cost system but there are risks to the environment from too many vessels using this method in one area and it was not suitable for some species. Little is known about trap fishing and its effectiveness for these species, and indeed the location of traps and their recovery from deep water may present technical difficulties which may not be solved easily. The recommendations from this workshop session were:

4.2. There is a urgent need to obtain as much knowledge as possible about the operating parameters of fishing vessels in deep water including engineering loads, performance of electronic equipment, vessel safety and stability .

4.3. It was recommended that, as a matter of priority, an economic assessment and comparative study be made of the various vessel types, fishing gears, and ancillary equipment currently used in deep water. It is necessary to establish cost benefits taking into account the value of the catch on one hand, and the cost of fuel and energy,the general operating expenses and assigning some value of the damage to the ecosytem on the other.

4.4. Gear effectiveness studies in relation to species of fish should be initiated. Some gears are not able to catch certain species, and others such as trawls can be highly destructive to both target and non-target species. Practical comparative sea trials are necessary to improve our understanding of these issues and to make recommendations as to the best choice of gear

4.5 Every effort should be made to incorporate or adapt the best selectivity techniques currently in use into deep-water gears. This is especially applicable to trawl fisheries in which mixed catches are inevitable and where the level of discards is unacceptably high.

4.6. Deep-water longline research should continue, and collaborative trials between institutes with a common interest in the subject undertaken as frequently as possible to achieve maximum benefit at least cost and in a short time.

4.7. Studies of the potential effects of ghost fishing by gill nets in deep water should be undertaken as soon as possible. The deep-water conditions may make this a more difficult problem than in shallow water. Research into the use of biodegradable materials for the complete net is needed, and the immediate use of this material for the floatlines of these nets should be investigated.

4.8. The study of the behaviour of deep-water fish and their reaction to different gears should be initiated.

5. Workshop IV. Joint research into deep-sea stocks

Chairman. John D.M.Gordon, Scotland
Rapporteur. Odd Aksel Bergstad, Norway

5.1. Throughout the Workshop delegates referred to the high cost of undertaking research into the resources of the deep ocean, and the general shortage of funding at national and international levels. It was agreed that collaboration and the pooling of results between countries would achieve much more than could be expected from any independent research.

It was generally recommended that collaborative effort should aim at assembling and collecting documented information on population biology and community ecology of the deep-sea species for the purpose of improving the assessment and management of these stocks.

The views expressed by delegates to the Workshop on future research have been summarised in the form of 16 specific action points. Some of these refer to the conduct of research into deep-water species, others suggest the research topics which should have priority. The sequence which follows is somewhat arbitrary.

5.2. In order to achieve consistency between areas, time periods and countries the data collection procedures should be standardised and documented for easy reference by all concerned, eg. fish length, survey methods, species identification.

5.3. The introduction of new common names for fish should be avoided and scientific names used as often as possible to avoid confusion and misinterpretation. They should be used as a matter of routine in all scientific/technical research.

5.4. Catch records of all species should be kept by using updated identification keys. In fisheries where the catches of several species are lumped together, or when discarding at sea occurs onboard sampling by trained observers will be necessary.

5.5. In studies of abundance and biomass the entire depth range of the species must be considered and not just in areas where exploitable concentrations occur. Interdisciplinary efforts between different sciences would appear to have considerable value, eg. collaboration between physical oceanographers and fish biologists is recommended especially in studies of distribution.

5.6. When studying abundance and biomass, several techniques of estimation should be employed, eg. trawl surveys, hydroacoustics and direct observation from an underwater vehicle. Efforts should also be made to develop and adapt new methods for abundance and biomass estimates in deep water such as hydroacoustics using towed vehicles.

5.7. In certain areas such as the northern Mid-Atlantic Ridge (Reykjanes Ridge) repeated stratified trawl surveys may be a possible means of obtaining information on trends in abundance in relation to exploitation.

5.8. Stock structure and migration should be studied by analysing distributions, age distributions, growth patterns, reproductive biology etc. Chemical methods should also be used such as DNA techniques and trace elements in otoliths.

5.9. Development and adoption of validated aging methods should have a high priority and it is strongly recommended that there should be intercalibration of age readings between different laboratories and countries.

5.10. Studies of reproductive biology of more species from different environments should be carried out in order to estimate reproductive capacities and probable regeneration times.

5.11. Early life history studies should be conducted for more species and areas. Laboratory studies of egg and larval development should be combined with field studies in relevant areas and depth ranges.

5.12 Analyses of probable responses to exploitation, such as the scope and time scale of compensatory growth and recruitment should be made based on information on life histories and physiology

5.13. Food webs and species associations should be described and the inter specific effects of exploitation analysed.

5.14. Widespread and intensified deep-water fishing may adversely affect populations and ecosystems which have so far remained untouched. To allow the assessment of possible impact on these from commercial fishing, records of assemblage structure, the by-catches of non-target species and invertebrates, and the physical habitat should be undertaken both before and once the fishery has started

5.15. The development of extensive gill net fisheries with the consequent risks from ghost fishing and destruction of non-target species calls for close monitoring of these fisheries to record both the species being caught and the scale of operations being employed. It is also important to know what steps are taken to recover lost gear and how much is lost each year. The use of bio-degradable material for netting is to be encouraged.

5.16. All available information on biology and ecology suggests that deep-water fishes are vulnerable to exploitation even when catches seem low by comparison with the initial abundance and the virgin biomass. In view of the increasing interest in the deep-water species an evaluation of the protective measures and precautions is urgently needed if these resources are survive in the future. Time is very short to take the necessary action.

5.17. The delegates considered that the ICES Study Group on the Biology and Assessment of Deep Sea Fisheries Resources which was to meet in August 1994 to be particularly timely and appropriate (see Annexe I). The delegates further re-affirmed the need for continued international co-operation and collaboration in the fields of aging, abundance estimation and the determination of stock and assemblage structures.

APPENDIX List of delegates

List of Delegates

Mr. Peter Chaplin, Chief Executive, Seafish, UK.
Mr. John Tumilty, Technical Director, Seafish, UK.
Mr. Philip MacMullen, Manager Marine Technology, Seafish, U.K.
Mr. Alan Hopper, U.K.
Prof. Alasdair McIntyre, U.K.
Dr. Stanislav Lisovsky, PINRO, Russia
Ms. Valentina Volkova, PINRO, Russia
Mr. Bruce Atkinson, Department of Fisheries & Oceans (DFO), Canada
Prof. Dick Haedrich, Ocean Sciences Centre, Canada
Dr. Rosemary Ommer, Institute for Social & Economic Research, Canada
Dr. Malcolm Clark, Ministry of Agriculture & Fisheries, New Zealand
Dr. Grimur Valdimarsson, Icelandic Fisheries Laboratories, Iceland
Mr. Halldor Peter Thorsteinsson, Icelandic Fisheries Laboratories, Iceland
Dr. Anatole Charuau, IFREMER, France
Dr. Jakup Reinert, Fiskirannsoknarstovan, Faroe Islands
Mr. Bjarti Thomsen, Fiskirannsoknarstovan, Faroe Islands
Dr. John Gordon, Scottish Association for Marine Science (SAMS), Scotland
Mr. J. E. Hunter, SAMS, Scotland
Dr. Nigel Merrett, Natural History Museum, U.K.
Dr. Jutta Magnusson, Marine Research Institute, Iceland
Dr. Jakob Magnusson, Marine Research Institute, Iceland
Dr. Odd Aksel Bergstad, Institute of Marine Research, Norway
Mr. Gudmunder Gunnarsson, Hampidjan Nets, Iceland
Mr. Hans Edvard Olsen, Directorate of Fisheries, Norway
Mr. Richard McCormick, BIM, Ireland
Mr. Nils Roar Hareide, Møreforsking, Norway
Mr. Per Stoknes, Møreforsking, Norway
Ms. Greta Garnes, Norway
Mr. James Hind, J. Marr, U.K.
Dr. Mike Pawson, Directorate of Fisheries Research, U.K.
Dr. John Hislop, SOAFD Marine Laboratory, U.K.
Mr. Roger Bailey, ICES, Denmark
Dr. Ole Jorgensen, Greenlands Fisheries Research Institute, Denmark
Dr. Luis Lopez-Abellan, Instituto Espanol de Oceanografia, Spain
Dr. Paul Connelly, Department of Marine, Ireland
Mr. Ciaran Kelly, Department of Marine, Ireland
Mr. Alistair Smith, University of Aberdeen
Mr. Andrew Duthie, DFO, Canada
Mr. W. R. Bowering, DFO, Canada
Mr. Alan Marsden, DFO, Canada

420

Mr. Willem Brugge, Commission of the European Communities, Brussels
Mr. Graham Patchell, Sealord Products, New Zealand
Mr. Sergio Iglesias, Instituto Espanol Oceanographico, Spain
Mr. David Symes, U.K.
Mr. Terry Thresh, Boyd Line, U.K.
Mr. Mike Tipple, Boyd Line, U.K.
Mr. Stephen Kilvington, J. Marr, U.K.
Mr. H. R. English, UK.
Mr. E. Oakshott, SOAFD, Scotland.
Mr. Carlos Sousa Reis - not attended, paper received

Steering Committee

John E. Tumilty	Technical Director, Sea Fish Industry Authority, England.
Dr. John D. M. Gordon	Scottish Association for Marine Science, Scotland.
Prof. Richard L. Haedrich	Department of Biology, Memorial University of Newfoundland, Canada.
Nils-Roar Hareide	Møreforsking, Ålesund, Norway.
Alan G. Hopper	Secretary to the Workshop, England.

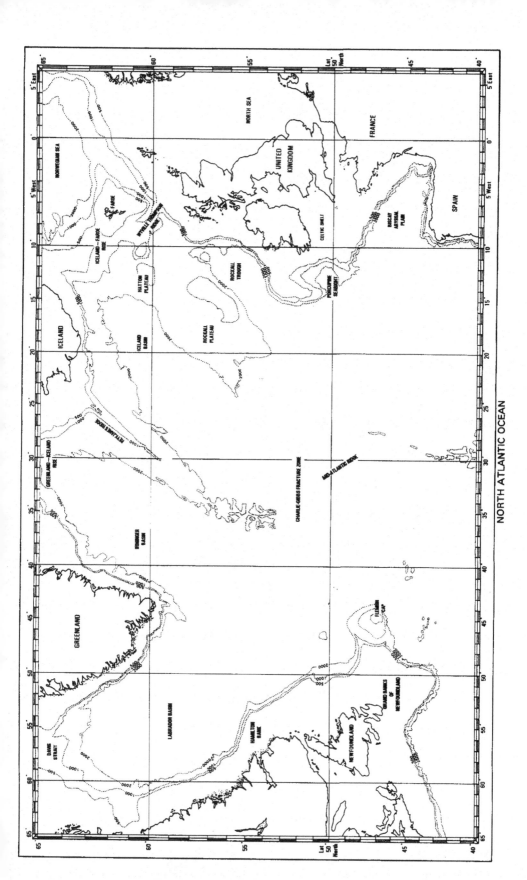

NORTH ATLANTIC OCEAN